Umbilical
C●rd Blood

A Future for
Regenerative Medicine?

Umbilical Cord Blood

A Future for Regenerative Medicine?

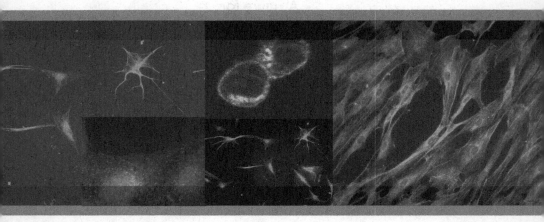

Editors

Suzanne Kadereit
University of Konstanz, Germany

Gerald Udolph
Institute of Medical Biology, Singapore

 World Scientific

NEW JERSEY · LONDON · SINGAPORE · BEIJING · SHANGHAI · HONG KONG · TAIPEI · CHENNAI

Published by

World Scientific Publishing Co. Pte. Ltd.

5 Toh Tuck Link, Singapore 596224

USA office: 27 Warren Street, Suite 401-402, Hackensack, NJ 07601

UK office: 57 Shelton Street, Covent Garden, London WC2H 9HE

British Library Cataloguing-in-Publication Data
A catalogue record for this book is available from the British Library.

UMBILICAL CORD BLOOD
A Future for Regenerative Medicine?

ISBN-13 978-981-283-329-7
ISBN-10 981-283-329-3

Typeset by Stallion Press
Email: enquiries@stallionpress.com

Printed in Singapore.

Contents

8   Immunological Properties of Mesenchymal Stem Cells   173
     Isolated from Bone Marrow and Umbilical Cord
     Mark F. Pittenger and Katarina LeBlanc

9   Expansion of Mesenchymal Stem Cells (MSCs) for   207
     Clinical Use
     Marie Prat-Lepesant, Marie-Jeanne Richard and
     Jean-Jacques Lataillade

Contributors

Finney, Marcie R.
Cleveland Cord Blood Center
25001 Emery Road, Suite 150
Cleveland, OH 44128, USA
E-mail: mfinney@clevelandcordblood.org

Giebel, Bernd
Institut für Transfusionsmedizin
Transplantationsdiagnostik und FuE
Universitätsklinikum Essen
Virchowstr. 179, 45147 Essen, Germany
E-mail: bernd.giebel@uk-essen.de

Görgens, Andre
Institut für Transfusionsmedizin
Transplantationsdiagnostik und FuE
Universitätsklinikum Essen
Virchowstr. 179, 45147 Essen, Germany
E-mail: andre.goergens@uk-essen.de

Hines, Pamela J.
American Association for the Advancement of Science
1200 New York Avenue, NW
Washington DC, WA 20005, USA
E-mail: phines@aaas.org

Hutmacher, Dietmar W.
Institute of Health and Biomedical Innovation
Queensland University of Technology
60 Musk Avenue, Kelvin Grove QLD 4059, Australia
E-mail: dietmar.hutmacher@qut.edu.au

Hwang, William Y. K.
Department of Haematology and Hematopoietic Stem Cell Laboratory
Singapore General Hospital
Outram Road, Singapore 169608
E-mail: william.hwang.y.k@sgh.com.sg/stemcelltransplants@gmail.com

Kadereit, Suzanne
Doerenkamp-Zbinden Chair for in vitro Toxicology and Biomedicine
University of Konstanz
Universitaetsstrasse 10, M657
78 457 Konstanz, Germany
E-mail: suzanne.kadereit@uni-konstanz.de

Klein, Travis J.
Institute of Health and Biomedical Innovation
Queensland University of Technology
60 Musk Avenue, Kelvin Grove QLD 4059, Australia
E-mail: t2.klein@qut.edu.au

Kögler, Gesine
Institute for Transplantation Diagnostics and Cell Therapeutics
University of Düsseldorf Medical School
Moorenstrasse 5, D-40225 Düsseldorf, Germany
E-mail: koegler@itz.uni-duesseldorf.de

Kuaha, Kunnika
Institute of Health and Biomedical Innovation
Queensland University of Technology
60 Musk Avenue, Kelvin Grove QLD 4059, Australia
E-mail: k.kuaha@qut.edu.au

Lataillade, Jean-Jacques
Hôpital d'Instruction des Armées Percy
Centre de Transfusion Sanguine des Armées
BP 410, 92141 Clamart cedex, France
E-mail: jjlataillade@ctsa-armees.fr

Laughlin, Mary J.
Department of Medicine, Division of Hematology/Oncology
Hematopoietic Cell Transplantation Program
University of Virginia Health System
1300 Jefferson Park Ave
Room 6263A, West Complex
Charlottesville, VA 22903, USA
E-mail: MJL5G@hscmail.mcc.virginia.edu

Le Blanc, Katarina
Karolinska Institutet
Division of Clinical Immunology
Karolinska University Hospital
Huddinge F79, 141 86 Stockholm, Sweden
E-mail: katarina.leblanc@ki.se

Lengerke, Claudia
Department of Hematology and Oncology
University of Tuebingen Medical Center II
Otfried-Müller-Str. 10, 72076 Tübingen, Germany
E-mail: claudia.lengerke@med.uni-tuebingen.de

McNiece, Ian K.
Stem Cell Institute
University of Miami
1501 NW 10th Ave, BRB room 908
Miami, FL 33136, USA
E-mail: imcniece@med.miami.edu

Nöth, Ulrich
Orthopädisches Zentrum für Muskuloskelettale
Forschung Orthopädische Klinik
König-Ludwig Haus
Universität Würzburg
Brettreichstrasse 11, 97074 Würzburg, Germany
E-mail: u-noeth.klh@uni-wuerzburg.de

Ong, Shin Y.
Duke-NUS Graduate Medical School
2 Jalan Bukit Merah
Singapore 169547
E-mail: shinyeu@gmail.com

Pittenger, Mark F.
Department of Surgery/Cardiac
University of Maryland School of Medicine
10 South Pine Street, MSTF Building-434
Baltimore, MD 21201, USA
E-mail: mfpittenger@comcast.net

Prat-Lepesant, Marie
Hôpital d'Instruction des Armées Percy
Centre de Transfusion Sanguine des Armées
BP 410, 92141 Clamart cedex, France
E-mail: mprat@ctsa-armees.fr

Radke, Teja F.
Institute for Transplantation Diagnostics and Cell Therapeutics
University of Düsseldorf Medical School
Moorenstrasse 5, D-40225 Düsseldorf, Germany
E-mail: radke@itz.uni-duesseldorf.de

Reichert, Johannes C.
Institute of Health and Biomedical Innovation
Queensland University of Technology
60 Musk Avenue, Kelvin Grove QLD 4059, Australia
E-mail: johannes.reichert@qut.edu.au

Richard, Marie-Jeanne
Dept. Biochimie, Toxico., Pharmaco. (DBTP)
Pavillon B Hôpital de la Tronche BP 217
38043 Grenoble Cedex 9, France
E-mail: MJRichard@chu-grenoble.fr

Rijckaert, Bart
Institute of Health and Biomedical Innovation
Queensland University of Technology
60 Musk Avenue, Kelvin Grove QLD 4059, Australia
E-mail: b.rijckaert@qut.edu.au

Seshareddy, Kiranbabu
Midwest Institute for Comparative Stem Cell Biology
Department of Anatomy and Physiology
Kansas State University
Manhattan, KS 66506, USA
E-mail: kiranseshareddy@gmail.com

Sorg, Rüdiger V.
Institute for Transplantation Diagnostics and Cell Therapeutics
University of Duesseldorf Medical School
Moorenstrasse 5, D-40225 Duesseldorf, Germany

Tanavde, Vivek M.
Genome and Gene Expression Analysis Division
Bioinformatics Institute
30 Biopolis Street #07-01
Singapore 138671
E-mail: vivek@bii.a-star.edu.sg

Udolph, Gerald
Institute of Medical Biology
8A Biomedical Grove #06-06
Singapore 138648
E-mail: gerald.udolph@imb.a-star.edu.sg

Weiss, Mark L.
Midwest Institute for Comparative Stem Cell Biology
Department of Anatomy and Physiology
Kansas State University
Manhattan, KS 66506, USA
E-mail: weiss@vet.k-state.edu

Weitzel, Patrick R.
Molecular and Clinical Hematology Branch
NHLBI/NIH
9000 Rockville Pike
Building 10 - MCC, 9N119
Bethesda, MD 20814, USA
E-mail: rpw5@case.edu

Preface

The Afterbirth: Once Discarded, Now Useful

Life is an ongoing symphony of building up, taking down, replacing, and remodeling tissues. The skin sheds and regenerates from within, the muscles build in response to exercise, the intestinal lining refreshes continuously. The cells needed for these processes are supplied by endogenous stem cells. Regenerative approaches that may repair sustainable biological tissues can capitalize on that strength of biology by using the cells from the umbilical cord, a tissue critical to gestation and seemingly useless after birth.

The 1950s saw experiments in which transplantation of one cell regenerated a frog, thereby exploring the malleability of cellular differentiation. Also in the 1950s, the first therapeutic bone marrow transplant was used to treat a patient with leukemia. Since then, many thousands of patients have been treated with transplants of blood-forming hematopoietic stem cells derived from bone marrow. Turning these pioneering efforts into successful medical treatments required advances in understanding the immunology of tissue rejection and the developmental biology of the blood-forming tissues, and even more advances in the technologies necessary to handle the cells and support the patients. About 30 years later, the first efforts to use hematopoietic cells from umbilical cord blood began with the treatment of a boy with Fanconi anemia.

This book, written for the research scientist, the medical specialist, and the interested student, gives an overview of current research about stem cells derived from umbilical cord blood and tissue. The book discusses the science surrounding a range of topics, including the various tissues in the umbilical cord, the current state of the research, the medical applications in place at the moment, and the potential for more advanced medical applications. The book also reflects on the complicated ethics concerning umbilical cord tissue collection, banking of tissue samples, and use of cord blood cells and tissues for medical treatments. The book is ultimately a story about cells, donors, patients, technologies, research insights, medical applications, and an emerging specialty.

In the context of regenerative medicine or tissue engineering, cells are the starting point. Cells could be used to replace damaged or degenerating tissues. Cells could be used to provide a helping environment to better sustain remaining endogenous cells. Cells could be used to learn more about development and cellular signaling, lessons that can be translated for application to normal tissues or to disease states. Cells could be used to deliver missing gene products or hormones. Stem cells are of particular utility because of their ability to proliferate and their undifferentiated status, characteristics that may be intricately intertwined.

Stem cells generally have a deceptively simple plan: maintain, proliferate, differentiate, signal, and die. Diversity in the details translates the general plan into the specific developmental pathways that form the various specialized tissues of the body from lineage-dedicated progenitor cells. Chapters on the basic biology of stem cells show that what is learned from the fruitfly *Drosophila* or the tiny worm *Caenorhabditis elegans* can help with understanding how stem cells function and how tissues are built.

The umbilical cord and placenta provide one source of stem cells. Cord blood can be collected from the umbilical cord and placenta after birth. Currently about one in five hematopoietic stem cell transplants use cord blood as a source. Some of the advantages of cord blood over bone marrow have to do with how cord blood is comparatively easy to collect after birth from tissues no longer needed and can be stored frozen in tissue banks. Other sources of stem cells that researchers are currently grappling with are complicated by limited availability, by extra interventions needed to make the cells useful, and by ethical constraints unique to each type of cell source.

Umbilical cord blood differs from bone marrow or peripheral blood as a source of hematopoietic stem cells. Umbilical cord blood is available in smaller quantities, but from a greater diversity of donors, than bone marrow. The complications attendant to transplantation may be fewer for umbilical cord blood, but bring a different suite of tag-along disease complications than bone marrow. Despite the advances already brought by hematopoietic stem cell transplantation from either source, considerable problems remain. Not all patients may find a match. Not all matches may result in success. The intervention is complicated and brings its own risks. This book considers the merits and problems of each approach.

Although the parallels with bone marrow transplant suggest that cord blood cells may have great utility for hematopoietic disorders, cells derived from umbilical cord may have other uses as well. The blood that the umbilical cord carries, some of which it still holds after birth, has been the main aspect of interest. However, the conduit itself, the umbilical cord, also offers some interesting cells that may turn out to be useful for therapeutic applications. Endothelial cells lining the umbilical vein and also mesenchymal-like cells that form the bulk of the tissue are already showing promise. The book explores some of the possible applications, such as support for vascularization or bone reconstruction. Even the cord blood cells are most likely not a homogeneous population. Subpopulations of cells identifiable by suites of proteins expressed on the cell surfaces seem to differ in their abilities to generate diverse sorts of new tissues. The book discusses outcomes of clinical trials, status of current research, and technologies most productively applied.

Other chapters look at the sorts of clinical settings in which cord blood transplantation might be a suitable option. How have various cancers and disorders of the blood responded to transplantation therapy based on cord blood, bone marrow, or peripheral (adult) blood? What sorts of complications have arisen? How are new technological refinements expanding the value of cord blood as a source of hematopoietic cell transplantation? What limits the number of stem cells in one sample? How does sample size per cord blood unit affect utility for some patients? What laboratory interventions are being developed to address these problems?

Treatments that depend on autologous transplantation, in which cells are derived from the same patient they will be used to treat, are currently quite limited. More often, interventions depend on donors. When the disease is rooted in a genetic anomaly, it may do little good to give the patient more of those same anomalous cells; donated cells with different genetics might do the job better. Both banking for autologous use and anonymous donation each carry their own ethical complications, as the book discusses. A particular advantage of cord blood over bone marrow as a tissue source is that cord blood is comparatively easy to collect and can be stored long-term. These factors cooperate to develop cord-blood collections with better representation of tissues suitable for patients with unusual tissue matching needs and for patients of ethnic minorities not well represented in the bone marrow registries.

Banking of cord blood samples also brings unique ethical challenges. Donors' rights need to be respected, and attempts to play on the heart-strings of parents either facing or anticipating a medical crisis need to be thwarted. Some of the ethical issues revolve around the angst and excitement of parents in the delivery room. With a newborn baby on their hands, and full of hopes and fears for the baby's future, parents may be susceptible to pressures that they should bank the cord blood for their own child's advantage, at a not inconsiderable cost. What costs are reasonable and whether better economic solutions exist, what cooperation between private and public interests is optimal, what situations are best treated by autologous transplant, are all challenging questions discussed here.

A heart-wrenching litany of diseases, disorders, and accidental damages plagues lives. Stem cells seem to offer promise of new medical therapies applied to treat a variety of diseases and disorders. However, we must be cautious in our expectations. Our understanding of stem cells and how they might best be used is still limited. Some of the current views will necessarily be revised as further research clarifies what stem cells can and cannot do. In the research labs, as reflected in this book, scientists are working hard to test the utility of umbilical cord cells as tools to address a variety of medical problems.

Moore's law describes the biannual doubling in capacity of integrated circuits, a pattern of technology growth that has driven similar growth in the memory space of desktop computers and the revolution by which everyday digital photography has overturned film photography. Similarly, we may expect to see considerable progress in the scientific understanding derived from cord blood tissues over the next few decades. What shape that advance takes remains to be seen. Although the unknown is less a barrier than a challenge to scientists, the translation of insights in regenerative medicine from basic science results to clinical applications must be guided by ethical and thoughtful respect for patients' circumstances.

Pamela J. Hines, Ph.D.
Senior Editor
Science
American Association for the Advancement of Science
1200 New York Ave. NW, Washington, DC 20005, USA
Email: phines@aaas.org

====================

Disclaimer: This preface represents solely my opinion on the current state of the field, and is not meant to represent the opinions, positions, or strategies of my employer or to be taken as advice. And just as the science in this field grows and changes over time, so too may my own opinions.

====================

PART I

Hematopoietic Stem Cells (HSCs)

1

Hematopoietic Stem and Progenitor Cells in Umbilical Cord Blood

Claudia Lengerke and Suzanne Kadereit

ABSTRACT

Hematopoietic stem cells (HSC) are among the best studied somatic stem cells. Studied for over 100 years and used in the clinical setting since over 50 years, HSCs are the only stem cells today that are transplanted in a routinely manner. In recent years it became apparent that in addition to bone marrow (BM), HSCs can also be obtained from umbilical cord blood (UCB). UCB-HSCs are thought to be more primitive than HSCs in the bone marrow, and the concentration of HSCs in UCB is higher than in BM. HSCs maintain the life long generation of blood cells of an organism and are essential for survival. And although HSCs are studied for so many years and are very well characterized, compared to other somatic stem cells, it is still difficult to grow HSCs in culture. Not surprisingly, UCB has become a major source of human HSCs for research, as UCB is more readily available than BM to most researchers. UCB has thus enabled the generation of a large body of research into the properties of these important human stem cells. The following chapter gives an introduction to the properties of HSCs and summarizes current knowledge about HSCs, particularly of UCB-HSCs, including phenotype, approaches to characterize their potential, and the specifics of UCB-HSCs.

INTRODUCTION

Hematopoiesis, or blood formation, is being studied since over 100 years.[1] Hematopoietic stem cells (HSCs) give rise to all differentiated blood cells in the body and have been characterized in great detail, particularly in the mouse. Here, the amazing potential of HSCs has been shown in single cell transplantation experiments documenting life-time repopulation capacity of the entire hematopoietic system by one single cell.[2] Importantly, however, for human health, the detailed analysis of hematopoiesis culminated 1957 in the first clinical application of stem cell transplantation, as Donnall Thomas intravenously infused for the first time bone marrow (BM) cells into irradiated patients with blood diseases, in an attempt to reconstitute a healthy hematopoietic system.[3] The first persistently engrafted patients succumbed however to graft-versus-host disease (GvHD),[4] a post-transplant complication which is now well characterized, but still not completely overcome. Over the following years, a better understanding of basic HSC biology and of immunological mechanisms active during allogeneic cell substitution resulted in continuous improvement of transplantation protocols (e.g., HLA-based selection of grafts and introduction of immunosuppressive drugs), making allogeneic HSCs transplantation the standard of care for a variety of malignant and non-malignant hematological diseases (e.g., acute leukemia, bone marrow failure syndromes). During allogeneic transplantation, HSCs are classically collected from the donor's BM by a surgical procedure or extracted from peripheral blood after stimulation with cytokines mobilizing HSCs from their bone marrow niche. Unfortunately, HLA-matched family donors are only available for about 30% of patients, and identification of HLA-matched unrelated donors can be challenging and time-consuming, particularly in the case of ethnic minorities. The report of successful hematopoietic reconstitution of a Fanconi anemia patient with umbilical cord blood (UCB)-derived cells introduced UCB in 1989 as a new resource providing transplantable HSCs.[5] This alternative strategy promised to substantially expand the potential HSC donor pool. An era of intense research into hematopoietic stem and progenitor cells from UCB ensued, and it became rapidly clear that UCB is not only a widely and readily available resource, but also very rich in HSCs and hematopoietic progenitor cell (HPC) activity. Cord blood has thus evolved into a mainstay source of human HSCs for basic and clinical research. Here, we will review the main characteristics of UCB-derived HSCs and compare them to conventionally isolated HSCs from adult bone marrow and mobilized peripheral blood.

FREQUENCY OF HSCs IN UCB

Transplantable HSCs and progenitors can be isolated from UCB, BM, and from peripheral blood of donors treated with cytokines to mobilize their stem cells from the bone marrow into the periphery, so-called peripheral blood stem cells (PBSCs). This latter source of HSCs is increasingly used to replace bone marrow aspirates. UCB units typically contain one to two logs fewer nucleated cells than classical bone marrow units.[6] Yet, the hematopoietic recovery after myeloablation depends not only on the number but also on the quality of transplanted HSCs. Studies over the last years have shown that HSCs undergo qualitative changes during ontogenetic development as well as later during life.[7] During development, HSCs are greatly expanded in the fetal liver, where they reside until shortly before birth. This is due to the fetal liver environment but also to intrinsical biological properties of fetal liver HSCs, which, in the mouse, have higher proliferation capacity than their postnatally collected counterparts.[8] Changes continue to occur also after birth: with age, the frequency of lympho-myeloid progenitors declines and less primitive progenitors capable of reconstituting just one of the two lineages predominate.[9] Thus, it is not surprising that newborn cord blood-derived HSCs show differences in proliferation, differentiation and reconstitution capacity compared to HSCs collected from the adult bone marrow or mobilized peripheral blood.[10,11]

IN VITRO COLONY ASSAYS
CFC, LTC-IC, CAFC and ML-IC

HSCs are defined by their ability to both self-renew and differentiate, giving rise to all blood lineages. HPCs develop from differentiating HSCs, and have more limited self-renewal as well as more restricted differentiation capacity than HSCs. More is known about the biology of human HPCs than about human HSCs, since progenitor function can be well assayed *in vitro*, while accurate analysis of stem cell properties requires an *in vivo* model.

The limited HPC self-renewal potential can be estimated by serial replating experiments,[12,13] while HPC differentiation capacity is classically analyzed in colony forming cell assays (CFC).[14] Here, multipotential progenitor cells termed CFU-GEMM (colony-forming unit-granulocyte, erythroid, macrophage, megakaryocyte) form colonies of mixed lineage blood cells in semisolid culture medium supplemented with supportive hematopoietic cytokines, such as erythropoietin (EPO), stem cell factor (SCF), granulocyte-macrophage colony-stimulating factor (GM-CSF), and interleukin-3 (IL-3), in the presence or absence of thrombopoetin (TPO). More committed but still immature granulocyte-macrophage progenitor cells termed CFU-GM (colony-forming unit-granulocyte, macrophage) give rise to colonies of granulocytes and macrophages when cultured in semisolid media supplemented with GM-CSF or IL-3 combined with SCF or Flt-3 ligand (FL). Granulocyte progenitors CFU-G (colony-forming unit-granulocyte) and macrophage progenitors CFU-M (colony-forming unit-macrophage) form granulocytic and macrophage colonies respectively only after stimulation with either granulocyte colony-stimulating factor (G-CSF) or macrophage colony stimulating factor (M-CSF), while generation of colonies deriving from erythroid progenitors BFU-E (burst-forming unit-erythroid) and megakaryocyte progenitors CFU-Meg (colony-forming unit-megakaryocyte) requires supplementation with EPO and TPO, respectively. More immature subsets of these lineage-restricted progenitors are detected by the positive colony growth response upon additional stimulation with either SCF or FL.[14]

HPCs quantified by the CFC assay represent short-term repopulating cells. Long-term engrafting HSCs are best analyzed in *in vivo* models, but surrogate assays have been developed to estimate HSC activity *in vitro*: the long-term culture-initiating cell (LTC-IC) assay,[15] the extended (E) LTC-IC[16] and the cobblestone area-forming cell (CAFC)[17] assay measure primitive myeloid cells, whereas the myeloid-lymphoid initiating cell (ML-IC) assay determines cells with both myeloid and lymphoid [natural killer (NK) cell] activity.[18] In LTC-IC and CAFC assays, putative stem cells are cultured under specific conditions (with or without stromal support) for five weeks. After this culture period, it is hypothesized that only long-term engrafting HSCs produce *de novo* colonies in secondary cultures (LT-CIC assay), or, alternatively, show presence of cobblestone areas in CAFC assay. The CAFC-assay has the advantage that appearance of cobblestone areas can be monitored over time, with occurrence at later time-points being indicative of a more primitive hematopoietic pheno-type. In the murine system, there is a good correlation between *in vitro* assays for progenitor/stem cell activity and the analyzed *in vivo* hematopoietic reconstitution capacity.[19] Whether this correlation also exists in the human system is less clear.

IN VIVO XENOGENEIC MOUSE MODELS

The SCID-Repopulating Cell (SRC)

Most of the functional studies on HSC characterization have been carried out in mice. Here, hematopoietic reconstitution of lethally irradiated animals has been achieved from a single HSC.[2] Experiments in dogs and nonhuman primate models suggest that murine studies may not always accurately predict HSC biology in larger animals. With human HSCs, any *in vivo* model is xenogeneic, making functional studies less exact and less amenable. Upon transplantation into fetal sheep, human HSCs successfully colonize the bone marrow, are able to differentiate into several lineages and persist over many years.[20] In mice, hematopoietic reconstitution with human HSCs has been disappointing until mice carrying the Prkdc mutation causing severe combined T- and B-cell immune deficiency (SCID) were used as recipients.[21,22] In these experi-ments, human HSCs were transplanted into sublethally irradiated SCID mice; the human cells repopulating the murine recipients were termed "human SCID-repopulating cells (SRCs)" and considered to be the engrafting human HSCs. However, chimerism rates were modest and primary immune response rarely detected in this model.[23] A further crucial step forward has been achieved 1995 by crossing SCID with NOD (non-obese diabetic)-mouse strains carrying additional defects in innate immunity (NK, macrophage and complement deficiency).[24] Human HSCs transplants into NOD/SCID mice showed five- to ten-fold higher engraftment as compared to SCID mice. Yet, human B- and T-cell maturation in NOD/SCID mice remained insufficient, and long-term follow-up was hampered by predisposition of these mice to develop lethal T-cell lymphomas. Further major improvement of the model has been achieved by administration of interleukin receptor 2 (IL-2R ß)[25] antibodies, or later by use of two new mouse strains deficient for the common cytokine-receptor gamma chain (NOD/SCID/IL2r$\gamma^{-/-}$)[26-28] and BALB/c/Rag2$^{-/-}$/IL2r$\gamma^{-/-}$.[29,30] These mice lack T, B and NK cells, and have a normal life span, with no limitations due to tumor development. Engraftment and differentiation of human HSCs and HPCs is best achieved when transplantation occurs into the growth and expansion supportive environment of the newborn, especially if administrated via intra-liver injection.[29,30] Upon transplantation, human cells of all lineages, including dendritic cells, T- and B-cells, develop and can be followed more than half a year later in the recipient mice. However, myelo-monocytic cells, NK cells, platelets and red cells are formed to a somewhat lesser extent. This is thought to be due to competition with residual mouse hematopoiesis and limited cross-reactivity between mouse and human cytokines and niche-factors, hampering the development of human hematopoiesis.[31-33] Approaches including non-lethal genetic deletion of insufficiently cross-reactive mouse growth factors and replacement with humans (that ideally are not acting on mouse receptors, e.g., GM-CSF and IL-3) may lead to further improvement of human cell differentiation in this xenogeneic model.[31]

Even though imperfect, the *in vivo* humanized mouse models are superior to the surrogate *in vitro* assays. It has become increasingly clear that the SRC is a more primitive cell than the LTC-IC, with only 0.1%–0.2% of HSCs UCB being SRC, compared to 10%–50% being LTC-IC.[34,35] These percentages drop to 0.01%–0.1% of SRC and 2%–10% of LTC-IC in bone marrow HSCs,[35] emphasizing the differences between UCB and adult HSC sources.

PHENOTYPING THE HUMAN HSC

CD34 Expression

The murine HSC has been well defined as expressing the markers Sca-1 and c-Kit (the receptor for stem cell factor), in absence of other lineage markers (Sca-1$^+$/c-Kit$^+$/Lin$^-$) and hematopoietic reconstitution has been achieved upon transplantation of one single cell.[2] Due to comparably less amenable transplant material and *in vivo* assays, markers clearly identifying the human HSC have not yet been established. It is however generally accepted that expression of CD34 correlates with stem cell phenotype and decreasing expression with differentiation towards lineage committed cells. The CD34 antigen is commonly used as selection marker for clinical HSC transplants, and the number of transplanted CD34$^+$ cells has been shown to correlate with engraftment.[36] Another hallmark of the human HSC is the absence of CD38 expression. While abundantly expressed in *in vitro* detectable clonogenic progenitors, CD38 seems to be a marker gained upon HSC differentiation. Within the CD34$^+$/CD38$^-$ cell population in UCB, one cell in 617 is capable of repopulating in the SRC assay, while no repopulating activity is contained within the CD38$^+$ cell population.[37] Similar data has been obtained with bone marrow CD34$^+$/CD38$^-$ cells, where one cell in 3.0×10^6 adult BM cells have repopulating capacity.[38]

Some data suggest that CD34 expression is not necessarily the ideal marker for isolating most primitive HSCs. For example, in cytokine-mobilized peripheral blood, CD34 expression is increased without an increase in SRC activity.[38,39] Moreover, long-term SRC activity has been shown in CD34$^-$ cells.[40,41] First reports identified a SRC in every 125,000 Lin$^-$/CD34$^-$/CD38$^-$ cells. This frequency could be increased to one SRC in 38,000 cells if the cells were first cultured for four days, suggesting a lack of homing molecules that could be increased through culture with cytokines.[42] Similarly, repopulating CD34$^-$ cells were also identified in human bone marrow in grafting experiments in fetal sheep.[43] The high repopulating capacity of CD34$^-$ cells was revealed when injecting the cells directly into the bone marrow as opposed to the classical tail vein injection. When transplanted through IBMI (intra-bone marrow injection) UCB CD34$^-$ cells engrafted and generated CD34$^+$ progeny, not only in the injected bone, but also in other bones following migration. The cells were moreover capable of engrafting secondary recipients, demonstrating their long-term repopulating capacity.[44] Further characterization revealed that within the CD34⁻ cells, the very primitive, long-term repopulating HSC is furthermore negative for the stem cell markers c-Kit and Flt3. When cultured *in vitro* on feeder cells, CD34⁻/Flt3⁻ cells produced CD34⁺/Flt3⁻ and CD34⁺/Flt3⁺cells, both populations with SRC activity. A hierarchy could be established, where the CD34⁻/Flt3⁻ cells had the highest SCID repopulating capacity, including repopulation of secondary recipients, followed by the CD34⁺/Flt3⁻ cells. The CD34⁺/Flt3⁺ cells contained SRCs, but had only weak secondary recipient repopulating capacity.[45] Importantly, human recipients of CD34⁺ BM grafts in which donor-derived multilineage hematopoiesis has been followed over more than seven years, are lacking donor-derived CD34⁻/Lin⁻ cells in their BM, thus the relevance of the CD34⁺/Lin⁻ cells for life-long maintenance and homeostasis of graft-derived hematopoiesis is not yet clear.[7]

CD133

Yin *et al.* have identified CD133 (AC133) expression to be largely restricted to the CD34⁺ subset of human hematopoietic cells.[46] AC133⁺/CD34⁺ cells have a higher frequency of LTC-IC and clonogenic potential than AC133⁻/CD34⁺ cells.[47,48] *In vivo* transplant experiments of human UCB CD34⁻/CD38⁻/Lin⁻ cells into NOD/SCID mice show CD133⁺ cells to have a 400-fold higher engraftment potential than AC133⁻ cells. Moreover, selectively CD133⁺ cells of the CD34⁻/CD38⁻/Lin⁻ compartment, and not CD133⁻, were able to give rise to CD34⁺ progenitors during *in vitro* culture.[49] More recently CD133 has been validated as a HSC selection marker in the human allogeneic transplant setting.[50]

Side Population

HSCs but not differentiated blood cells are capable of excluding the toxic dye Hoechst33342,[51] which is a functional property that is shared with stem cells of other tissues. After incubation of the cells with Hoechst33342 stem cells appear as a small dye-excluding side population (SP) in FACS (fluorescence activated cell sorting) and can be isolated from a majority of non-stem cells. Being able to purify stem cells by this technique has been a great advance in the field of stem cell research allowing, besides HSCs, also identification and characterization of stem cells from non-hematopoietic tissues (such as muscle and testis[52,53]). In HSCs, the capability to exclude Hoechst dye could be related to expression of ABC transporters, specifically Bcrp1 (also known as Abgc2 murine/ABCG2 human). Bcrp1 mRNA is expressed at high levels in human HSCs and immature progenitors and is sharply down-regulated upon differentiation.[54,55] Moreover, overexpression of Bcrp1 conferred SP phenotype to murine bone marrow cells, reducing the maturing progeny.[54] Recent results using ABCG2 retroviral overexpression in UCB-derived cells suggest that ABCG2 enhances the most primitive human progenitor pool as assayed by limiting dilution competitive repopulation assays in NOD/SCID mice, and moreover, surprisingly, affects B-cell development, inversing the myeloid/lymphoid differentiation ratio of the transplanted UCB cells.[56]

Aldehyde Dehydrogenase Activity

Another marker enabling functional identification of the HSC is the high expression of the detoxifying enzyme aldehyde dehydrogenase (ALDH).[57] In cord blood, 0.09% of the total mononuclear cell (MNC) fraction have high ALDH activity. These cells express CD34 and in majority also CD133. Interestingly, all *in vivo* SRC activity was contained within the ALDH^{hi} population, although the ALDH^{lo} population also included CD34$^+$/CD38$^-$ cells.[58] Further analysis showed that purification of cells based on high ALDH activity and coexpression of CD133 increased the repopulating proportions of the cells. While transplantation of 10^3 ALDH^{hi}/CD133$^+$ cells resulted in engraftment of 67% of the mice, five times more cells were required to engraft 50% of the mice when transplanting cells selected for CD133 expression.[59] Moreover, the ALDH^{hi} population also contained CD34$^-$ cells with long-term SRC activity.[40]

SPECIFICS OF UCB-DERIVED HSCs

Hematopoietic reconstitution is now routinely achieved with BM, mobilized peripheral blood, as well as cord blood-derived HSCs, yet considerable differences have been found in numbers of CD34$^+$ cells needed for recon-stitution. In the adult, higher numbers of CD34$^+$ cells are needed from mobilized peripheral blood in comparison to bone marrow, suggesting that the exposure to G-CSF and mobilization into the periphery may favor development of more differentiated CD34$^+$ progenitors. This would also explain the more rapid myeloid engraftment reported with mobilized peripheral blood grafts.

UCB-derived CD34$^+$ cells show a more potent engraftment capacity. Up to 20-fold less CD34$^+$ cells were needed to achieve comparably reliable engraftment when using UCB-derived grafts.[6,60] Surprisingly, no statistical difference could be noted between the numbers of primary LTC-IC and NK-IC generated from CD34$^+$/CD38$^-$/Lin$^-$ cells from the three sources (BM, MPB, UCB), arguing that the much higher regenerative potential of UCB-derived HSCs/HPCs seen in transplant assays is due to qualitative differences with adult HSCs, and not to a higher HSC frequency within the CD34$^+$/CD38$^-$/Lin$^-$ compartment. Purification to CD34$^+$/CD38$^-$/Lin$^-$ HSCs/HPCs and studies on a single cell level showed that indeed UCB-derived cells generate significantly more cells per colony in methylcellulose CFC and NK-cell assays, and more often give rise to two instead of one daughter LTC-IC and NK-IC.[61]

HSCs undergo great expansion during fetal development and early postnatal life, the period of time when UCB-HSCs are collected. Mouse fetal liver HSCs show greater proliferation capacity than HSCs derived postnatally from BM of young or older mice.[8,62] Moreover, qualitative changes are known to occur in HSCs throughout development and life. In the mouse, there is an age-dependent decline in frequency of more prim-itive HPCs capable to reconstitute both the myeloid and the lymphoid fraction, a feature accentuated following hematopoietic stress (e.g., *in vitro* manipulation, serial transplantation).[9,63] These data support the notion that UCB-derived HSCs differ from their adult (BM)-derived counterparts.

HOMING AND ENGRAFTMENT

When transplanted intravenously, HSCs actively navigate through the blood, cross the endothelial vasculature and enter their niche in the bone marrow. Particularly in response to stress, such as total body irradiation,

HSCs can colonize other organs (e.g., spleen) first. This coordinated, multistep process is called homing and represents a prerequisite for successful engraftment and repopulation. While short-term repopulating $CD34^+/CD38^+$ progenitors and mature, specialized T-cells and neutrophils can also home to the BM,[64,65] long-term repopulation requires HSCs.

Homing depends on intrinsic properties of the cells, but also on factors enhancing survival, migration and angiogenesis, such as the chemokine stromal derived factor-1 (SDF-1). *In vivo*, as well as *in vitro*, HSCs and progenitors migrate towards a gradient of the chemokine SDF-1 produced by stromal cells. This is mediated by the receptor CXCR4. It has been shown that engraftment and repopulating capacity of human $CD34^+/CD38^-$ cells is dependent on CXCR4, and that stem cell factor (SCF) and interleukin-6 (IL-6)-mediated increase of CXCR4 expression potentiated engraftment in primary and secondary recipients.[66] Furthermore, it appears that SDF-1 has also an effect on survival and cycling of $CD34^+$ cells, with higher survival and cycling in the presence of SDF-1.[67]

Neonatal, UCB-derived HSCs exhibit impaired homing capacity as compared to BM-derived HSCs. CB-$CD34^+$ cells demonstrate reduced rolling on endothelium and thus transendothelial extravasation, due to partial expression of a non-functional form of the P-selectin ligand, P-selectin glycoprotein ligand-1 (PSGL-1).[68] Interestingly, patients transplanted with 13 times more BM- as compared to UCB-derived $CD34^+$ cells, showed faster neutrophil and platelet reconstitution but also lower levels of progenitors one year after transplant. The data again demonstrate qualitative differences between the different HSC sources, and suggest that UCB-derived HSCs may not fully respond to differentiation signals and may be prone to maintaining an immature state. Lentiviral CXCR4 overexpression in UCB- or BM-derived $CD34^+$ cells enhanced cell motility and SDF-1 increased migration as well as SRC repopulation.[69,70] A similar stimulatory effect on engraftment was obtained by *in vivo* treatment with SDF-1.[71]

In an attempt to better understand the differences in homing and repopulation capacity, conventional intravenous transplantation has been compared to intra-osseous (IO), intra-bone marrow (IBM) and respectively intra-femoral (IF) transplantation. As expected, IBM injections significantly improved repopulation capacity of UCB-derived cells, leading to a

15-fold enhancement of SRC-frequency within transplanted CD34$^+$/CD38$^-$ cells. Notably, CXCR4 neutralization abolished engraftment of both intra-venously and intra-bone marrow transplanted cells, demonstrating an important role for this receptor beyond the homing process, perhaps in colonization and proliferation within the bone marrow.[72]

HSC DIFFERENTIATION AND SELF-RENEWAL

HSCs can differentiate to form all distinct mature blood cells, but can also reproduce themselves, which is known as self-renewal. Retaining the capacity of both self-renewal and differentiation is essential for preserv-ing continuous supply of functional HSCs in the hematopoietic system throughout life. These highly orchestrated biological processes are driven by cell intrinsic transcriptional mechanisms but also influenced by extrin-sic cues (e.g. growth factors) provided by the HSC environment, the so-called "niche".

During mammalian development, HSCs reside in distinct niches. Hematopoiesis is initiated in the yolk sac of the embryo, slightly later shifts to the aorta-gonado-mesonephros region and afterwards to the placenta, the fetal liver, and shortly after birth to the bone marrow. Cellular differentiation and homeostasis during these different stages is governed by cross-talk with growth factors and morphogen gradi-ents provided by the niche [such as NOTCH ligands, WNT proteins and bone morphogenetic proteins (BMP)] that activate specific intra-cellular signalling pathways.[73] Later in the adult, such pathways often regulate the stem cell self-renewal process. For example, addition of WNT3a to BM cell culture medium significantly increased self-renewal of the cells as measured by transplant and repopulation assays.[74] At the molecular level, WNT proteins stabilize beta-catenin, which then translocates to the nucleus and interacts with lymphoid-enhancer binding factor (LEF) and T-cell factor (TCF) family members, driving expression of genes such as Myc and cyclin D. SMAD4, a component of the BMP and transforming growth factor-beta (TGF-β) downstream pathway, has been also shown to be required during developmental hematopoiesis as well as for self-renewal of adult HSCs following transplantation.[75]

Of note, HSCs from different developmental stages can show differences in self-renewal properties. Fetal HSCs intrinsically cycle faster and perform symmetric divisions, more efficiently giving rise to stem daughter cells. A recent study on serially cultured and transplanted HSCs reports a new checkpoint marking development into an adult HSC: based on their expression of Sox17, fetal HSCs have an intrinsic ability to self-renew and differentiate, which is maintained until one week after birth, when they switch to quiescent, adult-type HSCs.[76,77]

In line with this observation, signaling pathways have been shown to have different effects at different stages of HSC development. Fetal and adult HSC responsiveness are differentially regulated by growth factors such as Kit ligand and thrombopoetin.[78] For example, whereas both HSC types were found to be Kit-dependent, fetal HSCs were stimulated at lower concentrations of Kit. While enhancing self-renewal, and thus being able to increase numbers of transplantable stem cells is a golden goal of HSC (and particularly of UCB-derived HSC) research, chronic activation of self-renewal pathways may lead to HSC depletion or force them into quiescence, or may result in uncontrolled growth.

Developmental morphogens act by activating transcriptional regulation of HSC self-renewal, and differentiation also involves target pathways of developmental morphogens, such as the classically studied *HOX* family genes.[73,79–81] *HOX* gene expression is directly regulated by caudal-type homeobox family genes such as *CDX1*, *CDX2* and *CDX4*,[80,82] as well as by HOX cofactors such as PBX or MEIS family members.[73,83] When ectopically expressed in adult BM-derived HSCs, *HOX* genes (e.g., *HOXA9*, *HOXB4*) as well as Hox-regulators (*CDX*, *PBX*, *MEIS*) have been shown to enhance self-renewal properties of the blood stem and progenitor cells.[84] There are however differences between the distinct *HOX* genes. For example, while *HOXB4* expands the progenitor pool, *HOXA9* promotes myelopoiesis and drives leukemogenesis[85] and *HOXA10* blocks erythroid and megakaryocytic cell fates.[86,87]

Transcriptional regulators such as *HOX* genes act in concert with epigenetic regulators (such as polycomb group genes), which mediate stimulation or repression of transcription factors by histone modifications. The best characterized epigenetic self-renewal factor is *Bmi1*, which is encoded by a polycomb group gene.[88–90] $Bmi1^{-/-}$ mice are born with an exhausted HSC pool. Conversely, ectopic *Bmi1* expression leads to enhanced self-renewal.[91] *Bmi1* has been shown to partly act by modulating the cell-cycle regulating proteins INK4A (p16) and ARF (p19),[91] but overall the downstream pathways of *HOX* and *HOX*-regulating polycomb genes in HSC self-renewal still need to be elucidated.

Another developmental gene that is epigenetically regulated and important for HSC self-renewal is *Mll*: inactivation in mice leads to decreased HSC self-renewal, while conversely, overexpression enhances their self-renewal capacity.[92] Moreover, Mll fusion proteins are found in leukemia.

The current view is that decisions between differentiation and self-renewal are based on competition of downstream transcription factors.[73] For example, *GATA1* drives erythroid differentiation, while *PU.1* normally drives myeloid differentiation from an HSC. The two factors physically bind to each other, limiting each other's ability to interact with binding sites in lineage specific genes. Small imbalances between transcriptional levels of such competing factors can dictate cell fate decisions.[93] There may be a similar competition between factors inducing self-renewal versus differentiation. As such, inactivation of various blood-specific genes has been linked to self-renewal (e.g., *GATA2*,[94] *GFI1*,[95] *SMAD4*[75]).

CONCLUSIONS

Since the first successful transplantation of a pediatric patient, umbilical cord blood has become the mainstay source of human HSCs for basic and translational research and enabled intensive exploration of the properties of these stem cells. It has become clear that cord blood HSCs are more immature than their bone marrow counterpart. The large availability of cord blood has allowed an in-depth characterization of this stem cell and allowed advances into characterization of self-renewal and differentiation properties that may hopefully one day enable *in vitro* culture and expansion of this life-saving stem cell population.

ACKNOWLEDGMENTS

This work was supported by grants from the Deutsche Krebshilfe (Max-Eder-Program), the DFG (SFB773), the University of Tübingen (fortüne-Program) to C.L. and funding from the Doerenkamp-Zbinden Foundation for S.K.

References

1. His W. Lecithoblast and angioblast der wirbelthiere: histogenetische studien. *Abhandl K S Ges Wiss Math-Phys* 1900;26:173–328.
2. Wagers AJ, Sherwood RI, Christensen JL, Weissman IL. Little evidence for developmental plasticity of adult hematopoietic stem cells. *Science* 2002;297(5590):2256–9.
3. Thomas ED, Lochte HL, Jr, Lu WC, Ferrebee JW. Intravenous infusion of bone marrow in patients receiving radiation and chemotherapy. *N Engl J Med* 1957;257(11):491–6.
4. Thomas ED. A history of haemopoietic cell transplantation. *Br J Haematol* 1999;105(2):330–9.
5. Gluckman E, Broxmeyer HA, Auerbach AD, Friedman HS, Douglas GW, Devergie A, *et al.* Hematopoietic reconstitution in a patient with Fanconi's anemia by means of umbilical-cord blood from an HLA-identical sibling. *N Engl J Med* 1989;321(17):1174–8.
6. Barker JN, Wagner JE. Umbilical cord blood transplantation: current practice and future innovations. *Crit Rev Oncol Hematol* 2003;48(1):35–43.
7. Verfaillie CM. Hematopoietic stem cells for transplantation. *Nat Immunol* 2002;3(4):314–7.
8. Rebel VI, Miller CL, Eaves CJ, Lansdorp PM. The repopulation potential of fetal liver hematopoietic stem cells in mice exceeds that of their liver adult bone marrow counterparts. *Blood* 1996;87(8):3500–7.
9. Sudo K, Ema H, Morita Y, Nakauchi H. Age-associated characteristics of murine hematopoietic stem cells. *J Exp Med* 2000;192(9):1273–80.
10. Lansdorp PM, Dragowska W, Mayani H. Ontogeny-related changes in proliferative potential of human hematopoietic cells. *J Exp Med* 1993;178(3):787–91.
11. Vaziri H, Dragowska W, Allsopp RC, Thomas TE, Harley CB, Lansdorp PM. Evidence for a mitotic clock in human hematopoietic stem cells: loss of telomeric DNA with age. *Proc Natl Acad Sci USA* 1994;91(21):9857–60.
12. Carow CE, Hangoc G, Broxmeyer HE. Human multipotential progenitor cells (CFU-GEMM) have extensive replating capacity for secondary CFU-GEMM: an effect enhanced by cord blood plasma. *Blood* 1993;81(4):942–9.

13. Carow CE, Hangoc G, Cooper SH, Williams DE, Broxmeyer HE. Mast cell growth factor (c-kit ligand) supports the growth of human multipotential progenitor cells with a high replating potential. *Blood* 1991;78(9):2216–21.

14. Broxmeyer HE, Srour E, Orschell C, Ingram DA, Cooper S, Plett PA, *et al.* Cord blood stem and progenitor cells. *Methods Enzymol* 2006;419:439–73.

15. Sutherland HJ, Eaves CJ, Eaves AC, Dragowska W, Lansdorp PM. Characterization and partial purification of human marrow cells capable of initiating long-term hematopoiesis *in vitro*. *Blood* 1989;74(5):1563–70.

16. Hao QL, Thiemann FT, Petersen D, Smogorzewska EM, Crooks GM. Extended long-term culture reveals a highly quiescent and primitive human hematopoietic progenitor population. *Blood* 1996;88(9):3306–13.

17. Ploemacher RE, van der Sluijs JP, Voerman JS, Brons NH. An *in vitro* limiting-dilution assay of long-term repopulating hematopoietic stem cells in the mouse. *Blood* 1989;74(8):2755–63.

18. Punzel M, Wissink SD, Miller JS, Moore KA, Lemischka IR, Verfaillie CM. The myeloid-lymphoid initiating cell (ML-IC) assay assesses the fate of multipotent human progenitors *in vitro*. Blood 1999;93(11):3750–6.

19. Weilbaecher K, Weissman I, Blume K, Heimfeld S. Culture of phenotypically defined hematopoietic stem cells and other progenitors at limiting dilution on Dexter monolayers. *Blood* 1991;78(4):945–52.

20. Zanjani ED, Almeida-Porada G, Flake AW. The human/sheep xenograft model: a large animal model of human hematopoiesis. *Int J Hematol* 1996; 63(3):179–92.

21. Mosier DE, Gulizia RJ, Baird SM, Wilson DB. Transfer of a functional human immune system to mice with severe combined immunodeficiency. *Nature* 1988;335(6187):256–9.

22. Lapidot T, Pflumio F, Doedens M, Murdoch B, Williams DE, Dick JE. Cytokine stimulation of multilineage hematopoiesis from immature human cells engrafted in SCID mice. *Science* 1992;255(5048):1137–41.

23. Greiner DL, Hesselton RA, Shultz LD. SCID mouse models of human stem cell engraftment. *Stem Cells* 1998;16(3):166–77.

24. Serreze DV, Leiter EH, Hanson MS, Christianson SW, Shultz LD, Hesselton RM, *et al.* Emv30null NOD-scid mice. An improved host for adoptive transfer of autoimmune diabetes and growth of human lymphohematopoietic cells. *Diabetes* 1995;44(12):1392–8.

25. Kerre TC, De Smet G, De Smedt M, Zippelius A, Pittet MJ, Langerak AW, *et al.* Adapted NOD/SCID model supports development of phenotypically and functionally mature T cells from human umbilical cord blood CD34(+) cells. *Blood* 2002;99(5):1620–6.

26. Ishikawa F, Yasukawa M, Lyons B, Yoshida S, Miyamoto T, Yoshimoto G, *et al.* Development of functional human blood and immune systems in NOD/SCID/IL2 receptor {gamma} chain(null) mice. *Blood* 2005;106(5): 1565–73.

27. Shultz LD, Lyons BL, Burzenski LM, Gott B, Chen X, Chaleff S, *et al.* Human lymphoid and myeloid cell development in NOD/LtSz-scid IL2R gamma null mice engrafted with mobilized human hemopoietic stem cells. *J Immunol* 2005;174(10):6477–89.

28. Ito M, Hiramatsu H, Kobayashi K, Suzue K, Kawahata M, Hioki K, *et al.* NOD/SCID/gamma(c)(null) mouse: an excellent recipient mouse model for engraftment of human cells. *Blood* 2002;100(9):3175–82.

29. Traggiai E, Chicha L, Mazzucchelli L, Bronz L, Piffaretti JC, Lanzavecchia A, *et al.* Development of a human adaptive immune system in cord blood cell-transplanted mice. *Science* 2004;304(5667):104–7.

30. Chicha L, Tussiwand R, Traggiai E, Mazzucchelli L, Bronz L, Piffaretti JC, *et al.* Human adaptive immune system Rag2-/-gamma(c)-/- mice. *Ann N Y Acad Sci* 2005;1044:236–43.

31. Manz MG. Human-hemato-lymphoid-system mice: opportunities and challenges. *Immunity* 2007;26(5):537–41.

32. Shultz LD, Ishikawa F, Greiner DL. Humanized mice in translational biomedical research. *Nat Rev Immunol* 2007;7(2):118–30.

33. Macchiarini F, Manz MG, Palucka AK, Shultz LD. Humanized mice: are we there yet? *J Exp Med* 2005;202(10):1307–11.

34. Larochelle A, Vormoor J, Hanenberg H, Wang JC, Bhatia M, Lapidot T, *et al.* Identification of primitive human hematopoietic cells capable of repopulating NOD/SCID mouse bone marrow: implications for gene therapy. *Nat Med* 1996;2(12):1329–37.

35. Coulombel L. Identification of hematopoietic stem/progenitor cells: strength and drawbacks of functional assays. *Oncogene* 2004;23(43): 7210–22.

36. Wagner JE, Barker JN, DeFor TE, Baker KS, Blazar BR, Eide C, *et al.* Transplantation of unrelated donor umbilical cord blood in 102 patients

with malignant and nonmalignant diseases: influence of CD34 cell dose and HLA disparity on treatment-related mortality and survival. *Blood* 2002;100(5):1611–8.

37. Bhatia M, Wang JC, Kapp U, Bonnet D, Dick JE. Purification of primitive human hematopoietic cells capable of repopulating immune-deficient mice. *Proc Natl Acad Sci USA* 1997;94(10):5320–5.

38. Wang JC, Doedens M, Dick JE. Primitive human hematopoietic cells are enriched in cord blood compared with adult bone marrow or mobilized peripheral blood as measured by the quantitative *in vivo* SCID-repopulating cell assay. *Blood* 1997;89(11):3919–24.

39. Hess DA, Levac KD, Karanu FN, Rosu-Myles M, White MJ, Gallacher L, *et al.* Functional analysis of human hematopoietic repopulating cells mobilized with granulocyte colony-stimulating factor alone versus granulo-cyte colony-stimulating factor in combination with stem cell factor. *Blood* 2002;100(3):869–78.

40. Pearce DJ, Bonnet D. The combined use of Hoechst efflux ability and alde-hyde dehydrogenase activity to identify murine and human hematopoietic stem cells. *Exp Hematol* 2007;35(9):1437–46.

41. Gotze KS, Schiemann M, Marz S, Jacobs VR, Debus G, Peschel C, *et al.* CD133-enriched CD34(-) (CD33/CD38/CD71)(-) cord blood cells acquire CD34 prior to cell division and hematopoietic activity is exclu-sively associated with CD34 expression. *Exp Hematol* 2007;35(9):1408–14.

42. Bhatia M, Bonnet D, Murdoch B, Gan OI, Dick JE. A newly discovered class of human hematopoietic cells with SCID-repopulating activity. *Nat Med* 1998;4(9):1038–45.

43. Zanjani ED, Almeida-Porada G, Livingston AG, Flake AW, Ogawa M. Human bone marrow CD34– cells engraft *in vivo* and undergo multilineage expression that includes giving rise to CD34+ cells. *Exp Hematol* 1998; 26(4):353–60.

44. Wang J, Kimura T, Asada R, Harada S, Yokota S, Kawamoto Y, *et al.* SCID-repopulating cell activity of human cord blood-derived CD34– cells assured by intra-bone marrow injection. *Blood* 2003;101(8):2924–31.

45. Kimura T, Asada R, Wang J, Morioka M, Matsui K, Kobayashi K, *et al.* Identification of long-term repopulating potential of human cord blood-derived CD34-flt3- severe combined immunodeficiency-repopulating cells by intra-bone marrow injection. *Stem Cells* 2007;25(6):1348–55.

46. Yin AH, Miraglia S, Zanjani ED, Almeida-Porada G, Ogawa M, Leary AG, *et al.* AC133, a novel marker for human hematopoietic stem and progenitor cells. *Blood* 1997;90(12):5002–12.

47. de Wynter EA, Buck D, Hart C, Heywood R, Coutinho LH, Clayton A, *et al.* CD34+AC133+ cells isolated from cord blood are highly enriched in long-term culture-initiating cells, NOD/SCID-repopulating cells and dendritic cell progenitors. *Stem Cells* 1998;16(6):387–96.

48. Gordon PR, Leimig T, Babarin-Dorner A, Houston J, Holladay M, Mueller I, *et al.* Large-scale isolation of CD133+ progenitor cells from G-CSF mobilized peripheral blood stem cells. *Bone Marrow Transplant* 2003;31(1):17–22.

49. Gallacher L, Murdoch B, Wu DM, Karanu FN, Keeney M, Bhatia M. Isolation and characterization of human CD34(-)Lin(-) and CD34(+)Lin(-) hematopoietic stem cells using cell surface markers AC133 and CD7. *Blood* 2000;95(9):2813–20.

50. Lang P, Bader P, Schumm M, Feuchtinger T, Einsele H, Fuhrer M, *et al.* Transplantation of a combination of CD133+ and CD34+ selected progenitor cells from alternative donors. *Br J Haematol* 2004;124(1):72–9.

51. Goodell MA, Brose K, Paradis G, Conner AS, Mulligan RC. Isolation and functional properties of murine hematopoietic stem cells that are replicating *in vivo*. *J Exp Med* 1996;183(4):1797–806.

52. Gussoni E, Soneoka Y, Strickland CD, Buzney EA, Khan MK, Flint AF, *et al.* Dystrophin expression in the mdx mouse restored by stem cell transplantation. *Nature* 1999;401(6751):390–4.

53. Goodell MA, Rosenzweig M, Kim H, Marks DF, DeMaria M, Paradis G, *et al.* Dye efflux studies suggest that hematopoietic stem cells expressing low or undetectable levels of CD34 antigen exist in multiple species. *Nat Med* 1997;3(12):1337–45.

54. Zhou S, Schuetz JD, Bunting KD, Colapietro AM, Sampath J, Morris JJ, *et al.* The ABC transporter Bcrp1/ABCG2 is expressed in a wide variety of stem cells and is a molecular determinant of the side-population phenotype. *Nat Med* 2001;7(9):1028–34.

55. Scharenberg CW, Harkey MA, Torok-Storb B. The ABCG2 transporter is an efficient Hoechst 33342 efflux pump and is preferentially expressed by immature human hematopoietic progenitors. *Blood* 2002;99(2):507–12.

56. Ahmed F, Arseni N, Glimm H, Hiddemann W, Buske C, Feuring-Buske M. Constitutive expression of the ATP-binding cassette transporter ABCG2

enhances the growth potential of early human hematopoietic progenitors. *Stem Cells* 2008;26(3):810–8.

57. Storms RW, Trujillo AP, Springer JB, Shah L, Colvin OM, Ludeman SM, *et al.* Isolation of primitive human hematopoietic progenitors on the basis of aldehyde dehydrogenase activity. *Proc Natl Acad Sci USA* 1999; 96(16):9118–23.

58. Hess DA, Meyerrose TE, Wirthlin L, Craft TP, Herrbrich PE, Creer MH, *et al.* Functional characterization of highly purified human hematopoietic repopulating cells isolated according to aldehyde dehydrogenase activity. *Blood* 2004;104(6):1648–55.

59. Hess DA, Wirthlin L, Craft TP, Herrbrich PE, Hohm SA, Lahey R, *et al.* Selection based on CD133 and high aldehyde dehydrogenase activity isolates long-term reconstituting human hematopoietic stem cells. *Blood* 2006;107(5):2162–9.

60. Gratwohl A, Baldomero H, Passweg J, Frassoni F, Niederwieser D, Schmitz N, *et al.* Hematopoietic stem cell transplantation for hematological malignancies in Europe. *Leukemia* 2003;17(5):941–59.

61. Theunissen K, Verfaillie CM. A multifactorial analysis of umbilical cord blood, adult bone marrow and mobilized peripheral blood progenitors using the improved ML-IC assay. *Exp Hematol* 2005;33(2):165–72.

62. Szilvassy SJ, Meyerrose TE, Ragland PL, Grimes B. Differential homing and engraftment properties of hematopoietic progenitor cells from murine bone marrow, mobilized peripheral blood, and fetal liver. *Blood* 2001;98(7):2108–15.

63. Ema H, Takano H, Sudo K, Nakauchi H. *In vitro* self-renewal division of hematopoietic stem cells. *J Exp Med* 2000;192(9):1281–8.

64. Broxmeyer HE, Kohli L, Kim CH, Lee Y, Mantel C, Cooper S, *et al.* Stromal cell-derived factor-1/CXCL12 directly enhances survival/antiapoptosis of myeloid progenitor cells through CXCR4 and G(alpha)i proteins and enhances engraftment of competitive, repopulating stem cells. *J Leukoc Biol* 2003;73(5):630–8.

65. Broxmeyer HE, Cooper S, Kohli L, Hangoc G, Lee Y, Mantel C, *et al.* Transgenic expression of stromal cell-derived factor-1/CXC chemokine ligand 12 enhances myeloid progenitor cell survival/antiapoptosis *in vitro* in response to growth factor withdrawal and enhances myelopoiesis *in vivo*. *J Immunol* 2003;170(1):421–9.

66. Peled A, Petit I, Kollet O, Magid M, Ponomaryov T, Byk T, *et al.* Dependence of human stem cell engraftment and repopulation of NOD/SCID mice on CXCR4. *Science* 1999;283(5403):845–8.

67. Lataillade JJ, Clay D, Dupuy C, Rigal S, Jasmin C, Bourin P, *et al.* Chemokine SDF-1 enhances circulating CD34(+) cell proliferation in synergy with cytokines: possible role in progenitor survival. *Blood* 2000;95(3):756–68.

68. Hidalgo A, Weiss LA, Frenette PS. Functional selectin ligands mediating human CD34(+) cell interactions with bone marrow endothelium are enhanced postnatally. *J Clin Invest* 2002;110(4):559–69.

69. Kahn J, Byk T, Jansson-Sjostrand L, Petit I, Shivtiel S, Nagler A, *et al.* Overexpression of CXCR4 on human CD34+ progenitors increases their proliferation, migration, and NOD/SCID repopulation. *Blood* 2004;103(8):2942–9.

70. Brenner S, Whiting-Theobald N, Kawai T, Linton GF, Rudikoff AG, Choi U, *et al.* CXCR4-transgene expression significantly improves marrow engraftment of cultured hematopoietic stem cells. *Stem Cells* 2004;22(7): 1128–33.

71. Cashman J, Dykstra B, Clark-Lewis I, Eaves A, Eaves C. Changes in the proliferative activity of human hematopoietic stem cells in NOD/SCID mice and enhancement of their transplantability after *in vivo* treatment with cell cycle inhibitors. *J Exp Med* 2002;196(9):1141–9.

72. Yahata T, Ando K, Sato T, Miyatake H, Nakamura Y, Muguruma Y, *et al.* A highly sensitive strategy for SCID-repopulating cell assay by direct injection of primitive human hematopoietic cells into NOD/SCID mice bone marrow. *Blood* 2003;101(8):2905–13.

73. Zon LI. Intrinsic and extrinsic control of haematopoietic stem-cell self-renewal. *Nature* 2008;453(7193):306–13.

74. Reya T, Duncan AW, Ailles L, Domen J, Scherer DC, Willert K, *et al.* A role for Wnt signalling in self-renewal of haematopoietic stem cells. *Nature* 2003;423(6938):409–14.

75. Karlsson G, Blank U, Moody JL, Ehinger M, Singbrant S, Deng CX, *et al.* Smad4 is critical for self-renewal of hematopoietic stem cells. *J Exp Med* 2007;204(3):467–74.

76. Bowie MB, Kent DG, Dykstra B, McKnight KD, McCaffrey L, Hoodless PA, *et al.* Identification of a new intrinsically timed developmental checkpoint that reprograms key hematopoietic stem cell properties. *Proc Natl Acad Sci USA* 2007;104(14):5878–82.

77. Kim I, Saunders TL, Morrison SJ. Sox17 dependence distinguishes the transcriptional regulation of fetal from adult hematopoietic stem cells. *Cell* 2007;130(3):470–83.

78. Bowie MB, Kent DG, Copley MR, Eaves CJ. Steel factor responsiveness regulates the high self-renewal phenotype of fetal hematopoietic stem cells. *Blood* 2007;109(11):5043–8.

79. Lengerke C, Schmitt S, Bowman TV, Jang IH, Maouche-Chretien L, McKinney-Freeman S, *et al.* BMP and Wnt specify hematopoietic fate by activation of the Cdx-Hox pathway. *Cell Stem Cell* 2008;2(1):72–82.

80. Lengerke C, McKinney-Freeman S, Naveiras O, Yates F, Wang Y, Bansal D, *et al.* The cdx-hox pathway in hematopoietic stem cell formation from embryonic stem cells. *Ann N Y Acad Sci* 2007;1106:197–208.

81. Lengerke C, Daley GQ. Patterning definitive hematopoietic stem cells from embryonic stem cells. *Exp Hematol* 2005;33(9):971–9.

82. McKinney-Freeman SL, Lengerke C, Jang IH, Schmitt S, Wang Y, Philitas M, *et al.* Modulation of murine embryonic stem cell-derived CD41+c-kit+ hematopoietic progenitors by ectopic expression of Cdx genes. *Blood* 2008;111(10):4944–53.

83. Lohnes D. The Cdx1 homeodomain protein: an integrator of posterior signaling in the mouse. *Bioessays* 2003;25(10):971–80.

84. Abramovich C, Humphries RK. Hox regulation of normal and leukemic hematopoietic stem cells. Curr *Opin Hematol* 2005;12(3):210–6.

85. Kroon E, Krosl J, Thorsteinsdottir U, Baban S, Buchberg AM, Sauvageau G. Hoxa9 transforms primary bone marrow cells through specific collaboration with Meis1a but not Pbx1b. *EMBO J* 1998;17(13):3714–25.

86. Thorsteinsdottir U, Sauvageau G, Hough MR, Dragowska W, Lansdorp PM, Lawrence HJ, *et al.* Overexpression of HOXA10 in murine hematopoietic cells perturbs both myeloid and lymphoid differentiation and leads to acute myeloid leukemia. *Mol Cell Biol* 1997;17(1):495–505.

87. Magnusson M, Brun AC, Miyake N, Larsson J, Ehinger M, Bjornsson JM, *et al.* HOXA10 is a critical regulator for hematopoietic stem cells and erythroid/megakaryocyte development. *Blood* 2007;109(9):3687–96.

88. Park IK, Qian D, Kiel M, Becker MW, Pihalja M, Weissman IL, *et al.* Bmi-1 is required for maintenance of adult self-renewing haematopoietic stem cells. *Nature* 2003;423(6937):302–5.

89. Lessard J, Sauvageau G. Bmi-1 determines the proliferative capacity of normal and leukaemic stem cells. *Nature* 2003;423(6937):255–60.

90. Lessard J, Schumacher A, Thorsteinsdottir U, van Lohuizen M, Magnuson T, Sauvageau G. Functional antagonism of the Polycomb-Group genes eed and Bmi1 in hemopoietic cell proliferation. *Genes Dev* 1999;13(20):2691–703.

91. Oguro H, Iwama A, Morita Y, Kamijo T, van Lohuizen M, Nakauchi H. Differential impact of Ink4a and Arf on hematopoietic stem cells and their bone marrow microenvironment in Bmi1-deficient mice. *J Exp Med* 2006; 203(10):2247–53.

92. Ernst P, Fisher JK, Avery W, Wade S, Foy D, Korsmeyer SJ. Definitive hematopoiesis requires the mixed-lineage leukemia gene. *Dev Cell* 2004; 6(3):437–43.

93. Galloway JL, Wingert RA, Thisse C, Thisse B, Zon LI. Loss of gata1 but not gata2 converts erythropoiesis to myelopoiesis in zebrafish embryos. *Dev Cell* 2005;8(1):109–16.

94. Rodrigues NP, Janzen V, Forkert R, Dombkowski DM, Boyd AS, Orkin SH, *et al.* Haploinsufficiency of GATA-2 perturbs adult hematopoietic stem-cell homeostasis. *Blood* 2005;106(2):477–84.

95. Zeng H, Yucel R, Kosan C, Klein-Hitpass L, Moroy T. Transcription factor Gfi1 regulates self-renewal and engraftment of hematopoietic stem cells. *EMBO J* 2004;23(20):4116–25.

2

Cord Blood and Cancer

Mary J. Laughlin and R. Patrick Weitzel

ABSTRACT

Umbilical cord blood is a compelling alternative to bone marrow necessitated by the lack of HLA-matched unrelated donors and made possible by the recent proliferation of banking facilities. In this chapter, we discuss the clinical features of umbilical cord blood as observed over 20 years of transplantation for the treatment of hematologic malignancies and summarize several important clinical studies in pediatric and adult recipients. These studies have revealed a particularly low risk of high grade (III–IV) acute graft-versus-host disorder in cord blood recipients, slower myeloid recovery and the importance of strategies for augmenting cell dose, particularly in adult recipients. Current treatment protocols therefore indicate double unit cord blood transplants, which have demonstrated safety following myeloablative and non-myeloablative conditioning regimens. Studies probing the various cellular and molecular bases for these advantages comprise a field of intense inquiry and continue to contribute to the attractiveness of cord blood as a source of allogeneic hematopoietic stem cells.

INTRODUCTION

Umbilical cord blood (UCB) for treatment of hematologic cancers has become widely accepted over the last decade. The number of UCB transplants has increased 200% in each of the past two years, and it is anticipated that the use of UCB for allogeneic transplantation in children and adults with hematologic disorders will continue to increase. UCB currently accounts for approximately 20% of allogeneic hematopoietic stem cell transplants (HSCT) performed. To date, over 10,000 children and adults worldwide have undergone UCB transplants. In addition to hematologic malignancies, UCB has been used to treat patients with marrow failure disorders, immunodeficiencies, sickle cell disease, beta-thalassemia, and inherited metabolic disorders.

HISTORY OF CORD BLOOD TRANSPLANTATION

The first demonstration of hematopoietic engraftment and reconstitution from an HLA-matched sibling UCB unit was shown in a patient with Fanconi anemia in 1988 with UCB collection, transportation, processing, and transplantation performed by an international group of collaborating faculty including Elaine Gluckman and Hal Broxmeyer.[1] Shortly thereafter, Joanne Kurtzberg and her team at Duke University, and J. Wagner and his collaborators at University of Minnesota, independently demonstrated the safety and feasibility of unrelated UCB grafting,[2–4] setting the stage for years of work confirming the therapeutic potential of UCB even in an HLA-mismatched setting.

UCB was believed to contain functional hematopoietic progenitors as early as 1974.[5] Two years earlier, a report by Ende and Ende[6] had demonstrated transient red cell chimerism following infusion of eight UCB donor samples into a 16-year-old male with acute lymphoblastic leukemia (ALL). Later reports by multiple groups[7] demonstrated multilineage hematopoietic precursor potential of UCB-derived cells by *in vitro* methylcellulose culture. In 1983, maintenance of these precursors' function in formation of multilineage colony-forming units was demonstrated following cryopreservation.[8] Thus, sufficient rationale was available by the early 1980s that UCB could comprise a suitable source of hematopoietic stem cells for clinical applications which, in addition to numerous other practical advantages, could potentially be stored frozen and recovered on-demand. In the ensuing decades, numerous large studies have confirmed the engraftment potential and hematopoietic reconstitution capabilities of UCB, and in addition have revealed multiple unique clinical advantages of UCB as an alternative source of stem cells in patients lacking an available histocompatible sibling allogeneic donor.

ADVANTAGES OF CORD BLOOD AS A SOURCE OF ALLOGENEIC STEM CELLS

UCB addresses several challenges inherent to allogeneic stem cell transplantation. Complete or partially HLA-compatible bone marrow (BM) units from adult donors remain difficult to match with US patients of certain ethnic backgrounds, particularly those of African, Asian, or Hispanic decent, as the adult donor registry is comprised primarily of Caucasians. However, as birthing accurately reflects the genetic background of a given population, development of local collection and banking strategies will effectively address this under-representation of ethnic minority populations. Indeed, compatible UCB grafts can now be located for 99% of patients for whom transplant physicians submit requests[9] via existing banks and registries.

Cord blood is collected fresh following delivery of a full-term infant, avoiding ethical implications associated with use of embryonic stem cells and eliminating risk or pain to the bone marrow donor. UCB may also be collected *in utero* during the final phase of labor or *ex utero*, and there are advantages and disadvantages to each approach with respect to invasiveness, number of cells collected, and incidence of microbial contamination. With use of either approach, collection of UCB stem cells does not interfere with the normal birthing process. UCB can be harvested following vaginal or caesarean delivery. Typically, following draining of the umbilical vein, the unit is depleted of red blood cells by density gradient separation, facilitating optimal conditions for hematopoietic stem cell viability during freezing and thaw and a reduction in unit volume, with advantages for storage and transport. UCB viability may be maintained in liquid nitrogen storage potentially indefinitely, with testing to date verifying viability after 15 years of liquid nitrogen storage.[10] Thus, units found to be clinically suitable can be stored for use whenever needed, without further involvement of the donor or family, permitting a shorter timeframe between donor identification and availability to the treating physician than with adult-derived cells, and eliminating donor attrition via death, illness, or relocation. This shorter timeframe is believed to be potentially life-saving for extremely high risk adult leukemia patients in need of immediate HSC transplant.

The use of a patient's own stem cells has historically been considered for cellular replacement therapies, but the use of autologous BM or mobilized peripheral blood stem cells (PBSCs) is limited due to diminishing stem cell number and function with age,[11-13] bone marrow involvement in certain types of leukemia, absence of immunologically mediated effects to control malignant cell re-growth, and invasiveness of the collection procedure. UCB effectively addresses these concerns.

In spite of the fact that HLA-matched adult bone marrow availability now approaches 75% for Caucasians, UCB carries several clinical advantages which may render it preferable to allogeneic BM or mobilized PBSCs in certain circumstances, particularly in a 4–5/6 HLA-matched setting. Perhaps the most notable of these advantages is the decrease in incidence and severity of acute graft-versus-host disease (GVHD) despite HLA disparity. The first evidence that UCB carries a lower risk of GVHD compared to HLA-matched bone marrow came about through a rather remarkable series of events in 1992. Wagner *et al.* transplanted a sibling UCB graft into a pediatric CML patient.[3] Shortly afterwards, the patient developed characteristic skin rash prompting histological analysis for acute GVHD, which was found to be negative. The patient later relapsed after seven months, and was then treated with a BM graft from the same donor sibling. This graft produced high-grade acute GVHD. Since these early observations, many clinical studies (discussed below) have confirmed that UCB grafts indeed elicit lower incidence and severity of acute GVHD.

Despite the clearly reduced incidence of acute GVHD, rates of malignant relapse have remained comparable to HLA-matched unrelated BM/PBSC transplants and the observed graft-versus-leukemia (GVL) effect has appeared robust in most trials, although the mechanisms underlying GVL activity following UCB engraftment remain unclear. Notably, earlier reports arose from trials involving patients with markedly advanced or non-responsive malignant disease, suggesting highly acceptable outcomes despite generally poor prognosis. The mechanisms by which the neonate maintains a naïve immune system, including reduced mature lymphocyte frequency, as well as impaired expression and function of activation co-receptors and signaling molecules associated with inflammation, are believed to contribute to reduced acute GVHD in the unrelated allogeneic setting and are discussed later in this chapter.

However, the UCB transplant field is challenged by the limitation of low graft cell dose. As UCB is a one-time donation with limited cell volume, the cell dose for reliable engraftment is sometimes insufficient for adults. In order to attain a 90% likelihood of engraftment, $2.5-3 \times 10^7$ of total nuclear cells (TNC) per kilogram of patient mass is generally used as a threshold. The kinetics of myeloid and platelet recovery after UCB transplantation have consistently shown to lag behind allo-BM/PBSCs (89% at a median of 26 days to neutrophil recovery versus 98% at a median of 18 days).[14] Rates and kinetics of donor-derived neutrophil recovery in UCB recipients have been shown in some trials to correlate with total nucleated and hematopoietic progenitor content, although there appears to be a point of diminishing returns where increased cell dose no longer correlates with further improvement in long-term engraftment outcome.[15] Remarkably, although engraftment rates are similarly favorable, the average total nucleated and CD34$^+$ cell content of UCB units is approximately ten times less than in adult-derived stem cell sources.[16] This rate of engraftment despite low UCB graft cell doses may be in part attributable to the significantly higher proliferative capacity of UCB-derived hematopoietic progenitors[17] and/or unique features of UCB graft T- and hematopoietic stem cells allowing adherence and transmigration across vasculature and occupancy of stem cell niches. Attempts to enhance hematopoietic progenitor cell dose *ex vivo* have to date been largely unsuccessful, but remain ongoing.[18] Additionally, T-cell depletion of UCB units may be detrimental to engraftment outcomes and GVL,[19] suggesting an important role for non-progenitor donor cells in facilitating engraftment. However, recent trials have indicated that co-transplantation of two allogeneic UCB units is safe and may overcome graft cell dose limitations, thereby permitting use of lower-intensity non-myeloablative conditioning regimens.[20–22]

Although neonatal immune tolerance underlies many of the chief clinical advantages of UCB, the chance of hematological abnormalities which may not yet be readily apparent in the newborn donor to be transmitted to the transplant recipient may comprise a potential risk of UCB compared to adult-derived allo-BM/PBSCs. Additionally, also potentially impacting on the kinetics of immune reconstitution is the risk of post-transplant infection at early time points (prior to day 100). The latter is relatively high, likely due to delayed neutrophil and platelet recovery, but also perhaps related to the naïve immunological phenotype of the graft T-cell population.[23]

CLINICAL TRIALS TO DATE — PEDIATRIC RECIPIENTS

Multi-institute outcomes for UCB initially reported on a cohort of HLA matched and mismatched sibling donors in 1995 and was updated in 1998.[4,24] Among 74 total patients receiving sibling UCB grafts (range 0.5–16.3, median 4.9 years), 56 patients received HLA 0-1 antigen mismatched grafts and 18 received two or three HLA-mismatched grafts. For recipients of 0-1 HLA-mismatched grafts, the probability of neutrophil recovery (absolute neutrophil count of $>5 \times 10^8$/L) was 91% ($\pm$2) at 60 days after transplantation at a median of 22 days (range: 9–46). Despite overall high engraftment rates, there was a trend toward a greater risk of graft failure in recipients with non-malignant hematologic disorders including marrow failure syndromes, hemoglobinopathies, or inherited metabolic disorders. Importantly, the probability of grade II–IV and grade III–IV GVHD was only 3% ($\pm$2) and 2% ($\pm$2), respectively. No extensive chronic GVHD was observed. At median follow-up of two years, survival was 61% ($\pm$12) in recipients of 0-1 HLA-antigen mismatched UCB grafts.

Gluckman et al. subsequently reported transplant outcomes in 74 recipients of related UCB.[25] The median age of the population was five years (range 0.2–20) with 46 patients having malignancy, 17 with BM failure syndromes, eight hemoglobinopathies and seven inborn errors of metabolism. Sixty of the 74 patients received HLA-identical grafts. In contrast to the prior report, the probability of neutrophil engraftment was only 79%. More favorable myeloid recovery and engraftment correlated with younger recipient age (p = 0.02), lower recipient body weight (p = 0.02), and HLA-identity (p = 0.04). Higher nucleated cell dose (p = 0.06) additionally may have been predictive of improved outcome, comprising the first dataset suggestive of a possible relationship between TNC dose and engraftment. The probability of grade II–IV GVHD was 9% in recipients of HLA-matched UCB, with chronic GVHD observed only in eight of 56 patients surviving beyond day 100 after transplantation. One-year survival was 63%, and was shown to statistically correlate to the same factors which favorably influenced engraftment.

In the absence of a prospective randomized trial, a retrospective analysis of registry data was performed by Rocha *et al.*[26,27] in order to assess the relative risks of delayed myeloid recovery or graft failure, GVHD, and mortality between UCB and adult donor HSC sources. One hundred and thirteen HLA-matched UCB recipients were compared with 2052 HLA-matched BM recipients. This report initially documented the delayed myeloid recovery in UCB recipients and also demonstrated for the first time that UCB recipients indeed had lower incidence and severity of GVHD than recipients of adult-derived grafts. Risk of grade II–IV acute GVHD was 14% in UCB recipients compared to 24% in BM recipients (p = 0.02). Survival rates after UCB and BM transplantation were similar at three years (64% versus 66%, p = 0.93). In this analysis, despite the lower demonstrated incidence of acute GVHD, relapse as a cause of death was no different in recipients of UCB compared to recipients of BM, suggesting that the GVL effect of the UCB graft was intact.

As a result of these promising findings with sibling donor UCB transplantation, cord blood banking programs were rapidly initiated and expanded throughout the US and Europe. These efforts ultimately facilitated the first two reports on the use of unrelated UCB transplantation in children in 1996.[28,29] These two studies clearly demonstrated that hematopoietic recovery and sustained engraftment could be achieved. Furthermore, incidence and severity of acute GVHD were surprisingly low, despite the fact that the majority of patients received one to two HLA-mismatched UCB units. Gluckman *et al.* later reported the outcomes in 65 patients (median age nine years) transplanted with unrelated UCB.[25] Similar to her observations in sibling UCB recipients, the probability of myeloid recovery was 87%. Notably, the UCB recipients who received a median cell dose of more than 3.7×10^7 nucleated cells/kg were more likely to have faster myeloid engraftment (median 25 days versus 35 days) and higher probability of myeloid recovery (94% versus 76%), demonstrating the importance of UCB nucleated cell dose for the first time. Cell dose may not have been identified as a key factor initially due to the more limited number of patients and the narrow range of TNC doses infused.

In a large landmark study by Rubinstein *et al.*,[30] patients with hematological malignancies (67%), genetic disease (24%) or acquired BM failure (9%) demonstrated an overall probability of myeloid recovery of 93% after unrelated UCB transplantation. This report confirmed the importance of UCB cell dose on engraftment, risk of adverse events and survival, although the incidence of myeloid recovery did not significantly improve once the nucleated cell dose exceeded 2.5×10^7/kg of body weight. HLA identity (HLA 0 versus >1 mismatch) was also associated with myeloid engraftment time (23 days as compared to 28 days, p = 0.0027). Notably, engraftment was similar in recipients of one versus two versus three antigen mismatched UCB grafts.

Correlation of acute GVHD risk with HLA-matching varied between reports involving unrelated donors. The study reported by Gluckman *et al.* and others failed to detect any difference in risk of GVHD in recipients of HLA 5/6 versus 4/6 matched UCB grafts. Rubinstein *et al.* found that risk of grade III–IV GVHD increased with increasing of mismatch (8% in 6/6 HLA-matched UCB, 19% in 5/6 HLA-matched UCB, and 28% in 4/6 HLA-matched UCB). However, it is now generally agreed that the risk of GVHD is considerably lower than anticipated at the level of HLA-mismatch in most UCB recipients compared with recipients of unrelated adult donor grafts.

In terms of survival, Gluckman *et al.* reported survival of 34% at one year which was adversely affected in recipients of older age, female gender, and advanced disease stage.[16] Notably, neither degree of HLA-mismatching nor cell dose was shown to influence survival in that study. Rubinstein *et al.* reported three-year survival at 48% in patients with genetic disease, and 27% in patients with hematological malignancy. Risk factors for adverse transplant-related events again included UCB graft low cell dose and HLA-mismatch. This comprised the first dataset where both engraftment and survival of the graft recipients was demonstrated to be influenced by cell dose and HLA disparity.

Eapen *et al.* recently conducted a comprehensive comparison study through collaborative efforts between the New York Blood Center (NYBC) and the Center for International Blood and Marrow Transplantation Research (CIBMTR) comparing the outcomes of children with acute leukemia who received HLA matched and mismatched UCB (n = 503) or 8/8 allele HLA-matched unrelated donor (MUD) marrow (n = 116).[31] All children undergoing myeloablative transplantation were younger than 16 years. The UCB recipients tended to be younger in age, more likely non-white, in relapse before transplantation, and to be those who received HLA-mismatched grafts. The median times to neutrophil and platelet recovery were slower in UCB recipients (25 days and 59 days respectively) compared to marrow recipients (19 days and 27 days). The probability of neutrophil recovery at day 42 was lower in UCB mismatched recipients regardless of the graft cell dose, compared to marrow recipients. However, the probability of neutrophil recovery was not different between matched UCB recipients and marrow recipients (HLA matched and mismatched). In comparison with allele-matched BM transplants, five-year leukemia-free survival (LFS) was similar to that after transplants with UCB HLA mismatched at one or two loci with potentially superior results in recipients of HLA-matched UCB. For recipients of 8/8 allele HLA-matched marrow, mismatched marrow, 6/6 antigen HLA-matched UCB, 5/6 antigen HLA-matched UCB (cell dose >3.0 × 10^7 NC/kg) and 4/6 antigen HLA-matched UCB (any cell dose), the adjusted LFS rates were 38%, 37%, 60%, 45%, and 33%, respectively. These data from the NYBC and the CIBMTR provocatively suggest that HLA-matched or high dose mismatched UCB transplantation can potentially be a front line therapy for pediatric acute leukemia patients even when HLA-matched marrow donors are available.

CLINICAL TRIALS TO DATE — ADULT RECIPIENTS

Considerably less data exists for adult recipients of UCB grafts compared to pediatric recipients. The first series of reports of unrelated UCB transplantation experience in adults were published together with pediatric populations in 1996. Reports specifically focused on the outcomes of UCB transplantation in adults were not published until 2000 and 2001. We reported the outcome of 68 adults (median age 31, range 17–58) with high-risk hematologic diseases who received 3–6/6 HLA-matched UCB grafts.[32] The probability of myeloid recovery was 90% at a median of 27 days. The incidence of grades II–IV and III–IV acute GVHD was higher than those previously reported in children (60% and 16%, respectively), and not associated with UCB graft HLA disparity. The incidence of chronic GVHD was 33% for those survivors beyond day 100. Event free survival was only 26%. Importantly, higher survival was observed in recipients who received CD34$^+$ cell dose $\geq 1.2 \times 10^5$/kg. Rocha et al. reported similar results in 108 adults (median age 26, range 15–53) with hematological malignancies, who underwent unrelated HLA-mismatched UCB transplantation.[27]

Across all trials including those with particularly high-risk patients with acute leukemia, event free survival (EFS) has ranged from 15% to 50%, with some small series reporting up to 70% EFS.[32–41] Patients with younger age, receiving higher UCB cell doses (nucleated cells, CD34, CFU-GM), with pre-transplant cytomegalovirus (CMV) serum negativity, and pre-transplant complete remission status had better outcomes. It is generally accepted that for many of these trials, data may be skewed unfavorably due to the poorer pre-transplant prognosis of patients, lower minimum cell dose compared with current standards (1×10^7/kg versus 2.5×10^7/kg), and higher degree of mismatching due to the lower initial availability of graft units with suitable cell dose for adult recipients.

We demonstrated lower incidence of acute GVHD in mismatched UCB grafts compared to mismatched BM (hazard ratio = 0.66; p = 0.04) in an analysis comparing adult recipients of UCB and unrelated bone marrow.[42] In this study, although UCB recipients received lower TNC doses and exhibited slower hematopoietic recovery, the rates of treatment-related mortality, graft failure, relapse, and overall mortality were statistically comparable between the two groups, and the adjusted probability of overall survival at three years was slightly higher in UCB (20% versus 26%, p = 0.62) (Figs. 1 and 2). Notably, no difference in outcome was observed comparing the recipients of 5/6 and 4/6 HLA-matched UCB.

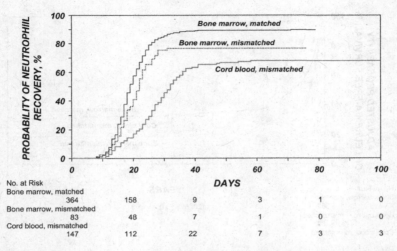

No. at Risk
Bone marrow, matched
| | 364 | 158 | 9 | 3 | 1 | 0 |

Bone marrow, mismatched
| | 83 | 48 | 7 | 1 | 0 | 0 |

Cord blood, mismatched
| | 147 | 112 | 22 | 7 | 3 | 3 |

Fig. 1. Cumulative incidence of neutrophil and platelet recovery after bone marrow and cord-blood transplantation. Despite early differences, the cumulative incidence of neutrophil recovery at day 100 was similar after the transplantation of mismatched bone marrow and of cord blood. Reprinted from Ref. 42, with kind permission from the Massachusetts Medical Society. Copyright © [2004] Massachusetts Medical Society. All rights reserved.

Thus, despite the early lack of large datasets, these studies suggested that HLA-mismatched cord blood should also be considered an acceptable graft source for adults in the absence of an HLA-matched adult donor.

DOUBLE UCB TRANSPLANTATION

The Minnesota group (Wagner and colleagues) has pioneered utilization of double UCB grafts in order to overcome the cell dose limitation in larger patients. Brunstein *et al.* recently updated their largest single institution experience in nearly 200 double UCB transplantation in the myeloablative and non-ablative settings.[22] The data included advanced stage hematologic malignancies in pediatric and adult patients and indicated that double UCB transplantation is technically feasible, clinically safe, and increased eligibility of adult patients for transplantation. The overall clinical outcomes (incidence of GVHD and survival) were comparable in patients that received double versus single UCB transplantation in a myeloablative setting. Interestingly, in analyzing the clinical

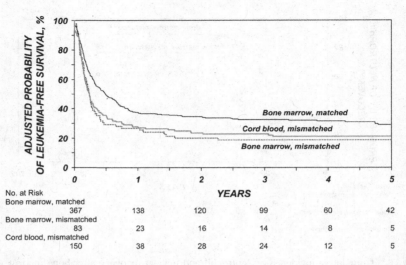

Fig. 2. Adjusted probability of leukemia-free survival after bone marrow and cord-blood transplantation. The adjusted probability of three-year survival without a recurrence of leukemia was 19% for recipients of mismatched marrow, 23% for recipients of cord blood, and 33% for recipients of HLA-matched marrow. Probabilities were adjusted for age, disease status at transplantation, and positivity for cytomegalovirus in the donor, recipient, or both. Reprinted from Ref. 42, with kind permission from the Massachusetts Medical Society. Copyright © [2004] Massachusetts Medical Society. All rights reserved.

outcomes in acute leukemia UCB recipients, the authors found that acute leukemia patients who received double UCB transplantation had a ten-fold decrease in the risk of leukemia relapse compared to patients who received single UCB transplantation. This might be attributed to a relatively higher degree of HLA-mismatch among double UCB recipients of whom 80% had at least one UCB unit with two antigens mismatched, promoting GVL effects. A randomized clinical trial has been designed to further confirm the clinical merit of double UCB transplantation in acute leukemia patients.

NON-MYELOABLATIVE UCB TRANSPLANTATION IN ADULTS

Elderly patients with significant co-morbidities and patients that have relapsed after extensive chemotherapy or prior stem cell transplantation may have unacceptable peri-transplantation risk. Again, Brunstein *et al.* from the Minnesota group recently reported on 110 adult patients with hematologic diseases.[22] The preparative regimens were largely low dose total body irradiation (TBI), Fludarabine and ATG based and the target UCB cell dose was 3.0×10^7/kg. About 85% patients needed to receive a second partially HLA-matched UCB. Neutrophil recovery was 92% at a median of 12 days. Incidences of grade III and IV acute and chronic GVHD were 22% and 23%, respectively. Transplantation-related mortality was 26% at three years. Overall and event-free survival (EFS) was 45% and 38%, respectively. Absence of high-risk clinical features and severe GVHD were favorable factors for overall survival, whereas absence of high-risk clinical feature and use of two UCB units were favorable factors for EFS.

Another study was reported by Miyakoshi *et al.* in mostly adult AML patients who were conditioned by non-myeloablative regimen (Fludarabine/melphalan/low-dose TBI). The one year overall survival was 33% with a relapse rate of 11%.[43] In general, the non-myeloablative UCB data are less well defined compared to myeloablative UCB transplantation.

CELLULAR AND MOLECULAR OBSERVATIONS UNDERLYING CLINICAL ADVANTAGES

Although CD34[+] cell quantification in UCB has not consistently been predictive of time to donor hematopoietic engraftment, some data may be confounded by quantification of CD34[+] cells in UCB grafts pre-freezing versus post-thaw and by reduced surface epitope density of CD34 expression on UCB progenitor cells. *In vitro* analyses of UCB CD34[+] cells point to a less mature phenotype compared to adult marrow and peripheral blood grafts.[44] Cobblestone area-forming cell (CAFC) assays show that UCB CD34[+] cells contain the highest frequency of CAFC at week 6 (3.6- to 10-fold higher than BM CD34[+] cells and peripheral blood stem cells, respectively), and the engraftment capacity *in vivo* in non-obese diabetic/severe combined immunodeficiency (NOD/SCID) mice is also significantly greater than BM CD34[+] cells.[45]

Engraftment facilitating activity of donor lymphocytes contained in the UCB transplant may also inhibit or eliminate residual recipient immune cells capable of graft rejection.[46] Indeed, T-cell depletion of UCB units is believed to be detrimental to engraftment outcomes and GVL function.[19] This may be of particular relevance in adults transplanted with low graft CD34$^+$ and nucleated UCB cell doses. Additionally, risk of post-transplant infection at early time points (prior to day 100) is high in some trials,[23] which may be potentially attributable to the naïve immunological phenotype of the graft T-cell population, impacting the early kinetics of immune reconstitution.[47,48] Larger retrospective studies, however, have indicated the risk of severe post-transplant infections following UCB grafts is similar to BM.[34] A complete understanding of the mechanisms by which the cord blood graft balances GVL activity with immunological tolerance remains unclear, but there is evidence to suggest that both the specific lymphocyte populations as well as impaired inflammatory lymphocyte signaling mechanisms within specific subsets play a role.

Compared to adult peripheral blood, UCB contains a significantly lower percentage of functionally cytotoxic CD8$^+$ T-cells, and UCB lymphocytes have a phenotype characteristic of T-cell immaturity.[49] The resting common leukocyte antigen CD45RA is co-expressed on >90% of UCB CD4$^+$ T-cells (compared to 40% of adult CD4$^+$ T-cells), and a greater proportion of UCB CD3$^+$ cells have been shown to express CD25/IL2-R (8% versus 18%) and the mature activation marker CD45RO (2% versus 10%) at lower levels relative to adult blood (AB).[50] UCB CD4$^+$/45RA$^+$ T-cell subsets have been demonstrated to show markedly decreased helper function and increased suppressive activity when compared to the same cells obtained from adult peripheral blood by several groups. Additional studies suggest higher proportion of NK-cell differentiation in UCB compared to AB.[51–53] Further mechanisms potentially underlying UCB immune tolerance may include altered toll-like receptors and adhesion molecule expression on donor graft antigen-presenting cells.[54] These observations suggest the "naïve" features of neonatal T-cells in UCB grafts might contribute to the observed immune tolerance in the allogeneic transplant setting and allow successful outcomes despite donor-recipient HLA disparity and reduced graft cellular content.

A relatively high incidence of CD4$^+$, CD25^{hi} and FoxP3$^+$ T-regulatory cells has been observed in UCB as well.[55] These cells have been confirmed to impair the activation and expansion of alloreactive helper and cytotoxic T-cells and contribute to self-tolerance and repression of GVHD in mouse xenograft models.[56–58] Likewise, graft T-cells in UCB produce significantly less IFN-γ, TNF-α, and more IL-12 and IL-10 than their AB counterparts,[59,60] perhaps due to impaired calcium signaling, impaired expression of lymphocyte receptors such as IL-2R and CTLA4,[61] and dramatically reduced expression of transcription factors including nuclear factor of activated T-cells (NFAT1) known to regulate their gene expression.[62]

CURRENT INDICATIONS

UCB is now the primary source of unrelated HSCs for pediatric recipients. For pediatric patients with ALL and AML, leukemia-free and long-term survival across studies has ranged from 30% to 50%. Survival for pediatric CML patients is considerably less (approximately 25%), although less data exists and patients selected for these trials in general exhibited more aggressive disease. Although many studies have correlated advanced disease status with increased risk of relapse, relapse rates for pediatric CML recipients of UCB remain low (15%). Multiple studies indicate similar rates of long-term overall survival for pediatric ALL patients when comparing recipients of UCB and BM. UCB has been confirmed by multiple groups as comparable or superior to mismatched BM in treatment of adult malignancies.

The main challenge in UCB transplantation however, particularly in adults, is graft cell dose. The number of hematopoietic progenitors in a single cord blood unit may be insufficient for adult recipients, and current unit selection strategies prioritize maximization of cell dose similarly with HLA-identity. One strategy employed to overcome potential drawbacks in cell dose which has exhibited much greater successes than *ex vivo* expansion efforts has been the infusion of multiple UCB units. The first trials in 2005[34] with fully-myeloablative double cord blood transplants (DCBT) reported a slightly earlier median time to myeloid engraftment without increase in GVHD or graft failure over single-unit therapy. Later studies from the same group have suggested a potential benefit in relapse rates when comparing double- and single-UCB treatments at high (>2.5 × 10^7/kg) cell doses. Disease-free survival across studies has shown to be over 50% in some studies between ten and 22 months.

DCBT may have particular advantages as a graft source option where non-myeloablative conditioning is required for an older patient with more advanced disease and increased co-morbidities. Numerous studies have confirmed relative safety of DCBT following myeloablative or non-myeloablative conditioning,[21,22,34,63] and data across studies indicates GVL functionality without increased rates of GVHD, although a statistical confirmation of DCBT benefits compared to single-unit therapy has yet to be fully realized in the non-myeloablative setting. Notably, DCBT trials have not revealed immunological complications characteristic of a "graft-versus-graft" reaction, although it is clear that only one cord will eventually go on to completely repopulate the recipient long-term. Studies to predict which cord will predominate have been fairly inconclusive, although higher nucleated and progenitor cell dose as well as order of infusion have been suggested.[64] A recent study of 96 patients has also suggested a reduced risk of relapse with DCBT[65] without a statistical association with cell dose. Unit-to-unit HLA matching may also be of importance and multiple transplant groups require that UCB units be at least 4/6 matched to each other in addition to the recipient.

Currently, two positions have emerged at North American institutions for UCB indications. Transplant teams at the University of Minnesota advocate 4–6/6 matched UCB as a front-line therapy for acute leukemia in the absence of a fully HLA-matched marrow donor, particularly in pediatric patients. Transplant physicians at Case Western Reserve University generally defer to 10/10 HLA-matched adult derived marrow or mobilized peripheral blood stem cell sources when available, although if a suitably matched UCB unit with a large ($>2.5 \times 10^7$/kg) cell count is available or the transplant is particularly urgent, UCB may be prioritized, thereby reducing the risk of adverse events arising from HLA-mismatch.

Both groups have stipulated protocols advocating use of two UCB units for transplant into patients with malignancies. The Minnesota position stipulates two units from partially HLA-matched donors due to the possible observed augmentation of the GVL effect for malignant diseases. Single UCB-units are still preferred for treatment of non-malignant conditions. For DCBT, both groups require the individual UCB units to be at least 4/6 matched to each other. However, the Case group takes into account higher-resolution HLA-typing, showing preference to genetic identity at the DR, B, and A loci for single-unit transplants. For patients requiring DCBT due to cell dose considerations, priority is given to units which share the greatest HLA-homology with the patient, given that they meet the 4/6 matching requirement to each other, and each have a minimum TNC count pre-cryopreservation of 1.5×10^7/kg.

SUMMARY AND CONCLUSIONS

Over the past two decades, UCB has become an increasingly attractive graft choice for hematopoietic stem cell transplantation. Numerous studies have confirmed its feasibility and safety for the treatment of malignant as well as non-malignant diseases. Although cell dose comprises a significant challenge for treatment of adults, innovative strategies such as *ex vivo* expansion, DCBT, and non-myeloablative conditioning regimens promise to improve outcomes and increase the flexibility of treatment options available to the physician. UCB brings additional advantages to the bedside, particularly reduced risk of grade III–IV GVHD. As availability of UCB units increases through national, regional, and local banking strategies, we anticipate that the hurdles to the broader application of allogeneic hematopoietic stem cell transplant posed by matched-unit availability will be effectively surpassed.

References

1. Gluckman E, Broxmeyer HA, Auerbach AD, Friedman HS, Douglas GW, Devergie A, *et al.* Hematopoietic reconstitution in a patient with Fanconi's anemia by means of umbilical-cord blood from an HLA-identical sibling. *N Engl J Med* 1989;321(17):1174–8.
2. Kurtzberg J, Graham M, Casey J, Olson J, Stevens CE, Rubinstein P. The use of umbilical cord blood in mismatched related and unrelated hemopoietic stem cell transplantation. *Blood Cells* 1994;20(2–3):275–83; discussion 84.
3. Wagner JE, Broxmeyer HE, Byrd RL, Zehnbauer B, Schmeckpeper B, Shah N, *et al.* Transplantation of umbilical cord blood after myeloablative therapy: analysis of engraftment. *Blood* 1992;79(7):1874–81.
4. Wagner JE, Kernan NA, Steinbuch M, Broxmeyer HE, Gluckman E. Allogeneic sibling umbilical-cord-blood transplantation in children with malignant and non-malignant disease. *Lancet* 1995;346(8969):214–9.

5. Knudtzon S. *In vitro* growth of granulocytic colonies from circulating cells in human cord blood. *Blood* 1974 Mar;43(3):357–61.
6. Ende M, Ende N. Hematopoietic transplantation by means of fetal (cord) blood. A new method. *Virginia Med Mon* 1972 Mar;99(3):276–80.
7. Nakahata T, Ogawa M. Hemopoietic colony-forming cells in umbilical cord blood with extensive capability to generate mono- and multipotential hemopoietic progenitors. *J Clin Invest* 1982 Dec;70(6):1324–8.
8. Koike K. Cryopreservation of pluripotent and committed hemopoietic progenitor cells from human bone marrow and cord blood. *Pediatr Int* 1983;25(3):275–83.
9. Rubinstein P. Why cord blood? *Human Immunol* 2006 Jun;67(6):398–404.
10. Broxmeyer HE, Srour EF, Hangoc G, Cooper S, Anderson SA, Bodine DM. High-efficiency recovery of functional hematopoietic progenitor and stem cells from human cord blood cryopreserved for 15 years. *Proc Natl Acad Sci USA* 2003 Jan 21;100(2):645–50.
11. Chambers SM, Goodell MA. Hematopoietic stem cell aging: wrinkles in stem cell potential. *Stem Cell Rev* 2007 Fall;3(3):201–11.
12. Chambers SM, Shaw CA, Gatza C, Fisk CJ, Donehower LA, Goodell MA. Aging hematopoietic stem cells decline in function and exhibit epigenetic dysregulation. *PLoS Biol* 2007 Aug;5(8):e201.
13. Rossi DJ, Bryder D, Weissman IL. Hematopoietic stem cell aging: mechanism and consequence. *Exp Gerontol* 2007 May;42(5):385–90.
14. Rocha V, Cornish J, Sievers E, Filipovich A, Locatelli F, Peters C, *et al.* Comparison of outcomes of unrelated bone marrow and umbilical cord blood transplants in children with acute leukemia. *Blood* 2001;97(10):2962–71.
15. Wagner JE, Barker JN, DeFor TE, Baker KS, Blazar BR, Eide C, *et al.* Transplantation of unrelated donor umbilical cord blood in 102 patients with malignant and nonmalignant diseases: influence of CD34 cell dose and HLA disparity on treatment-related mortality and survival. *Blood* 2002 Sep 1;100(5):1611–8.
16. Gluckman E, Rocha V, Arcese W, Michel G, Sanz G, Chan KW, *et al.* Factors associated with outcomes of unrelated cord blood transplant: guidelines for donor choice. *Exp Hematol* 2004 Apr;32(4):397–407.
17. Theunissen K, Verfaillie CM. A multifactorial analysis of umbilical cord blood, adult bone marrow and mobilized peripheral blood progenitors using the improved ML-IC assay. *Exp Hematol* 2005 Feb;33(2):165–72.

18. Hofmeister CC, Zhang J, Knight KL, Le P, Stiff PJ. *Ex vivo* expansion of umbilical cord blood stem cells for transplantation: growing knowledge from the hematopoietic niche. *Bone Marrow Transplant* 2007 Jan;39(1):11–23.

19. Martin PJ, Hansen JA, Torok-Storb B, Durnam D, Przepiorka D, O'Quigley J, *et al.* Graft failure in patients receiving T cell-depleted HLA-identical allogeneic marrow transplants. *Bone Marrow Transplant* 1988 Sep;3(5): 445–56.

20. Ballen KK, Hicks J, Dharan B, Ambruso D, Anderson K, Bianco C, *et al.* Racial and ethnic composition of volunteer cord blood donors: comparison with volunteer unrelated marrow donors. *Transfusion* 2002 Oct;42(10):1279–84.

21. Barker JN, Davies SM, DeFor T, Ramsay NK, Weisdorf DJ, Wagner JE. Survival after transplantation of unrelated donor umbilical cord blood is comparable to that of human leukocyte antigen-matched unrelated donor bone marrow: results of a matched-pair analysis. *Blood* 2001 May 15;97(10):2957–61.

22. Brunstein CG, Barker JN, Weisdorf DJ, DeFor TE, Miller JS, Blazar BR, *et al.* Umbilical cord blood transplantation after nonmyeloablative conditioning: impact on transplantation outcomes in 110 adults with hematologic disease. *Blood* 2007 Oct 15;110(8):3064–70.

23. Safdar A, Rodriguez GH, De Lima MJ, Petropoulos D, Chemaly RF, Worth LL, *et al.* Infections in 100 cord blood transplantations: spectrum of early and late post-transplant infections in adult and pediatric patients 1996–2005. *Medicine* 2007 Nov;86(6):324–33.

24. Wagner JE, Kurtzberg J. Banking and transplantation of unrelated donor umbilical cord blood: status of the National Heart, Lung, and Blood Institute-sponsored trial. *Transfusion* 1998 Sep;38(9):807–9.

25. Gluckman E, Rocha V, Boyer-Chammard A, Locatelli F, Arcese W, Pasquini R, *et al.* Outcome of cord-blood transplantation from related and unrelated donors. Eurocord Transplant Group and the European Blood and Marrow Transplantation Group. *N Engl J Med* 1997;337(6):373–81.

26. Rocha V, Arcese W, Sanz G, *et al.* Prognostic factors of outcome after unrelated cord blood transplant (UCBT) in adults with hematologic malignancies. *Blood* 2000;96(Suppl. 1):587a.

27. Rocha V, Wagner JE, Jr, Sobocinski KA, Klein JP, Zhang MJ, Horowitz MM, *et al.* Graft-versus-host disease in children who have received a cord-blood or

bone marrow transplant from an HLA-identical sibling. Eurocord and International Bone Marrow Transplant Registry Working Committee on Alternative Donor and Stem Cell Sources. *N Engl J Med* 2000 Jun 22;342(25):1846–54.

28. Kurtzberg J, Laughlin M, Graham ML, Smith C, Olson JF, Halperin E, *et al.* Placental blood as a source of hematopoietic stem cells for transplantation into unrelated recipients. *N Engl J Med* 1996;335(3):157–66.

29. Wagner JE, Rosenthal J, Sweetman R, Shu XO, Davies SM, Ramsay NK, *et al.* Successful transplantation of HLA-matched and HLA-mismatched umbilical cord blood from unrelated donors: analysis of engraftment and acute graft-versus-host disease. *Blood* 1996;88(3):795–802.

30. Rubinstein P, Carrier C, Scaradavou A, Kurtzberg J, Adamson J, Migliaccio A, *et al.* Outcomes among 562 recipients of placental-blood transplants from unrelated donors. *New Engl J Med* 1998;339(22):1565–77.

31. Eapen M, Rubinstein P, Zhang MJ, Stevens C, Kurtzberg J, Scaradavou A, *et al.* Outcomes of transplantation of unrelated donor umbilical cord blood and bone marrow in children with acute leukaemia: a comparison study. *Lancet* 2007 Jun 9;369(9577):1947–54.

32. Laughlin MJ, Barker J, Bambach B, Koc O, Rizzieri D, Wagner J, *et al.* Hematopoietic engraftment and survival in adult recipients of umbilical cord blood from unrelated donors. *N Engl J Med* 2001;344(24):1815–22.

33. Arcese W, Rocha V, Labopin M, Sanz G, Iori AP, de Lima M, *et al.* Unrelated cord blood transplants in adults with hematologic malignancies. *Haematologica* 2006 Feb;91(2):223–30.

34. Barker JN, Weisdorf DJ, DeFor TE, Blazar BR, McGlave PB, Miller JS, *et al.* Transplantation of 2 partially HLA-matched umbilical cord blood units to enhance engraftment in adults with hematologic malignancy. *Blood* 2005 Feb 1;105(3):1343–7.

35. Cornetta K, Laughlin M, Carter S, Wall D, Weinthal J, Delaney C, *et al.* Umbilical cord blood transplantation in adults: results of the prospective Cord Blood Transplantation (COBLT). *Biol Blood Marrow Transplant* 2005 Feb;11(2):149–60.

36. Lori AP, Cerretti R, De Felice L, Screnci M, Mengarelli A, Romano A, *et al.* Pre-transplant prognostic factors for patients with high-risk leukemia undergoing an unrelated cord blood transplantation. *Bone Marrow Transplant* 2004 Jun;33(11):1097–105.

37. Konuma T, Ooi J, Takahashi S, Tomonari A, Uchiyama M, Fukuno K, *et al.* Unrelated cord blood transplantation after myeloablative conditioning in patients with acute leukemia aged between 50 and 55 years. *Bone Marrow Transplant* 2006 Apr;37(8):803–4.

38. Long GD, Laughlin M, Madan B, Kurtzberg J, Gasparetto C, Morris A, *et al.* Unrelated umbilical cord blood transplantation in adult patients. *Biol Blood Marrow Transplant* 2003 Dec;9(12):772–80.

39. Ooi J, Iseki T, Nagayama H, Tomonari A, Ito K, Shirafuji N, *et al.* Unrelated cord blood transplantation for adult patients with myelodysplastic syndrome-related secondary acute myeloid leukaemia. *Br J Haematol* 2001;114(4):834–6.

40. Sanz G, Saavedra S, Planelles D, Senent L, Cervera J, Barragan E, *et al.* Standardized, unrelated donor cord blood transplantation in adults with hematologic malignancies. *Blood* 2001;98(8):2332–38.

41. Tomonari A, Takahashi S, Ooi J, Nakaoka T, Takasugi K, Uchiyama M, *et al.* Cord blood transplantation for acute myelogenous leukemia using a conditioning regimen consisting of granulocyte colony-stimulating factor-combined high-dose cytarabine, fludarabine, and total body irradiation. *Eur J Haematol* 2006 Jul;77(1):46–50.

42. Laughlin MJ, Eapen M, Rubinstein P, Wagner JE, Zhang MJ, Champlin RE, *et al.* Outcomes after transplantation of cord blood or bone marrow from unrelated donors in adults with leukemia. *N Engl J Med* 2004 Nov 25;351(22):2265–75.

43. Miyakoshi S, Yuji K, Kami M, Kusumi E, Kishi Y, Kobayashi K, *et al.* Successful engraftment after reduced-intensity umbilical cord blood transplantation for adult patients with advanced hematological diseases. *Clin Cancer Res* 2004 Jun 1;10(11):3586–92.

44. Gomi S, Hasegawa S, Dan K, Wakabayashi I. A comparative analysis of the transplant potential of umbilical cord blood versus mobilized peripheral blood stem cells. *Nippon Ika Daigaku Zasshi* 1997;64:307–13.

45. Ng YY, van Kessel B, Lokhorst HM, Baert MR, van den Burg CM, Bloem AC, *et al.* Gene-expression profiling of CD34+ cells from various hematopoietic stem-cell sources reveals functional differences in stem-cell activity. *J Leukoc Biol* 2004 Feb;75(2):314–23.

46. Hiruma K, Hirsch R, Patchen M, Bluestone JA, Gress RE. Effects of anti-CD3 monoclonal antibody on engraftment of T-cell-depleted bone marrow

allografts in mice: host T-cell suppression, growth factors, and space. *Blood* 1992 Jun 1;79(11):3050–8.

47. Cohen G, Carter SL, Weinberg KI, Masinsin B, Guinan E, Kurtzberg J, *et al.* Antigen-specific T-lymphocyte function after cord blood transplantation. *Biol Blood Marrow Transplant* 2006 Dec;12(12):1335–42.

48. Komanduri KV, St John LS, de Lima M, McMannis J, Rosinski S, McNiece I, *et al.* Delayed immune reconstitution after cord blood transplantation is characterized by impaired thymopoiesis and late memory T cell skewing. *Blood* 2007 Dec 15;110(13):4543–51.

49. Rainaut M, Pagniez M, Hercend T, Daffos F, Forestier F. Characterization of mononuclear cell subpopulations in normal fetal peripheral blood. *Human Immunol* 1987 Apr;18(4):331–7.

50. Hannet I, Erkeller-Yuksel F, Lydyard P, Deneys V, DeBruyere M. Developmental and maturational changes in human blood lymphocyte subpopulations. *Immunol Today* 1992 Jun;13(6):215, 218.

51. Brahmi Z, Hommel-Berrey G, Smith F, Thomson B. NK cells recover early and mediate cytotoxicity via perforin/granzyme and Fas/FasL pathways in umbilical cord blood recipients. *Human Immunol* 2001 Aug;62(8):782–90.

52. El Marsafy S, Dosquet C, Coudert MC, Bensussan A, Carosella E, Gluckman E. Study of cord blood natural killer cell suppressor activity. *Eur J Haematol* 2001 Apr;66(4):215–20.

53. Gaddy J, Broxmeyer HE. Cord blood Cd16(+)56(–) cells with low lytic activity are possible precursors of mature natural killer cells. *Cell Immunol* 1997;180(2):132–42.

54. Liu E, Law HK, Lau YL. Tolerance associated with cord blood transplantation may depend on the state of host dendritic cells. *Br J Haematol* 2004 Aug;126(4):517–26.

55. Godfrey WR, Spoden DJ, Ge YG, Baker SR, Liu B, Levine BL, *et al.* Cord blood CD4(+)CD25(+)-derived T regulatory cell lines express FoxP3 protein and manifest potent suppressor function. *Blood* 2005 Jan 15;105(2):750–8.

56. Cohen JL, Trenado A, Vasey D, Klatzmann D, Salomon BL. CD4(+)CD25(+) immunoregulatory T Cells: new therapeutics for graft-versus-host disease. *J Exp Med* 2002 Aug 5;196(3):401–6.

57. Hoffmann P, Ermann J, Edinger M, Fathman CG, Strober S. Donor-type CD4(+)CD25(+) regulatory T cells suppress lethal acute graft-versus-host

disease after allogeneic bone marrow transplantation. *J Exp Med* 2002 Aug 5;196(3):389–99.

58. Taylor PA, Lees CJ, Blazar BR. The infusion of *ex vivo* activated and expanded CD4(+)CD25(+) immune regulatory cells inhibits graft-versus-host disease lethality. *Blood* 2002 May 15;99(10):3493–9.

59. Kaminski B, Kadereit S, Miller R, Leahy P, Stein K, Topa D, *et al.* Reduced expression of NFAT-associated genese in UCB versus adult CD4 T lymphocytes during primary stimulation. *Blood* 2003;102(13):4608–17.

60. Scott ME, Kubin M, Kohl S. High level interleukin-12 production, but diminished interferon-gamma production, by cord blood mononuclear cells. *Pediatr Res* 1997;41(4 Part 1):547–53.

61. Miller RE, Fayen JD, Mohammad SF, Stein K, Kadereit S, Woods KD, *et al.* Reduced CTLA-4 protein and messenger RNA expression in umbilical cord blood T lymphocytes. *Exp Hematol* 2002 Jul;30(7):738–44.

62. Kadereit S, Mohammad S, Miller R, Woods K, Listrom C, McKinnon K, *et al.* Reduced NFAT1 protein expression in human umbilical cord blood T lymphocytes. *Blood* 1999;94(9):3101–7.

63. Ballen KK, Spitzer TR, Yeap BY, McAfee S, Dey BR, Attar E, *et al.* Double unrelated reduced-intensity umbilical cord blood transplantation in adults. *Biol Blood Marrow Transplant* 2007 Jan;13(1):82–9.

64. Haspel RL, Kao G, Yeap BY, Cutler C, Soiffer RJ, Alyea EP, *et al.* Preinfusion variables predict the predominant unit in the setting of reduced-intensity double cord blood transplantation. *Bone Marrow Transplant* 2008 Mar;41(6):523–9.

65. Verneris MR, Brunstein C, DeFor TE, Barker J, Weisdorf DJ, Blazar BR, *et al.* Risk of relapse (REL) after umbilical cord blood transplantation (UCBT) in patients with acute leukemia: marked reduction in recipients of two units. *ASH Annu Meet Abstr* 2005 Nov 16;106(11):93a(Abstract 305).

3

Self-Renewal of Primitive Hematopoietic Cells: A Focus on Asymmetric Cell Division

Andre Görgens and Bernd Giebel

ABSTRACT

The different mammalian blood cell types originate from hematopoietic stem cells (HSCs). HSCs are required in order to sustain the daily production of hundreds of millions of different blood cells fulfilling different functions. Since HSCs have been used for clinical applications for more than three decades, they represent the best studied somatic stem cell entity so far. However, the mechanisms controlling HSC maintenance or differentiation remain largely unknown. Focussing mainly on the process of asymmetric cell division, this chapter will review some of the current knowledge about the mechanisms controlling the decision "self-renewal *versus* differentiation" of primitive human hematopoietic cells.

INTRODUCTION

Somatic stem cells are undifferentiated cells which have been identified in many tissues and organs. They are required for tissue homeostasis i.e., by creating differentiating daughter cells they provide the fundament for the regenerative capabilities of somatic tissues and organs. To fulfil this function during the entire life span of an organism, somatic stem cells not only have to replenish cells lost by natural turnover or by tissue damaging events, but they also have to maintain the stem cell pool constant i.e., they need to retain their self-renewal capacity. As depicted in Fig. 1, loss of stem cells or of stem cell activities would be as detrimental as their uncontrolled expansion. On the one hand reduced somatic stem cell activities could decrease regenerative capabilities, possibly resulting in tissue degeneration, as for example observed during aging of multicellular

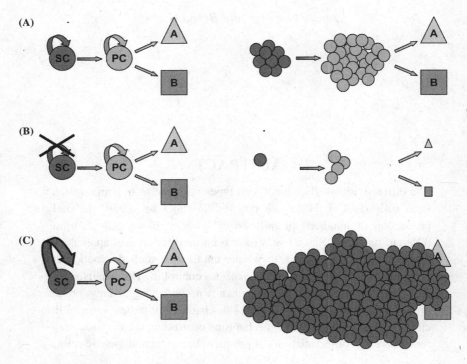

Fig. 1.   The decision of somatic stem cells (SC) between self-renewal and differentiation needs to be tightly controlled. **(A)** Adequately controlled self-renewal and differentiation will maintain tissue homeostasis. **(B)** A reduction of the self-renewal capability will decrease the regenerative potential of any given tissue and finally might result in their degeneration. **(C)** In contrast, increased self-renewal capacities might result in tumor formation. PC: Progenitor cell, A: differentiated cell type A, B: differentiated cell type B.

organisms.[1-3] On the other hand an uncontrolled expansion could result in tumor formation.[4-6] Thus, it becomes evident that the decision between self-renewal and differentiation requires tightly controlled mechanisms. In this respect, it is of fundamental interest in stem cell biology to uncover such underlying mechanisms. Understanding such mechanisms might not only provide new avenues to regenerative medicine, but might also reveal important knowledge that could be helpful to develop new antitumor strategies which could help to increase or decrease the self-renewal capacities of endogenous stem cells or of tumor cells, respectively (Fig. 1).

PRINCIPAL MECHANISMS FOR CONTROLLING THE DECISION "SELF-RENEWAL VERSUS DIFFERENTIATION" OF SOMATIC STEM CELLS

There are different possible strategies for controlling self-renewal *versus* differentiation. For example, following cell division, somatic stem cells could give rise to daughter cells which primarily have identical developmental capabilities. Depending on the activity of extrinsic factors that are provided by the surrounding environment, the initial cell fate could be modulated. A special set of extrinsic factors for example might be required to maintain stem cell activity, whereas other combinations could be responsible to drive the cells into differentiation. By limiting the environmental areas providing stem cell supporting factors, the stem cell pool can, at least theoretically, be kept relatively constant (Fig. 2A). Indeed, research on model organisms such as the fruit fly *Drosophila melanogaster* led to the discovery of stem cell supporting areas in the gonads of these animals.[7-9] In analogy to special ecological environments, the ecological niches, these areas are named stem cell niches.

In another scenario, the decision whether a newly formed stem cell daughter retains its stem cell capacity or whether it loses this capacity might depend solely on intrinsic factors. In case a certain concentration of a given intrinsic factor acting as a cell fate determinant is required to maintain the stem cell fate, the distribution of such determinants during mitotic stages could control the presumptive cell fates of both daughter cells. Both daughter cells would inherit different amounts of such cell fate determinants, thus obtaining different developmental capacities, one

(A) **(B)**

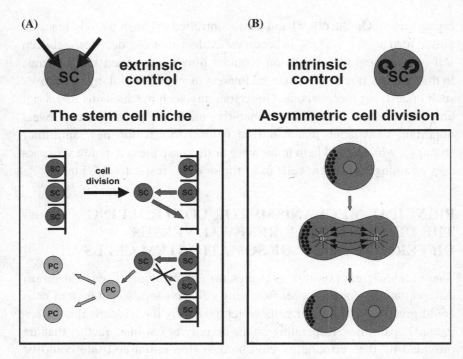

Fig. 2. Extrinsic or intrinsic control of somatic stem cells (SC) to either self-renew or to differentiate. **(A)** The model of the stem cell niche predicts that a particular combination of extrinsic factors being present within the stem cell niches is required to maintain stem cell fate. Stem cells which do not enter an available niche will differentiate. **(B)** Intrinsic cell fate determinants segregate differently into arising daughter cells during asymmetric cell division and control stem cell maintenance versus differentiation.

maintaining the stem cell capacity and one being specified to differentiate (Fig. 2B). Indeed, evidence for such a model has been provided from model organisms, too. In the case of the neural stem cells in *Drosophila melanogaster*, the neuroblasts, it has been shown that during mitotic division proteins acting as cell fate determinants segregate differently into the two daughter cells. As a result of this asymmetric cell division only one of the two daughter cells retains the stem cell capacity whereas the other becomes committed and differentiates.[10,11]

In the following two paragraphs both models will be discussed in more detail.

Extrinsic Control

The stem cell niche in Drosophila

Studies on germ line stem cells (GSCs) of *Drosophila* contributed largely to the understanding of the concept of stem cell niches. In male flies, six to 12 GSCs reside at the tip of the testis, adjacent to the tightly packed somatic hub cells.[12,13] In females, two to three GSCs are localized in the tip of each ovariole, the germarium, in close contact to four to seven somatic cap cells.[14,15] In both sexes, the direct cell-cell contact between the GSC and the hub cells or the cap cells is essential to maintain the GSC fate.[9,12,13,16] Apart from this cellular contact, female GSCs depend on the secretion of a bone morphogenic protein (BMP) homologue protein Decapentaplegic (DPP) and Glassbottomed Boat (GBB) from the closely attached terminal filament cells.[8,9,17] Their signaling results in repression of the differentiation factor *bag-of-marbles* (*bam*), whose activity is both necessary and sufficient to drive differentiation of immediate GSC daughters.[17,18]

Even though a BMP dependence of male GSCs has also been described, the cytokine-like protein Unpaired which is expressed by the hub cells activates the JAK-STAT signaling pathway and, thereby controls GSC self-renewal.[12,13,19]

In most of their cell divisions GSCs create an anterior located daughter cell maintaining the contact with the cap cells and a posterior located cell losing this contact. While the anterior cell maintains the GSC capacity, the posterior cell normally becomes committed to differentiate. However, if one GSC is lost, both offspring of the remaining GSC can form cellular contacts with the cap cells and thus both are maintained as GSCs.[9] Experiments in which all GSCs were ablated revealed an alternative strategy for GSC replacement. It was found that depending on the activity of DPP-committed progenitor cells, the cystocytes, can de-differentiate and develop into functional GSCs.[20,21]

These examples reveal that within a stem cell niche a special combination of extrinsic signals, some of them being provided in a direct cellular contact and others being secreted, act in combination to control the fate of the stem cells and the size of the stem cell pool. In case individual stem cells are lost, they can be replaced either by offspring of remaining stem cells or by progenitor cells that are induced to de-differentiate.

Intrinsic Control

Asymmetric cell division in model organisms

Much of our current knowledge about mechanisms controlling asymmetric cell divisions has come from studies analyzing the development of the zygote of the nematode *Caenorhabditis elegans*. Like in many organisms, the zygote of *C. elegans* displays an asymmetric distribution of determinants that are differentially partitioned to the new blastomers. Here, after fertilization and before the first cleavage, certain vesicles containing germ line specifying determinants, the so-called P-granules, become localized at the posterior pole. Since the zygote divides along the anterior-posterior axis, the P-granules completely segregate into the smaller, posterior blastomere, the so-called P1 cell.[22] As a result, the cell inheriting P-granules will form the germline, while the anterior located blastomere, the AB cell, loses this capacity.[23–25]

The asymmetric segregation of the P-granules depends on a variety of different factors. Initially, a number of such factors have been identified in genetic screens as genes whose activity is required for the proper asymmetric development of the two cell-stage blastomers. As defects in certain genes were found to cause an equal developmental potential of both two cell-stage blastomers, corresponding genes were named *partitioning defect* (*par*) genes.[26] It turned out that different PAR proteins i.e., Par-3 and Par-6, form a complex with an atypical protein kinase C (aPKC) that concentrates in the anterior half of the embryo. By reorganizing the actinomyosin network this complex has been found to be required to establish the anterior-posterior polarity axis in the *C. elegans* zygote.[27–29] Therefore, it became evident that the establishment of the anterior-posterior axis in the zygote is an essential prerequisite for its asymmetric cell division.

Remarkably, during recent years and as extensively reviewed elsewhere,[30,31] it turned out that ranging from worm to man the evolutionary conserved PAR/aPKC complex controls establishment and maintenance of cell polarity in a variety of different cell types. In this context, the PAR/aPKC complex also organizes the cell polarity and asymmetric cell division of the *Drosophila* neuroblasts.[32–36]

Depending on the activity of the PAR/aPKC complex in mitotic stages, the cell fate determinants Prospero and Numb, are localized to the basal pole of the neuroblast.[10,11,33] The PAR/aPKC complex also coordinates the binding of Inscuteable (Insc),[32,33,35,36] a protein which is recruited to the apical membrane in premitotic neuroblasts.[37] During mitotic stages Insc controls rotation of the mitotic spindle apparatus in a manner that neuroblast divide perpendicular to the apicobasal cell axis. Depending on this rotation, the apical pole is inherited by one and their basal pole including the cell fate determinants Prospero and Numb by the other arising daughter cell, which then becomes committed to differentiate.[38,39]

In conclusion, both examples, the *C. elegans* zygote and the *Drosophila* neuroblasts, reveal that asymmetrically dividing cells need to comply with three major prerequisites: firstly, cells need to establish a cell polarity axis, secondly, cell fate determinants need to be localized asymmetrically into the daughter cells, and thirdly, the spindle apparatus needs to be orientated in a manner that the cell fate determinants segregate differently into the arising daughter cells.

STEM CELL NICHES AND ASYMMETRIC CELL DIVISION ARE NOT MUTUALLY EXCLUSIVE

Although the discussed models are based on either extrinsic or intrinsic cell mechanisms, there are well-investigated cellular systems in which combinations of intrinsic and extrinsic mechanisms orchestrate cell fate decision processes of developing structures. The development of the peripheral nervous system of *Drosophila* provides an excellent model system to study such mechanisms.

In the developing peripheral nervous system, so-called sensory organ precursor cells (SOPs) repetitively divide asymmetrically to form the four different cells of an individual peripheral mechano-sensory organ in a stereotypical manner.[40]

In this context, the cell surface receptor Notch plays a decisive role.[41,42] Upon binding of its ligands Delta and Serrate the Notch receptor is activated.[43–46] The ligand-binding induces a γ-secretase dependent proteolytic cleavage and a release of the Notch intracellular domain (NICD).[29,47–52] This domain subsequently enters the nucleus and turns a transcriptional repressor into a transcriptional activator.[53–60]

It was found that during SOP development, the Notch-ligands serve as extrinsic signals that are presented to both SOP daughter cells.[44] Even though both of these cells express Notch and thus are capable to receive the extrinsic signal, only one of them transduces the signal to its nucleus.[61] It turned out that SOPs express the cell fate determinant Numb which, during mitosis, segregates mainly into one of the two daughter cells, where it antagonizes Notch function.[62–64] Therefore, the intrinsic cell fate determinant Numb antagonizes the extrinsically provided signal that is required to activate the Notch signaling pathway.

THE HEMATOPOIETIC STEM CELL COMPARTMENT

Like any other somatic cell type with stem cell properties, HSCs are able to self-renew and differentiate into specialized cell types. In contrast to other somatic stem cells, HSCs are successfully used in clinical trials for more than 30 years.[65] Initially, a potential clinical usage of human HSCs was indicated when lethally irradiated mice were rescued by the transplantation of cell fractions containing HSCs; the transplanted HSCs led to a complete reconstitution of the murine blood system.[66,67] Since then basic research and clinical trials improved our knowledge about HSC biology as well as their clinical applications, so that nowadays HSC containing cell fractions of different sources are routinely used to transplant different cohorts of patients.[65]

The Hematopoietic Stem Cell Niche

Triggered by their enormous clinical relevance, many attempts have been performed to formulate conditions allowing *ex vivo* HSC expansion. As extensively reviewed elsewhere,[68] almost all of these attempts failed to expand HSCs in larger amounts. However, differentiation of primitive hematopoietic cells is slowed down upon culture on certain stromal cells and the primitive potential of such HSC-enriched cell fractions can be maintained for at least several weeks in culture.[69–72] In contrast, it seems that HSCs can expand *in vivo*. As already mentioned, myeloablated patients can be cured by HSC-containing grafts for their normal lasting life time.[65] Moreover, small numbers of HSCs can be transplanted sequentially in animal models without losing their stem cell potential, leading to successful regeneration in each generation.[73] These observations demonstrate that the environment is important in controlling the development of HSCs and their progeny. In this context, Raymond Schofield proposed as early as in 1978 the stem cell niche hypothesis, in which the stem cell is seen in association with other cells which determine its and its progeny behavior. As long as stem cells can occupy stem cell niches they can be maintained as stem cells, otherwise they get committed to differentiate.[74]

Indeed, evidence for HSC-niches has recently been reported, with osteoblasts forming putative bone marrow (BM) niches in the endosteum, and sinusoidal endothelial cells the vascular HSC-niches in the spleen and BM.[75–77] It seems that the receptor tyrosine kinase Tie2 which is expressed on HSCs and its cognate ligand Angiopoietin-1 produced by the osteoblasts play a crucial role in these niches.[78] Furthermore, osteoblasts express the Notch ligand Jagged-1 which can activate the Notch signaling pathway. This, together with the fact that in cocultures the long-term supportive effect of osteoblasts is strongly reduced by blocking Notch activation, suggests that another important function of HSC niches is the activation of the Notch signaling pathway in the stem cell.[75] More recent data that are reviewed elsewhere suggest additional roles of the Wnt signaling pathway, and of HSC-supporting factors such as the angiopoietin-like (Angptl) proteins which might be involved in maintaining the stem cell pool.[79,80]

ASYMMETRIC CELL DIVISION OF PRIMITIVE HEMATOPOIETIC CELLS

Before the concept of the HSC niche was improved, it was widely assumed that HSCs divide asymmetrically to give rise to one daughter cell maintaining the stem cell capability and to another cell being committed to differentiate. Observations supporting this hypothesis are presented in the next paragraph.

Primitive Human Hematopoietic Cells Can Adopt Different Cell Fates

First evidence that HSCs might divide asymmetrically came from studies of the group of Ogawa.[81–83] They separated daughter cells of primitive hematopoietic cells and compared their proliferation rate and differentiation capacity. It was observed that in some cases sister cells developed colonies of different types and/or of different sizes. They assumed that these differences might be the result of stochastic processes that decide whether stem cells self-renew or differentiate.[81–83] Differences in the development of separated daughter cells were also found by the group of Landsdorp.[84] They separated progenies of human umbilical cord blood derived $CD34^+/CD45RA^{low}/CD71^{low}$ cells, a cell fraction highly enriched for primitive hematopoietic cells, and cultured them under identical or different culture conditions. Similar to the group of Ogawa, they found that a small fraction of initially deposited cells gave rise to siblings that formed different colony types. In their subsequent studies the development of individual fetal liver-derived $CD34^+/CD38^-/CD71^{low}/CD45RA^{low}$ cells were analyzed. In this context Brümmendorf and colleagues[84] recognized that in some cases progenies of deposited cells followed different proliferation kinetics i.e., some of the cells showed fast and others slow proliferation kinetics. Since slow dividing cells showed higher expansion rates than fast dividing cells, they were judged to be the more primitive daughter cells. Next, the group subcloned $CD34^+/CD38^-$ cells of slow dividing cell clones and observed that arising daughter cells again could be discriminated in terms of their proliferation kinetics. The authors assumed that differences in the cell fate of arising daughter cells might continuously and intrinsically be generated by asymmetric cell divisions, in which similarly to asymmetric cell divisions in *Drosophila*, molecules acting as cell fate determinants are differently distributed during mitosis. However, they also discussed that there are alternative explanations for their observations, such as postmitotic mechanisms.[84]

In another set of experiments the group of Ho[29] labelled $CD34^+/CD38^-$ cells derived from human fetal liver, umbilical cord blood and adult bone marrow with the membrane dye PKH-26. Studying the proliferation kinetics of these cells over time, they observed that depending on the cytokines used, deposited cells performed their first *in vitro* cell division after 36 to 38 hours (in the presence of late acting cytokines) or after 48 to 50 hours (in the presence of early acting cytokines) post-seeding. However, in a cytokine independent, but cell source dependent manner, approximately 20%–40% of the CD34$^+$/CD38$^-$ cells gave rise to daughter cells with asynchronous proliferation kinetics.[29] In this context, functional studies on umbilical cord blood-derived CD34$^+$/CD38$^-$ cells revealed that a higher content of more primitive hematopoietic cells gives rise to daughter cells with asynchronous proliferation kinetics than more mature progenitor cells. Confirming the data of Brümmendorf and colleagues (1998), slow dividing cells were found to be more primitive than fast dividing cells.[85]

In collaboration with Ho's group, our group then performed cell separation studies.[86] Since umbilical cord blood-derived primitive hematopoietic cells contain higher regenerative abilities than corresponding adult cells derived from BM or from the peripheral blood of G-CSF mobilized stem cell donors,[87,88] mainly umbilical cord blood-derived cells were used for these studies. To specifically analyze more primitive hematopoietic cells it was first tested whether, apart from cell surface antigens, cell culture conditions might also help to enrich for more primitive hematopoietic cells. Indeed, it was shown that in the presence of late acting cytokines mainly more mature CD34$^+$ cells are enforced to perform their first *in vitro* cell divisions within the first five days of culture, whereas most of the more primitive CD34$^+$ cells remain quiescent for at least the first five days. In this context it should be mentioned that within the presence of early acting cytokines, almost all CD34$^+$ cells are induced to perform their first *in vitro* cell division between a 48- to 72-hour culture.[86]

For cell separation studies, individual CB-derived CD34$^+$/CD38$^-$ cells were sorted into individual wells of 96-well plates and were cultured in the presence of late acting cytokines. Within 24 hours, siblings of cells which performed their first cell division between culture day 5 and day 10 were separated from each other and individually expanded on stromal cells (AFT024), which support maintenance of primitive hematopoietic cell fates.[71] Next, the offspring of each of the separated cells were tested for their capacity to form myeloid as well as lymphoid cell types. In case initially deposited or separated daughter cells gave rise to myeloid as well as to lymphoid offspring they were retrospectively termed as myeloid-lymphoid initiating cells or briefly as ML-IC.[71] More than 80% of the initially deposited cells which were retrospectively judged as ML-ICs gave rise to daughter cells, one of which also revealed the ML-IC potential, and one of which showed a reduced developmental capacity. Therefore, these data suggest that at the point of time the ML-IC siblings were separated, they were differently specified, one still retaining the ML-IC capacity, whereas the other was more committed.[86]

Even though these data are compatible with the model of asymmetric cell division as it was initially discussed by Brümmendorf and colleagues, the observed differences might also result from post-mitotically acting processes i.e., immediately after mitosis arising daughter cells might contain identical developmental capacities that might be altered during the time between the initially deposited cell had finished its mitosis and before the daughter cells got separated.[84,86] Since both possibilities, asymmetric cell division and post-mitotic mechanisms, could explain the outcome of the delineated studies, it remained an open question whether primitive hematopoietic cells indeed could divide asymmetrically.[86]

Cell Polarity of Primitive Hematopoietic Cells

As discussed before, an important prerequisite for asymmetrically dividing cells in model organisms is that they are polarized. In the case that observed differences in the development of primitive hematopoietic cell siblings are indeed established by asymmetric cell divisions, they should obey similar principles than other asymmetrically dividing cells and thus should contain a cell polarity axis. Indeed, while freshly isolated CD34$^+$ cells regularly appear as small round cells without any apparent indication of cellular polarity, cultured CD34$^+$ cells adopt an elongated morphology, forming a leading edge on one end and a structure named uropod at the other end.[89] This morphology resembles the migrational phenotype of immunoreactive leukocytes. Moreover, certain drugs such as the

PI3K-inhibitor Ly294002 that interfere with processes required to establish polarized cell shapes of cultured CD34$^+$ cells, also interfere with their *in vitro* migratory capabilities.[89]

Since several proteins were known to adopt a polarized distribution in migrating leukocytes and several of these proteins are expressed on primitive hematopoietic cells, their distribution was studied on cultured CD34$^+$ cells. In this context it could be shown that the stem cell marker CD133/Prominin-1 colocalizes with known uropod markers, the cell adhesion molecules CD43, CD44, CD50 and CD54, at the tip of the uropod. In contrast, the chemokine receptor CXCR4 and the ganglioside GM3 are highly enriched at the opposite pole, the leading edge, whereas the cell surface proteins, CD34 and CD45 show an even distribution over the entire cell surface of the CD34$^+$ cells.[89]

Next, using some of these antigens, the cell polarity axis in mitotic cells was studied. CD43, CD44, CD50 and CD54 together with CD133 become highly enriched in the cleavage furrow and at later mitotic stages on the cytoplasmic bridge connecting the arising daughter cells, on the so-called midbody.[90] This set of data demonstrates the existence of a cell polarity axis in dividing CD34$^+$ cells.

However, since in dividing CD34$^+$ cells none of the investigated antigens revealed an asymmetric distribution, it still remained an open question whether primitive hematopoietic cells can divide asymmetrically. Either the cells do not divide asymmetrically or the investigated cell polarity markers are not inherited differently in asymmetrically dividing CD34$^+$ cells.

Identification of Asymmetrically Segregating Antigens

Studying the proliferation kinetics of the cultured CD34$^+$ cells, it was realized that initially most of the CD34$^+$ cells also express CD133. However, with the onset of cell divisions between culture day 2 and day 3, a novel cell population appears whose cells have largely lost their cell surface expression of CD133, but not that of CD34 (Fig. 3). Remarkably, the proportion of this population increases over time and the underlying kinetics are compatible to a model in which — according to the functional studies described above — approximately 70% of the more primitive cells divide

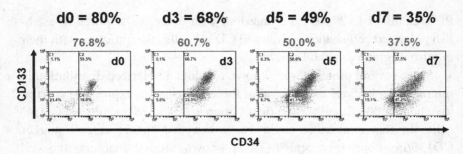

Fig. 3. Hypothetical and measured content of primitive hematopoietic cells within the fraction of cultured CD34$^+$ cells. As pointed out in the text functional analyses revealed that approximately 30% of primitive hematopoietic cells give rise to daughter cells following different proliferation kinetics and/or realizing different cell fates. Assuming that (i) 30% of the primitive hematopoietic cells (CD34$^+$/CD133$^+$) divide asymmetrically, giving rise to a primitive (CD34$^+$/CD133$^+$) and a more specified daughter cell (CD34$^+$/CD133$^{low/-}$), (ii) starting at culture day 2 all CD34$^+$ cells divide once a day, and (iii) approximately 80% of the freshly isolated CD34$^+$ cells express CD133, the content of CD34$^+$/CD133$^+$ cells at a given culture day *n* can be calculated with the formula: **content of CD34$^+$/CD133$^+$ cells = 0.8 × (0.7 + 0.15)$^{n-2}$**. The theoretical content of CD34$^+$/CD133$^+$ cells within a CD34$^+$ cell population at a given culture day is given in black letters, that of the experimentally measured content in red letters. Corresponding plots are shown below.

symmetrically to create two CD34$^+$/CD133$^+$ cells and approximately 30% asymmetrically to create a more primitive CD34$^+$/CD133$^+$ daughter cell and a more mature CD34$^+$/CD133$^{low/neg}$ cell.[86,90] Since in none of the dividing cells studied CD133 revealed an asymmetric cell surface distribution, CD133 itself was not a candidate for an asymmetrical segregating protein. However, due to the intriguing population kinetics it was hypothesized that any other cell surface molecule being differently expressed on CD34$^+$/CD133$^+$ and CD34$^+$/CD133$^{low/neg}$ cells, in principle provides a candidate for an asymmetrically segregating molecule.

In this context, a flow cytometry based screen identified several cell surface antigens such as CD53, CD62L, CD63 and CD71 being differently expressed on CD133$^+$/CD34$^+$ and CD133$^{low/-}$/CD34$^+$ cells. On dividing CD34$^+$ cells the tetraspanins CD53 and CD63 as well as the transferrin receptor (CD71) and the L-selectin (CD62L) showed an asymmetric distribution on 20%–30% of the mitotic CD34$^+$ cells studied,[90] strongly suggesting that human primitive hematopoietic cells indeed can divide asymmetrically. Even though it is not known yet, whether the identified molecules are causally linked to the process of asymmetric cell division, three of them (CD53, CD63 and CD71) provide physical links to the endosomal compartment.[90] This coupled with the more recent finding that a subset of endosomes, the ones expressing the early endosomal antigen-1 (EEA-1), segregate mainly into the anterior part of the asymmetrically dividing zygote of *C. elegans*,[91] suggests that asymmetric segregation of endosomes might provide a more general mechanism in asymmetrically dividing cells.[90] Supporting this notion, it just recently has been published that an asymmetric segregation of certain endosomes can also be observed in asymmetrically dividing sensory organ precursor cells of *Drosophila melanogaster*.[92]

Asymmetric Cell Division and CD133

According to previous findings, endosomes can segregate asymmetrically in dividing cells. Since endosomes are required for molecular trafficking within the cells, it appears likely that several proteins, including proteins with extracellular epitopes, are differentially distributed during the course of asymmetric cell divisions. In this context, the asymmetrical distribution of certain proteins might occur on the intracellular level rather than on the cell surface. Since asymmetrically segregating proteins were identified by comparing the cell surface expression of different proteins on $CD34^+/CD133^+$ and $CD34^+/CD133^{low/neg}$ cells,[90] it is an interesting question how the differences in the CD133 content itself are established. In principle, this might be mediated by post-mitotic mechanisms in which only the more primitive $CD34^+$ cells can re-express and/or maintain CD133 expression. Alternatively, differences might also be established by the asymmetric segregation of the CD133 protein itself. Even though we did not find any indication for an asymmetric segregation of the CD133 on the cell surface, it cannot be excluded that an intracellular fraction of CD133 segregates differently into the daughter cells to establish the observed differences. Indeed, Fonseca *et al.* (2008) analyzed the intracellular CD133 distribution in dividing primitive hematopoietic cells and observed a co-localization of CD133 and CD63 within the intracellular compartment. Confirming previous data, they found that in approximately 20% of the mitotic cells studied, both CD63 and CD133 mainly segregate into one of the arising daughter cells.[90,93] Therefore, in asymmetrically dividing primitive hematopoietic cells CD133 can indeed segregate asymmetrically.

Endosomes are Involved in Cell Fate Specification Processes

Since asymmetrically segregating proteins were identified by comparing the CD34$^+$/CD133$^+$ and CD34$^+$/CD133$^{low/neg}$ cell fractions,[90] the question was whether the asymmetric segregation of the identified proteins and most likely that of certain endosomes might correlate with the cell fate of corresponding cells. To get some insight, the potential of the cells was compared in long-term culture initiating cell (LTC-IC) assays, an *in vitro* read out system to estimate the content of primitive myeloid cells. Almost all cells with LTC-IC capacities were enriched within the CD34$^+$/CD133$^+$ cell fraction.[90] According to these results, the asymmetric distribution of the proteins identified seems to correlate with the cell fate the cells finally adopt.

Evidence that endosomes and endosomal proteins are involved in cell fate specification processes has been obtained from model organisms. Within the developing wing of *Drosophila melanogaster* an anterior-posterior gradient of the TGF-β homologue protein DPP controls the fate of the wing epithelial cells in a concentration-dependent manner. Upon binding to its receptor Thickveins (Tkv), a type I TGF-β receptor, DPP induces the phosphorylation of the R-Smad transcription factor Mad and recruits a common Smad to become an active transcription factor that then translocates into the nucleus and induces the expression of target genes. The endosomal protein Smad-anchor-for-receptor-activation (SARA) is required for the recruitment of R-Smads to type I TGF-β receptors.[94] In mammals as well as in *Drosophila*, Sara accumulates in early endosomes; which in the *Drosophila* wing epithelial cells also contain DPP and its receptor Tkv.[95,96] During mitosis of normally developing wing epithelial cells, these endosomes equally segregate into the arising daughter cells.[95] However, even though in Sara mutant wing epithelial cells DPP and Tkv containing endosomes still exist, their distribution into arising daughter cells is randomized. Frequently, daughter cells obtain different amounts of Tkv containing endosomes, and in correlation with that they adopt different cell fates. Cells with a higher content of Tkv containing endosomes adopt more anterior and those with a lower content more posterior cell fates.[95]

Homeostasis Requires Orchestration of Extrinsic and Intrinsic Cues

In addition to asymmetric division of primitive hematopoietic cells, data suggests that extrinsic signals may play a role as well. The observation that a higher percentage of primitive hematopoietic cells cultured on the stromal cell line AFT024 create daughter cells with different proliferation kinetics than corresponding cells that were cultured under stroma-free conditions, suggests that extrinsic signals can modulate intrinsic cell fate specification processes.[97] Hence, special combinations of extrinsic and intrinsic signals seem to regulate cell fate specification processes within the primitive hematopoietic compartment. Thus, it becomes an important question of how extrinsic and intrinsic cues cooperate to regulate mammalian hematopoietic homeostasis. It remains a major challenge to define culture conditions allowing the expansion of HSCs. Therefore a better understanding of the processes and mechanisms controlling asymmetric cell divisions within the primitive hematopoietic system may help us to improve and control cell culture conditions allowing optimized expansion of HSCs *ex vivo*.

ACKNOWLEDGMENTS

We thank Julia Beckmann, Peter Horn and Michael Punzel for discussion and critical review of the manuscript. Our studies are supported by grants from the Deutsche Forschungsgemeinschaft (SPP1109 GI 336/1-4, GI 336/4-1).

References

1. Ruzankina Y, Brown EJ. Relationships between stem cell exhaustion, tumour suppression and ageing. *Br J Cancer* 2007 Nov 5;97(9):1189–93.
2. Sharpless NE, DePinho RA. How stem cells age and why this makes us grow old. *Nat Rev Mol Cell Biol* 2007 Sep;8(9):703–13.

3. Rossi DJ, Jamieson CH, Weissman IL. Stems cells and the pathways to aging and cancer. *Cell* 2008 Feb 22;132(4):681–96.
4. Visvader JE, Lindeman GJ. Cancer stem cells in solid tumours: accumulating evidence and unresolved questions. *Nat Rev Cancer* 2008 Oct;8(10):755–68.
5. Al-Hajj M. Cancer stem cells and oncology therapeutics. *Curr Opin Oncol* 2007 Jan;19(1):61–4.
6. Savona M, Talpaz M. Getting to the stem of chronic myeloid leukaemia. *Nat Rev Cancer* 2008 May;8(5):341–50.
7. Tran J, Brenner TJ, DiNardo S. Somatic control over the germline stem cell lineage during Drosophila spermatogenesis. *Nature* 2000 Oct 12;407(6805): 754–7.
8. Xie T, Spradling AC. Decapentaplegic is essential for the maintenance and division of germline stem cells in the Drosophila ovary. *Cell* 1998 Jul 24;94(2):251–60.
9. Xie T, Spradling AC. A niche maintaining germ line stem cells in the Drosophila ovary. *Science* 2000 Oct 13;290(5490):328–30.
10. Spana EP, Doe CQ. The prospero transcription factor is asymmetrically localized to the cell cortex during neuroblast mitosis in Drosophila. *Development* 1995 Oct;121(10):3187–95.
11. Hirata J, Nakagoshi H, Nabeshima Y, Matsuzaki F. Asymmetric segregation of the homeodomain protein Prospero during Drosophila development. *Nature* 1995 Oct 19;377(6550):627–30.
12. Kiger AA, Jones DL, Schulz C, Rogers MB, Fuller MT. Stem cell self-renewal specified by JAK-STAT activation in response to a support cell cue. *Science* 2001 Dec 21;294(5551):2542–5.
13. Tulina N, Matunis E. Control of stem cell self-renewal in Drosophila spermatogenesis by JAK-STAT signaling. *Science* 2001 Dec 21;294(5551): 2546–9.
14. Lin H, Spradling AC. Germline stem cell division and egg chamber development in transplanted Drosophila germaria. *Dev Biol* 1993 Sep;159(1):140–52.
15. Wieschaus E, Szabad J. The development and function of the female germ line in *Drosophila melanogaster*: a cell lineage study. *Dev Biol* 1979 Jan;68(1):29–46.
16. Song X, Xie T. DE-cadherin-mediated cell adhesion is essential for maintaining somatic stem cells in the Drosophila ovary. *Proc Natl Acad Sci USA* 2002 Nov 12;99(23):14813–8.

17. Song X, Wong MD, Kawase E, Xi R, Ding BC, McCarthy JJ, *et al.* Bmp signals from niche cells directly repress transcription of a differentiation-promoting gene, bag of marbles, in germline stem cells in the Drosophila ovary. *Development* 2004 Mar;131(6):1353–64.

18. Chen D, McKearin D. Dpp signaling silences bam transcription directly to establish asymmetric divisions of germline stem cells. *Curr Biol* 2003 Oct 14;13(20):1786–91.

19. Kawase E, Wong MD, Ding BC, Xie T. Gbb/Bmp signaling is essential for maintaining germline stem cells and for repressing bam transcription in the Drosophila testis. *Development* 2004 Mar;131(6):1365–75.

20. Kai T, Spradling A. Differentiating germ cells can revert into functional stem cells in *Drosophila melanogaster* ovaries. *Nature* 2004 Apr 1;428(6982):564–9.

21. Kai T, Spradling A. An empty Drosophila stem cell niche reactivates the proliferation of ectopic cells. *Proc Natl Acad Sci USA* 2003 Apr 15;100(8):4633–8.

22. Strome S, Wood WB. Immunofluorescence visualization of germ-line-specific cytoplasmic granules in embryos, larvae, and adults of *Caenorhabditis elegans*. *Proc Natl Acad Sci USA* 1982 Mar;79(5):1558–62.

23. Hird SN, Paulsen JE, Strome S. Segregation of germ granules in living *Caenorhabditis elegans* embryos: cell-type-specific mechanisms for cyto-plasmic localisation. *Development* 1996 Apr;122(4):1303–12.

24. Sulston JE, Schierenberg E, White JG, Thomson JN. The embryonic cell lineage of the nematode *Caenorhabditis elegans*. *Dev Biol* 1983 Nov;100(1):64–119.

25. Priess JR, Thomson JN. Cellular interactions in early *C. elegans* embryos. *Cell* 1987 Jan 30;48(2):241–50.

26. Kemphues KJ, Priess JR, Morton DG, Cheng NS. Identification of genes required for cytoplasmic localization in early *C. elegans* embryos. *Cell* 1988 Feb 12;52(3):311–20.

27. Watts JL, Etemad-Moghadam B, Guo S, Boyd L, Draper BW, Mello CC, *et al.* par-6, a gene involved in the establishment of asymmetry in early *C. elegans* embryos, mediates the asymmetric localization of PAR-3. *Development* 1996 Oct;122(10):3133–40.

28. Tabuse Y, Izumi Y, Piano F, Kemphues KJ, Miwa J, Ohno S. Atypical protein kinase C cooperates with PAR-3 to establish embryonic polarity in *Caenorhabditis elegans*. *Development* 1998 Sep;125(18):3607–14.

29. Huang S, Law P, Francis K, Palsson BO, Ho AD. Symmetry of initial cell divisions among primitive hematopoietic progenitors is independent of ontogenic age and regulatory molecules. *Blood* 1999 Oct 15;94(8):2595–604.

30. Suzuki A, Ohno S. The PAR-aPKC system: lessons in polarity. *J Cell Sci* 2006 Mar 15;119(Pt 6):979–87.

31. Gonczy P. Mechanisms of asymmetric cell division: flies and worms pave the way. *Nat Rev Mol Cell Biol* 2008 May;9(5):355–66.

32. Schober M, Schaefer M, Knoblich JA. Bazooka recruits Inscuteable to orient asymmetric cell divisions in Drosophila neuroblasts. *Nature* 1999 Dec 2; 402(6761):548–51.

33. Wodarz A, Ramrath A, Kuchinke U, Knust E. Bazooka provides an apical cue for Inscuteable localization in Drosophila neuroblasts. *Nature* 1999 Dec 2; 402(6761):544–7.

34. Rolls MM, Albertson R, Shih HP, Lee CY, Doe CQ. Drosophila aPKC regulates cell polarity and cell proliferation in neuroblasts and epithelia. *J Cell Biol* 2003 Dec 8;163(5):1089–98.

35. Petronczki M, Knoblich JA. DmPAR-6 directs epithelial polarity and asymmetric cell division of neuroblasts in Drosophila. *Nat Cell Biol* 2001 Jan;3(1):43–9.

36. Wodarz A, Ramrath A, Grimm A, Knust E. Drosophila atypical protein kinase C associates with Bazooka and controls polarity of epithelia and neuroblasts. *J Cell Biol* 2000 Sep 18;150(6):1361–74.

37. Kraut R, Campos-Ortega JA. Inscuteable, a neural precursor gene of Drosophila, encodes a candidate for a cytoskeleton adaptor protein. *Dev Biol* 1996 Feb 25;174(1):65–81.

38. Kraut R, Chia W, Jan LY, Jan YN, Knoblich JA. Role of inscuteable in orienting asymmetric cell divisions in Drosophila. *Nature* 1996 Sep 5;383(6595):50–5.

39. Kaltschmidt JA, Davidson CM, Brown NH, Brand AH. Rotation and asymmetry of the mitotic spindle direct asymmetric cell division in the developing central nervous system. *Nat Cell Biol* 2000 Jan;2(1):7–12.

40. Gho M, Bellaiche Y, Schweisguth F. Revisiting the Drosophila microchaete lineage: a novel intrinsically asymmetric cell division generates a glial cell. *Development* 1999 Aug;126(16):3573–84.

41. de Celis JF, Mari-Beffa M, Garcia-Bellido A. Cell-autonomous role of Notch, an epidermal growth factor homologue, in sensory organ differentiation in Drosophila. *Proc Natl Acad Sci USA* 1991 Jan 15;88(2):632–6.

42. Hartenstein V, Posakony JW. A dual function of the Notch gene in *Drosophila sensillum* development. *Dev Biol* 1990 Nov;142(1):13–30.

43. Fehon RG, Kooh PJ, Rebay I, Regan CL, Xu T, Muskavitch MA, *et al.* Molecular interactions between the protein products of the neurogenic loci Notch and Delta, two EGF-homologous genes in Drosophila. *Cell* 1990 May 4;61(3):523–34.

44. Zeng C, Younger-Shepherd S, Jan LY, Jan YN. Delta and Serrate are redundant Notch ligands required for asymmetric cell divisions within the Drosophila sensory organ lineage. *Genes Dev* 1998 Apr 15;12(8):1086–91.

45. Heitzler P, Simpson P. The choice of cell fate in the epidermis of Drosophila. *Cell* 1991 Mar 22;64(6):1083–92.

46. Rebay I, Fleming RJ, Fehon RG, Cherbas L, Cherbas P, Artavanis-Tsakonas S. Specific EGF repeats of Notch mediate interactions with Delta and Serrate: implications for Notch as a multifunctional receptor. *Cell* 1991 Nov 15; 67(4):687–99.

47. Struhl G, Greenwald I. Presenilin is required for activity and nuclear access of Notch in Drosophila. *Nature* 1999 Apr 8;398(6727):522–5.

48. De Strooper B, Annaert W, Cupers P, Saftig P, Craessaerts K, Mumm JS, *et al.* A presenilin-1-dependent gamma-secretase-like protease mediates release of Notch intracellular domain. *Nature* 1999 Apr 8;398(6727):518–22.

49. Song W, Nadeau P, Yuan M, Yang X, Shen J, Yankner BA. Proteolytic release and nuclear translocation of Notch-1 are induced by presenilin-1 and impaired by pathogenic presenilin-1 mutations. *Proc Natl Acad Sci USA* 1999 Jun 8;96(12):6959–63.

50. Rebay I, Fehon RG, Artavanis-Tsakonas S. Specific truncations of Drosophila Notch define dominant activated and dominant negative forms of the receptor. *Cell* 1993 Jul 30;74(2):319–29.

51. Lieber T, Kidd S, Alcamo E, Corbin V, Young MW. Antineurogenic phenotypes induced by truncated Notch proteins indicate a role in signal transduction and may point to a novel function for Notch in nuclei. *Genes Dev* 1993 Oct;7(10):1949–65.

52. Struhl G, Fitzgerald K, Greenwald I. Intrinsic activity of the Lin-12 and Notch intracellular domains *in vivo*. *Cell* 1993 Jul 30;74(2):331–45.

53. Kitagawa M, Oyama T, Kawashima T, Yedvobnick B, Kumar A, Matsuno K, *et al.* A human protein with sequence similarity to Drosophila mastermind coordinates the nuclear form of notch and a CSL protein to build a transcriptional activator complex on target promoters. *Mol Cell Biol* 2001 Jul;21(13):4337–46.

54. Petcherski AG, Kimble J. Mastermind is a putative activator for Notch. *Curr Biol* 2000 Jun 29;10(13):R471–3.
55. Wu L, Aster JC, Blacklow SC, Lake R, Artavanis-Tsakonas S, Griffin JD. MAML1, a human homologue of Drosophila mastermind, is a transcriptional co-activator for NOTCH receptors. *Nat Genet* 2000 Dec;26(4):484–9.
56. Schroeter EH, Kisslinger JA, Kopan R. Notch-1 signalling requires ligand-induced proteolytic release of intracellular domain. *Nature* 1998 May 28;393(6683):382–6.
57. Jarriault S, Brou C, Logeat F, Schroeter EH, Kopan R, Israel A. Signalling downstream of activated mammalian Notch. *Nature* 1995 Sep 28;377(6547): 355–8.
58. Kidd S, Lieber T, Young MW. Ligand-induced cleavage and regulation of nuclear entry of Notch in *Drosophila melanogaster* embryos. *Genes Dev* 1998 Dec 1;12(23):3728–40.
59. Struhl G, Adachi A. Nuclear access and action of notch *in vivo*. *Cell* 1998 May 15;93(4):649–60.
60. Hsieh JJ, Zhou S, Chen L, Young DB, Hayward SD. CIR, a corepressor linking the DNA binding factor CBF1 to the histone deacetylase complex. *Proc Natl Acad Sci USA* 1999 Jan 5;96(1):23–8.
61. Rhyu MS, Jan LY, Jan YN. Asymmetric distribution of numb protein during division of the sensory organ precursor cell confers distinct fates to daughter cells. *Cell* 1994 Feb 11;76(3):477–91.
62. Guo M, Jan LY, Jan YN. Control of daughter cell fates during asymmetric division: interaction of Numb and Notch. *Neuron* 1996 Jul;17(1):27–41.
63. Spana EP, Doe CQ. Numb antagonizes Notch signaling to specify sibling neuron cell fates. *Neuron* 1996 Jul;17(1):21–6.
64. Frise E, Knoblich JA, Younger-Shepherd S, Jan LY, Jan YN. The Drosophila Numb protein inhibits signaling of the Notch receptor during cell-cell inter-action in sensory organ lineage. *Proc Natl Acad Sci USA* 1996 Oct 15; 93(21):11925–32.
65. Baron F, Storb R. Allogeneic hematopoietic cell transplantation as treatment for hematological malignancies: a review. *Springer Semin Immunopathol* 2004 Nov;26(1–2):71–94.
66. Jacobson LO, Simmons EL, Marks EK, Eldredge JH. Recovery from radia-tion injury. *Science* 1951 May 4;113(2940):510–11.

67. Lorenz E, Uphoff D, Reid TR, Shelton E. Modification of irradiation injury in mice and guinea pigs by bone marrow injections. *J Natl Cancer Inst* 1951 Aug;12(1):197–201.

68. Hofmeister CC, Zhang J, Knight KL, Le P, Stiff PJ. *Ex vivo* expansion of umbilical cord blood stem cells for transplantation: growing knowledge from the hematopoietic niche. *Bone Marrow Transplant* 2007 Jan;39(1): 11–23.

69. Moore KA, Ema H, Lemischka IR. *In vitro* maintenance of highly purified, transplantable hematopoietic stem cells. *Blood* 1997 Jun 15;89(12): 4337–47.

70. Nolta JA, Thiemann FT, Arakawa-Hoyt J, Dao MA, Barsky LW, Moore KA, *et al.* The AFT024 stromal cell line supports long-term *ex vivo* maintenance of engrafting multipotent human hematopoietic progenitors. *Leukemia* 2002 Mar;16(3):352–61.

71. Punzel M, Moore KA, Lemischka IR, Verfaillie CM. The type of stromal feeder used in limiting dilution assays influences frequency and maintenance assessment of human long-term culture initiating cells. *Leukemia* 1999 Jan;13(1):92–7.

72. Shih CC, Hu MC, Hu J, Medeiros J, Forman SJ. Long-term *ex vivo* maintenance and expansion of transplantable human hematopoietic stem cells. *Blood* 1999 Sep 1;94(5):1623–36.

73. Iscove NN, Nawa K. Hematopoietic stem cells expand during serial transplantation *in vivo* without apparent exhaustion. *Curr Biol* 1997 Oct 1;7(10):805–8.

74. Schofield R. The relationship between the spleen colony-forming cell and the haemopoietic stem cell. *Blood Cells* 1978;4(1–2):7–25.

75. Calvi LM, Adams GB, Weibrecht KW, Weber JM, Olson DP, Knight MC, *et al.* Osteoblastic cells regulate the haematopoietic stem cell niche. *Nature* 2003 Oct 23;425(6960):841–6.

76. Kiel MJ, Yilmaz OH, Iwashita T, Terhorst C, Morrison SJ. SLAM family receptors distinguish hematopoietic stem and progenitor cells and reveal endothelial niches for stem cells. *Cell* 2005 Jul 1;121(7):1109–21.

77. Zhang J, Niu C, Ye L, Huang H, He X, Tong WG, *et al.* Identification of the haematopoietic stem cell niche and control of the niche size. *Nature* 2003 Oct 23;425(6960):836–41.

78. Arai F, Hirao A, Ohmura M, Sato H, Matsuoka S, Takubo K, *et al.* Tie2/Angiopoietin-1 signaling regulates hematopoietic stem cell quiescence in the bone marrow niche. *Cell* 2004 Jul 23;118(2):149–61.

79. Hutton JF, D'Andrea RJ, Lewis ID. Potential for clinical *ex vivo* expansion of cord blood haemopoietic stem cells using non-haemopoietic factor supplements. *Curr Stem Cell Res Ther* 2007 Sep;2(3):229–37.

80. Blank U, Karlsson G, Karlsson S. Signaling pathways governing stem-cell fate. *Blood* 2008 Jan 15;111(2):492–503.

81. Leary AG, Strauss LC, Civin CI, Ogawa M. Disparate differentiation in hemopoietic colonies derived from human paired progenitors. *Blood* 1985 Aug;66(2):327–32.

82. Suda T, Suda J, Ogawa M. Disparate differentiation in mouse hemopoietic colonies derived from paired progenitors. *Proc Natl Acad Sci USA* 1984 Apr;81(8):2520–4.

83. Suda J, Suda T, Ogawa M. Analysis of differentiation of mouse hemopoietic stem cells in culture by sequential replating of paired progenitors. *Blood* 1984 Aug;64(2):393–9.

84. Brummendorf TH, Dragowska W, Zijlmans J, Thornbury G, Lansdorp PM. Asymmetric cell divisions sustain long-term hematopoiesis from single-sorted human fetal liver cells. *J Exp Med* 1998 Sep 21;188(6):1117–24.

85. Punzel M, Zhang T, Liu D, Eckstein V, Ho AD. Functional analysis of initial cell divisions defines the subsequent fate of individual human CD34(+) CD38(–) cells. *Exp Hematol* 2002 May;30(5):464–72.

86. Giebel B, Zhang T, Beckmann J, Spanholtz J, Wernet P, Ho AD, *et al.* Primitive human hematopoietic cells give rise to differentially specified daughter cells upon their initial cell division. *Blood* 2006 Mar 1;107(5):2146–52.

87. Ng YY, van Kessel B, Lokhorst HM, Baert MR, van den Burg CM, Bloem AC, *et al.* Gene-expression profiling of CD34+ cells from various hematopoietic stem-cell sources reveals functional differences in stem-cell activity. *J Leukoc Biol* 2004 Feb;75(2):314–23.

88. Theunissen K, Verfaillie CM. A multifactorial analysis of umbilical cord blood, adult bone marrow and mobilized peripheral blood progenitors using the improved ML-IC assay. *Exp Hematol* 2005 Feb;33(2):165–72.

89. Giebel B, Corbeil D, Beckmann J, Hohn J, Freund D, Giesen K, *et al.* Segregation of lipid raft markers including CD133 in polarized human hematopoietic stem and progenitor cells. *Blood* 2004 Oct 15;104(8):2332–8.

90. Beckmann J, Scheitza S, Wernet P, Fischer JC, Giebel B. Asymmetric cell division within the human hematopoietic stem and progenitor cell compartment: identification of asymmetrically segregating proteins. *Blood* 2007 Jun 15;109(12):5494–501.

91. Andrews R, Ahringer J. Asymmetry of early endosome distribution in *C. elegans* embryos. *PLoS ONE* 2007;2(6):e493.

92. Coumailleau F, Furthauer M, Knoblich JA, Gonzalez-Gaitan M. Directional Delta and Notch trafficking in Sara endosomes during asymmetric cell division. *Nature* 2009 Apr 23;458(7241):1051–5.

93. Fonseca AV, Bauer N, Corbeil D. The stem cell marker CD133 meets the endosomal compartment — new insights into the cell division of hematopoietic stem cells. *Blood Cells Mol Dis* 2008 Sep–Oct;41(2):194–5.

94. Tsukazaki T, Chiang TA, Davison AF, Attisano L, Wrana JL. SARA, a FYVE domain protein that recruits Smad2 to the TGFbeta receptor. *Cell* 1998 Dec 11;95(6):779–91.

95. Bokel C, Schwabedissen A, Entchev E, Renaud O, Gonzalez-Gaitan M. Sara endosomes and the maintenance of Dpp signaling levels across mitosis. *Science* 2006 Nov 17;314(5802):1135–9.

96. Panopoulou E, Gillooly DJ, Wrana JL, Zerial M, Stenmark H, Murphy C, *et al.* Early endosomal regulation of Smad-dependent signaling in endothelial cells. *J Biol Chem* 2002 May 17;277(20):18046–52.

97. Punzel M, Liu D, Zhang T, Eckstein V, Miesala K, Ho AD. The symmetry of initial divisions of human hematopoietic progenitors is altered only by the cellular microenvironment. *Exp Hematol* 2003 Apr;31(4):339–47.

4

Ex Vivo Expansion of Cord Blood Hematopoietic Cells

Ian K. McNiece

ABSTRACT

Umbilical cord blood (UCB) provides an alternate source for patients undergoing high dose chemotherapy for treatment of cancer or genetic diseases. In particular, UCB has become a standard therapeutic option for selected patients with hematologic malignancies. Several studies have reported on the use of UCB for transplantation in adult patients, however, low cell doses per cord blood unit have limited the use of CB in this setting due to subsequent delays in engraftment. *Ex vivo* expansion is one approach that is being explored as a means to provide larger cell numbers from individual UCB products. In addition, recent clinical trials have demonstrated the potential to combine two UCB products to increase the cell dose delivered, however the majority of UCB products with sufficient cell dose, have a one or more antigen mismatch to each other or the recipient. Better matched products are available and *ex vivo* expansion may increase the cell doses for these products enabling better matched grafts.

This chapter will present an overview of past clinical trials and the approaches currently being explored.

INTRODUCTION

The cellular content of UCB products has been evaluated in relationship to the time to engraftment and the total nucleated cell dose, CD34$^+$ cell content and myeloid progenitor content (CFU-GM) have been proposed as predictive of outcome.[1-3] UCB products contain similar cell populations to bone marrow (BM) and mobilized peripheral blood progenitor cell products (PBPCs), including hematopoietic stem cells (HSCs), primitive progenitor cells, mature progenitor cells and mature functional cells. However, the total cell number and progenitor cells are much lower in UCB compared to BM and PBPCs. For example, BM and PBPC grafts contain approximately 10^8 CD34$^+$ cells while UCB contains approximately 5×10^6 CD34$^+$ cells. In contrast the frequency of HSCs, as determined by NOD/SCID engraftment is enriched in the CD34$^+$ cell population of UCB compared to BM or PBPCs. As few as 100,000 UCB CD34$^+$ cells can engraft NOD/SCID mice, while approximately one million BM CD34$^+$ cells and five million PBPC CD34$^+$ cells are required for engraftment of human cells.

These numbers would suggest that UCB contains similar levels of HSCs to BM and PBPCs but significantly lower levels of committed progenitor cells. Clinical data are consistent with this theory. Patients transplanted with UCB grafts have delayed neutrophil and platelet engraftment compared to patients transplanted with BM or PBPC products, however there does not appear to be any long term engraftment problems in patients transplanted with UCB grafts. This suggests that UCB products contain sufficient long term engrafting cells (i.e., HSCs), but less short term engrafting cells (mature progenitor cells).

Therefore a simple goal of *ex vivo* expansion would be to generate more committed progenitor cells that have the potential to provide faster short term engraftment. This can be achieved by *ex vivo* culture in hematopoietic growth factors (HGF), however we must consider the potential negative impact of depleting HSCs by driving their differentiation to mature progenitor cells. Therefore the ideal protocol for evaluating *ex vivo* expanded UCB cells involves the use of two products, one component which has been *ex vivo* expanded and the second that has not been manipulated.

The cell dose transplanted into recipients following ablative chemotherapy has been shown to be the major predictor of time to neutrophil and platelet engraftment. The minimum dose proposed for adult patients is 1×10^7 total nuclear cells (TNC) per kg of body weight, while 3.7×10^7 TNC/kg has been proposed for pediatric patients.[4,5]

EX VIVO EXPANDED CELLS PROVIDE RAPID ENGRAFTMENT

The potential enhancement of engraftment by *ex vivo* expanded cells has been demonstrated in clinical trials. Several studies[6–9] have been reported using *ex vivo* expanded PBPC CD34$^+$ cells in myeloablated patients. In these studies the use of *ex vivo* expanded cells resulted in faster neutrophil engraftment, with patients having minimal days of neutropenia compared to patients receiving unexpanded PBPC products. Our own study conducted at the University of Colorado,[6] resulted in neutrophil engraftment as early as four days post-transplant. Analysis of the patient data demonstrated minimal correlation of the time to engraftment to CD34$^+$ cell dose, but demonstrated a highly significant correlation to the dose of total nucleated cells per kg of body weight of the recipient.[6] Evaluation of cytospins prepared of the expanded cells demonstrated a high percentage of mature neutrophil cells. Based upon these data we have focused our experimental protocols on driving differentiation of UCB cells to produce a cellular product that contains a high proportion of mature neutrophil cells. In addition, these conditions drive the production of mature progenitor cells.[6,10] However, these conditions also appear to deplete products of long term engrafting cells as demonstrated by engraftment of fetal sheep[11] again indicating the need for developing clinical protocols that utilize two graft components, one expanded and the second unmanipulated.

IMMUNOMAGNETIC SELECTION OF UCB PRODUCTS FOR *EX VIVO* EXPANSION

A number of approaches have been explored for *ex vivo* expansion of UCB products from liquid culture in bags to expansion in bioreactors. A number of groups have demonstrated that selection of CD34$^+$ cells or

CD133$^+$ cells was necessary for optimal expansion. In 1997 we reported that culture of UCB mononuclear cells (MNCs) in a HGF cocktail including stem cell factor (SCF), granulocyte colony stimulating factor (G-CSF) and thrombopoietin (TPO), resulted in only a 1.4-fold expansion of total cells, 0.8-fold of mature progenitor cells (GM-CFC) and 0.3-fold expansion of erythroid progenitors (BFU-E).[12] In contrast, culture of CD34$^+$ selected UCB cells resulted in 113-fold expansion of total cells, 73-fold expansion of GM-CFC and 49-fold expansion of BFU-E. Based upon these results we initiated expansion cultures in clinical trials with CD34$^+$-selected UCB cells. Processing of clinical products has led us to two conclusions:

(1) Although we can expand significantly the TNC and committed progenitor cells from CD34$^+$ cells we rarely reach pre-selection total cell numbers. For a typical UCB product starting with a cell dose of 1×10^9 TNC and containing 0.5% CD34$^+$ cells, we would obtain a maximum of 5×10^6 CD34$^+$ cells post-selection. Therefore after culture for ten to 14 days we would require a minimum of 200-fold expansion of TNC to obtain preprocessing levels.

(2) The use of unrelated UCB grafts in the clinical setting requires the use of frozen UCB products. However, thawing and selection of frozen UCB products results in significant losses of CD34$^+$ cells (50% or greater loss of CD34$^+$ cells) and often results in low purities.[13] With a 50% recovery of CD34$^+$ cells after selection we now require at least a 400-fold cell expansion to obtain equivalent TNC as we started with. Again, in our experience with clinical studies, the purity of the CD34-selected product impacted the level of expansion achieved. The median purity of CD34$^+$ cells was 47.5% (range 14% to 81%) and the median expansion was 56-fold of TNC.[13] Products with a purity greater than 50% resulted in a median of 139-fold expansion, while products with a purity of less than 50% resulted in only 32-fold expansion. Therefore, in our experience, the use of CD34 selected products has rarely resulted in increased cell doses of *ex vivo* expanded cells compared to the starting unmanipulated product.

AVAILABILITY OF CLINICAL GRADE REAGENTS

A number of approaches have been evaluated for *ex vivo* expansion of UCB products including various culture media, HGF cocktails and various culture vessels (flasks, bags, etc). Most protocols utilize a ten- to 21-day culture in 5% CO_2 incubators. The development of closed culture systems using clinical grade HGF and media is important in order to comply with regulatory requirements. In our culture system we have used three HGF, namely, SCF, G-CSF and TPO as these HGF have been manufactured to GMP (Good Manufacturing Practices) standards. It is most likely that addition of other HGF could enhance the expansion potential of these cultures; however the current lack of GMP-grade HGF inhibits their use in clinical trials. Similarly, media must be manufactured to GMP and there are currently only limited options available. In our initial clinical trials we used a defined medium that was manufactured by Amgen for clinical use. However, Amgen discontinued the production of this medium and in subsequent expansion trials we have been using a proprietary formulation from Sigma (Stemline II expansion medium). Sigma manufactures this defined medium to GMP grade and it has been used in two expansion clinical trials previously conducted at Johns Hopkins University.

Although the use of GMP grade reagents is preferred by the FDA, it is possible to conduct phase I and phase II studies with non-GMP grade reagents, however the progression to a phase III study may be complicated. The regulatory requirements in other countries differ significantly, with some countries requiring all GMP grade reagents.

CLINICAL EXPERIENCE WITH *EX VIVO* EXPANDED CELLS

Despite hundreds of reports of preclinical studies evaluating *ex vivo* expansion of UCB products, only a small number of clinical trials have been conducted to evaluate the clinical potential of *ex vivo* expanded UCB cells. Kurtzberg *et al.* (n = 21 patients) and Stiff *et al.* (n = 9 patients) used the Aastrom system for expansion of UCB cells, however, no significant effects on engraftment kinetics were observed in these patients.[14,15] We have reported the results of a clinical trial we conducted at the University of Colorado (n = 43 patients) and again the conclusion was that the rate of engraftment was not significantly increased by the use of expanded cells.[16]

Recent and Ongoing Trials of *Ex Vivo* Expanded UCB

ViaCell — Selective amplification

A clinical trial was performed by Viacell to evaluate the use of selective amplification in which lineage negative cells (Lin⁻) were isolated and then expanded *in vitro*. A subsequent selection was performed to isolate Lin⁻ cells from the expanded population followed by further expansion *in vitro*. The results of ten patients treated on this study were released in February 2007.[17] In the patients that achieved neutrophil recovery, the median time to engraftment was 24 days. Platelet engraftment was demonstrated in seven patients with a time to recovery of 54 days. While the presence of *ex vivo* expanded cells was detected in some patients at early time points, chimerism from the expanded cells was not detected at 21 to 28 days. Although the *ex vivo* expanded cells were safe, Viacell decided not to advance this approach in future clinical trials.

Gamida Cell — The use of a copper chelator

Preclinical studies performed by Gamida Cell[18] have demonstrated that the copper chelator tetraethylenepentamine (TEPA) can attenuate the differentiation of *ex vivo* cultures of UCB cells, resulting in a preferential expansion of primitive progenitor cells. A phase I/II study was conducted at MD Anderson[19] to test the feasibility and safety of transplantation of CD133⁺ UCB cells cultured in media supplemented with SCF, TPO, Flt-3 ligand (Flt-3L), interleukin-6 (IL-6) and TEPA. Ten patients with advances hematological malignancies were transplanted with a UCB unit which was banked in two aliquots. The smaller fraction was culture-expanded *ex vivo* for 21 days and transplanted 24 hours after infusion of the larger unmanipulated fraction. The median expansion was 100-fold for TNC, 20-fold for CD34⁺ cells and 85-fold for GM-CFC. Nine patients engrafted with a median time to neutrophil and platelet engraftment of

30 (range 16–46) and 48 (range 35–105) days, respectively. These engraftment rates are equivalent to studies using unmanipulated UCB products. Subsequently, the first patient has been enrolled in a pivotal registration study. This study is a single arm, multi-center study to evaluate the efficacy and safety of StemEx (the *ex vivo* expanded UCB cells generated using the culture conditions described above). The study is planned to enrol 100 patients with high-risk hematological malignancies.

MD Anderson randomized trial

deLima and colleagues reported at the 2007 meeting of the American Society for Hematology (ASH), the results of a randomized study of double UCB transplantation comparing two unmanipulated products to one unmanipulated product plus an *ex vivo* expanded product.[20] Forty-eight patients were enrolled on this study. The CD133[+] fraction was isolated from *one of* the products and cultured for 14 days in media containing SCF, G-CSF and TPO. The time to engraftment for neutrophils and platelets was 20 and 41 days, respectively for the cohort receiving expanded cells and 23.5 and 54, respectively for the cohort receiving only unmanipulated UCB products. Chimerism showed that one UCB unit dominated in all patients. Among 19 evaluable patients receiving *ex vivo* expanded cells, in 63% the unmanipulated unit provided 100% of hematopoiesis from day 30. The expanded unit *was* predominant in three cases (for 12, two, and two months) and present but not dominant in four patients (5%–25%, for 12, seven, three, and three months). The conclusion from this study was that the use of expanded cells was safe, and that the fold-expansion was highly variable.

These studies suggest that the culture conditions currently being used are not capable of expanding the appropriate cell population or that insufficient numbers are being generated to impact the time to recovery of neutrophils or platelets. Our conclusion from our own experience and data is that the requirement for selection of CD34[+] cells or CD133[+] cells from frozen UCB products greatly minimizes the potential of generating a suitable expanded UCB product enhancing the rate of engraftment. Therefore, in recent studies we have evaluated methods for expanding UCB products without an initial selection for CD34[+] or CD133[+] cells.

Ex vivo expansion of cord blood MNCs on mesenchymal stem cells (MSCs)

We have developed a coculture system which is capable of expanding UCB MNCs by culturing the UCB MNCs on confluent MSC layers. The literature contains many reports of the ability of MSC to support the growth of hematopoietic cells (e.g., HSCs). It has been demonstrated that MSCs produce a number of HGF and adhesion molecules that may stimulate growth of hematopoietic cells.[13] Our initial data reproducibly demonstrated a ten- to 20-fold expansion of TNC with 18-fold expansion of GM-CFC and 16- to 37-fold expansion of CD34$^+$ cells.

In previous studies we have evaluated the potential of *ex vivo* expansion of frozen UCB products using the coculture on MSCs. Five UCB units were thawed and washed, resulting in a median recovery of 3.3×10^8 TNC (range 1.4 to 3.6×10^8, n = 5). For a 50 kg recipient, these UCB products would provide only 0.73×10^7 TNC/kg of body weight with none of the five thawed products reaching the minimal target dose of 1×10^7 TNC.

Each unit was expanded by culturing the MNC fraction on preformed layers of MSCs. Ten T162 cm^2 flasks were used for each UCB such that each flask contained 10% of the UCB MNC. After *ex vivo* culture for 14 days, in a cocktail of SCF, G-CSF and TPO in Stemline II media, a median of nine-fold expansion of TNC was obtained with a range of 6.5- to 24-fold. The median TNC post-expansion was 21.6×10^8 (range 11 to 79×10^8 TNC). A median expansion of mature progenitor cells (GM-CFC) of 46-fold was also obtained in the coculture. For a 50 kg recipient, the expanded UCB product would be equivalent to 4.3×10^7 TNC/kg (range 2.2 to 16×10^7), with all five expanded products reaching the minimal target of 1×10^7 TNC/kg. In fact all expanded products contained more the 1×10^7 TNC/kg based upon a 100 kg recipient.

We would propose two potential advantages to the use of coculture for expansion based upon these results. Firstly, the possible enhanced engraftment and secondly, the ability to use better matched UCB products that may have a low cell dose. Wagner and colleagues have described the use of two UCB products to provide an increased cell dose, however, the majority of patients receive a two antigen miss-matched UCB unit.[21] Better matched UCB units are routinely identified but are not suitable due to low cell doses. The expansion of the better matched UCB units could potentially decrease the graft versus host disease that can result.

A clinical trial to evaluate the potential of *ex vivo* expanded cells generated using this coculture approach, is currently being conducted at MD Anderson by Dr. E. J. Shpall. The preliminary data from these studies show a median time to neutrophil engraftment for six patients of 13 days (personal communication from E. J. Shpall).

Notch-mediated expansion of UCB products

Bernstein and colleagues[22] reported the results of a trial conducted at the Fred Hutchinson Cancer Research Center using the Notch ligand, Delta 1 to expand UCB CD34$^+$ cells. Patients received the expanded UCB product plus a second unexpanded UCB product. To date, six patients have been treated on this protocol and the median time to neutrophil engraftment was 14 days (range seven to 34 days) for patients receiving expanded cells compared to 25 days (range 16 to 48 days), in patients (n = 17) undergoing an identical transplant regimen, but with two non-cultured UCB units.

SUMMARY

Despite more than a decade of research, the clinical results for *ex vivo* expanded UCB products have not provided a major impact on time to engraftment of recipients. New approaches are currently being developed and recent data from two separate trials described above, using different culture approaches, have demonstrated faster neutrophil engraftment. These initial results need to be repeated in larger clinical trials but provide the first evidence that *ex vivo* expanded UCB cells can enhance engraftment. Based upon these clinical results it will be possible to better characterize the expanded cells products and potentially identify the cell population within the expanded products that is responsible for the faster neutrophil engraftment. This in turn may provide a basis for optimization of expansion methodologies leading to further enhanced clinical outcomes.

Disclosure Statement:

Under a licensing agreement between ViaCell Inc., and the Johns Hopkins University, Ian McNiece is entitled to a share of royalty received by the University on sales of products described in this article. The terms of this arrangement are being managed by the Johns Hopkins University in accordance with its conflict of interest policies.

References

1. Gluckman E, Rocha V. Donor selection for unrelated cord blood transplants. *Curr Opin Immunol* 2006 Oct;18(5):565–70.
2. Laughlin MJ, Barker J, Bambach B, Koc O, Rizzieri D, Wagner J, *et al.* Hematopoietic engraftment and survival in adult recipients of umbilical cord blood from unrelated donors. *N Engl J Med* 2001;344(24):1815–22.
3. Migliaccio AR, Adamson JW, Stevens CE, Dobrila NL, Carrier CM, Rubinstein P. Cell dose and speed of engraftment in placental/umbilical cord blood transplantation: graft progenitor cell content is a better predictor than nucleated cell quantity. *Blood* 2000 Oct 15;96(8):2717–22.
4. Gluckman E, Rocha V, Boyer-Chammard A, Locatelli F, Arcese W, Pasquini R, *et al.* Outcome of cord-blood transplantation from related and unrelated donors. Eurocord Transplant Group and the European Blood and Marrow Transplantation Group. *N Engl J Med* 1997;337(6):373–81.
5. Takahashi S, Ooi J, Tomonari A, Konuma T, Tsukada N, Tojo A, *et al.* Posttransplantation engraftment and safety of cord blood transplantation with grafts containing relatively low cell doses in adults. *Int J Hematol* 2006;84(4):359–62.
6. McNiece I, Jones R, Bearman SI, Cagnoni P, Nieto Y, Franklin W, *et al.* *Ex vivo* expanded peripheral blood progenitor cells provide rapid neutrophil recovery after high-dose chemotherapy in patients with breast cancer. *Blood* 2000 Nov 1;96(9):3001–7.
7. Paquette RL, Dergham ST, Karpf E, Wang H-J, Slamon DJ, Souza L, *et al.* *Ex vivo* expanded unselected peripheral blood: progenitor cells reduce post-transplantation neutropenia, thrombocytopenia, and anemia in patients with breast cancer. *Blood* 2000 Oct 1;96(7):2385–90.
8. Prince HM, Paul JS, Genevieve W, Dominic PW, Lesley B, Guy CT, *et al.* Improved haematopoietic recovery following transplantation with *ex vivo* expanded mobilized blood cells. *Br J Haematol* 2004;126(4):536–45.

9. Reiffers J, Cailliot C, Dazey B, Attal M, Caraux J, Boiron J. Abrogation of post-myeloablative chemotherapy neutropenia by *ex vivo* expanded autologous CD34-positive cells. *Lancet* 1999;354(9184):1092–3.

10. McNiece I, Kubegov D, Kerzic P, Shpall EJ, Gross S. Increased expansion and differentiation of cord blood products using a two-step expansion culture. *Exp Hematol* 2000;28(10):1181–6.

11. McNiece I, Almeida-Porada G, Shpall E, Zanjani E. *Ex vivo* expanded cord blood cells provide rapid engraftment in fetal sheep but lack long-term engrafting potential. *Exp Hematol* 2002;30(6):612–6.

12. Briddell R, Kern B, Zilm K, Stoney G, McNiece I. Purification of CD34$^+$ cells is essential for optimal *ex vivo* expansion of umbilical cord blood cells. *J Hematother* 1997;6(2):145–50.

13. McNiece I, Harrington J, Turney J, Kellner J, Shpall EJ. *Ex vivo* expansion of cord blood mononuclear cells on mesenchymal stem cells. *Cytotherapy* 2004;6(4):311–7.

14. Jaroscak J, Goltry K, Smith A, Waters-Pick B, Martin PL, Driscoll TA, *et al.* Augmentation of umbilical cord blood (UCB) transplantation with *ex vivo* expanded UCB cells: results of a phase 1 trial using the AastromReplicell System. *Blood* 2003 June 15;101(12):5061–7.

15. Pecora A, Stiff P, LeMaistre C, Bayer R, Bachier C, Goldberg S, *et al.* A phase II trial evaluating the safety and effectiveness of the AastromReplicell system for augmentation of low-dose blood stem cell transplantation. *Bone Marrow Transplant* 2001 Aug;28(3):295–303.

16. Shpall EJ, Quinones R, Giller R, Zeng C, Baron AE, Jones RB, *et al.* Transplantation of *ex vivo* expanded cord blood. *Biol Blood Marrow Transplant* 2002;8(7):368–76.

17. Viacell website, www.viacellinc.com. Press release, February 6, 2007.

18. Peled T, Mandel J, Goudsmid RN, Landor C, Hasson N, Harati D, *et al.* Pre-clinical development of cord blood-derived progenitor cell graft expanded *ex vivo* with cytokines and the polyamine copper chelator tetraethylenepentamine. *Cytotherapy* 2004;6(4):344–55.

19. Shpall E, Robinson S, de Lima M. Transplantation of *ex vivo* expanded cord blood. In: *Conference on Biology and Clinical Applications of Cord Blood Cells* 2007 Oct; 19–21.

20. de Lima M, McMannis J, Komanduri K, *et al.* Randomized study of double cord blood transplantation (CBT) with versus without *ex vivo* expansion (EXP). *Blood* 2007;118(11):599a.

21. Barker JN, Weisdorf DJ, DeFor TE, Blazar BR, McGlave PB, Miller JS, *et al.* Transplantation of 2 partially HLA-matched umbilical cord blood units to enhance engraftment in adults with hematologic malignancy. *Blood* 2005 Feb 1;105(3):1343–7.

22. Delaney C, Varnum-Finney B, Aoyama K, Brashem-Stein C, Bernstein ID. Dose-dependent effects of the Notch ligand Delta1 on *ex vivo* differentiation and *in vivo* marrow repopulating ability of cord blood cells. *Blood* 2005 Oct 15;106(8):2693–9.

PART II

Mesenchymal Stem Cells (MSCs)

5

Mesenchymal Stem Cells

Mark F. Pittenger

ABSTRACT

Mesenchymal stem cells (MSCs), sometimes referred to as marrow stromal cells or multipotential stromal cells, represent a class of adult progenitor cells capable of differentiation to several mesenchymal lineages. A number of different names for MSCs describe similar but perhaps not identical multipotential cells of mesenchymal origin isolated in different labs — mesenchymal stem cells, mesenchymal progenitor cells, multipotential mesenchymal stromal cells, and others — but the abbreviation MSCs continues to be used by each of these laboratories. They can be isolated from many tissues although bone marrow has been used most often. MSCs isolated from different species from rat through man have very similar properties. MSCs may prove useful for repair and regeneration of a variety of mesenchymal tissues such as bone, cartilage, muscle, marrow stroma, and MSCs produce beneficial growth factors and cytokines that may help repair additional tissues. There is also evidence for their differentiation to non-mesenchymal lineages. MSCs interact with cells of the immune system in ways that allows them to be transplanted across immunological barriers and this is the subject of subsequent chapters. This chapter will provide a brief background on the origins of MSC research.

INTRODUCTION

Mesenchymal stem cells (MSCs) are being studied by an ever increasing number of laboratories, leading to a rich and complex research field. These cells have the potential to teach us much about mesenchymal cell biology, control of gene expression, cellular differentiation, cellular responses to tissue damage, and initiation of repair. As models of stem cells, they can teach us about control of proliferation, differentiation signals, and the genes that regulate endodermal, mesenchymal, and ectodermal decisions and maintenance. As building blocks for tissue engineering, they can help us design strategies for creating complex tissues. As players in the body's immune system, MSCs are a useful tool to help dissect complex innate and adaptive responses, and understand whether MSCs can be used successfully for allogeneic as well as autologous transplantation. While the field of MSC research and applications is expanding, it is useful to look back and review earlier reports of these cells as there is continuing complexity that can confound experts as well as researchers new to the field.

TISSUE REPAIR AND REGENERATION

Healing damaged tissue takes place more rapidly in children than in adults and this is likely due to a number of factors. One of those factors appears to be the abundance of stem and progenitor cells including MSCs in the developing tissues of the child. As we reach adulthood, these cells are less necessary for tissue growth and appear to diminish over time, perhaps some differentiating to adult cell types while some are likely retained as resident tissue stem cells.

Over the years, the number of tissue resident stem cells further diminishes as they are called upon for normal tissue repair and maintenance, and normal cellular senescence. Still, there appear to be small numbers of stem cells or progenitor cells such as MSCs that can be isolated from many tissues at later stages of life. These cells offer a wonderful opportunity, indeed, a responsibility, to understand important aspects of human biology involving tissue repair and regeneration. One of these adult stem cells that can be isolated throughout life and propagated in culture was termed the mesenchymal stem cell or MSC by Arnold Caplan of Case Western University.[1] A key element in the potential acceptance of a preparation of MSCs as a cellular therapeutic was the early demonstration of safety in humans by Caplan and colleagues of University Hospitals of Cleveland, who first tested MSCs as support cells during hematopoietic stem cell (HSC) transplantations.[2] Caplan and colleagues were the first to systematically pursue the isolation of the MSC from human tissue. The reasoning was that a human source of MSCs should be sought that could be harvested as a simple procedure in the doctor's office under local anesthetic, without sacrificing or harming the tissue that was to be repaired and regenerated. In this case, bone marrow was chosen to be a good choice as it is remarkably renewable, harbors MSCs, is a known source of HSCs, and marrow can be isolated from the marrow cavity of the hipbone in a simple procedure under local anesthetic.

Our efforts on the isolation of MSCs and examination of their multipotential nature resulted in evidence for the controlled *in vitro* multilineage differentiation of human MSCs to adipo-, chondro- and osteogenic fates (Fig. 1). Our key technical contributions at this time were the demonstration of reproducible isolation techniques, clonal analysis and *in vitro* assays for the differentiation of MSCs and this opened up the field to many more investigators.[3]

MSCs ACROSS SPECIES AND TISSUES

Probably the first descriptions of fibroblastic cells that could be isolated and grown from bone marrow samples and that retained the ability to differentiate to bone tissue was presented by Dr. Alexander Friedenstein of the Gamalaya Institute in Moscow in the 1960s, using guinea pig bone marrow as the source.[4,5] When these cells were culture expanded *ex vivo*, and then placed in capsules under the skin of a recipient syngeneic animal, new bone and cartilage tissue was identified when histology was performed. Maureen Owen and colleagues performed similar experiments using rabbit bone marrow as a source of cells.[6,7] Although the same type of cell, or a close homologue of the MSC, can be found in many tissues, including adipose tissue,[8] three sources including bone marrow, the endosteal surface of bone, and bone itself, have proven to be reproducible

(A)

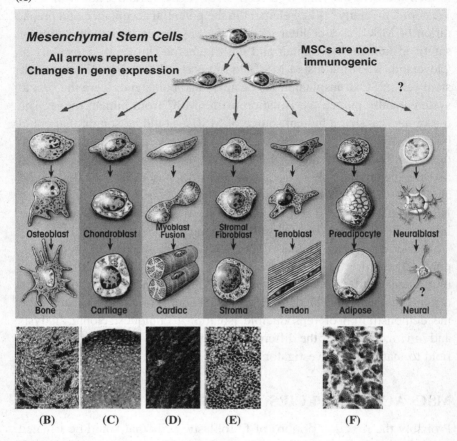

Fig. 1. Differentiation of MSCs to mesenchymal lineages (**A**) and examples of osteogenic differentiation as shown by alkaline phosphatase reaction product in pink and von Kossa staining for calcium deposits (**B**), chondrogenic differentiation as shown be collagen II immunostaining (**C**), MSCs (labeled red) engrafted in the heart showing desmin counter staining (**D**), cobblestone formation by HSCs on a bed of MSCs as a stromal feeder layer (**E**) and adipogenic differentiation as revealed by lipid vacuoles stained with oil red O (**F**).

and convenient sources of MSCs from all species tested. MSCs also have been isolated from mouse,[9] rat,[10] rabbit,[11] dog,[12] goat,[13] pig,[14] and non-human primates.[15] With mouse MSCs extra care must be taken to eliminate mouse HSCs that survive in the MSC culture conditions. With other species, the HSCs do not survive the standard MSC culture conditions. There are also a number of papers that describe MSC-like cells derived from microvascular pericytes, or mural cells, from several species. If these cells are also MSCs, then MSCs are present in all tissues in small numbers.

CHARACTERIZATION OF MSCs

MSCs are quite rare in the body of adults, although they may be abundant at early stages of development. Because they are rare, about 100 times more rare than hematopoietic stem cells, MSCs are difficult to isolate and study as primary cells. Thus, it is the cultured MSCs that have been expanded ~1000-fold or more, that are most often used in laboratory studies. Part of the ongoing debate is whether the MSCs that are grown from tissue biopsies represent a true cell that is present in the body, or whether they are a result of the culture conditions and therefore a "culture artifact." Perhaps this question will never be completely answered. Cells from the body may be affected by methods used to isolate them. All cultured cells are a product of the culture conditions used to produce them and primary and secondary cultures of multipotent cells are especially prone to reflecting their culture conditions. That is, multipotent cells of a single genotype cultured under different conditions in the same laboratory will have a different phenotype when analyzed. However, it is clear that laboratories thousands of miles apart can produce MSCs of quite similar phenotype from different donors (different genotype) when the culture conditions and assay methods are carefully standardized. This is the one of the central tenets of the MSC therapies now undergoing clinical trials.

MSCs are frequently characterized by flow cytometry methods to detect the surface molecules on the cells. This methodology is useful to detect contaminating cells as well. For this, the MSCs can be treated with trypsin to detach them from the surface without interfering with subsequent staining by most antibodies. A list of surface molecules found on MSCs is shown in Table 1. These markers represent receptors for interaction of MSCs with other cell types, extracellular matrix, etc. They do not identify MSCs as stem cells — only biological assays can do that — but the markers are useful to characterize the expanded cells.

Table 1.　Surface antigen detection on MSCs.

CD11a,b	–
CD13	+
CD14	–
CD18 (Integrin β2)	–
CD29	+
CD31 (PECAM)	–
CD34	–
CD44	+
CD45	–*
CD49b (Integrin α2)	+
CD49d (Integrin α4)	–
CD49e (Integrin α5)	+
CD50 (ICAM3)	–
CD54 (ICAM1)	+
CD62E (E-Selectin)	–
CD71 (transferrin R.)	+
CD73 (SH-2)	+
CD90 (Thy-1)	+
CD105 (Endoglin)	+
CD106 (VCAM)	+
CD117	–
CD133	–
CD166 (ALCAM)	+
Galectin-1	+
Nestin	+
CD271	+
HLA-A,B,C	+
HLA-DR	Inducible with IFN
SSEA 3, 4	+
Oct 4	+
Trk A,B,C	+
Differentiation to lineages	
Adipogenic	+
Osteogenic	+
Chondrogenic	+
Stromal	+
Neural	(+)
Myoblastic	(+)
Endothelial	(+)

Notes: + Positive; – Negative; (+) Detection varied; *Positive at isolation, lost in culture.

FUNCTIONAL CHARACTERISTICS OF MSCs

In addition to their ability to differentiate to mesenchymal cell types, MSCs secrete growth factors and cytokines that have autocrine and paracrine activities. The MSCs produce vascular endothelial growth factor (VEGF), stem cell factor (SCF-1), leukemia inhibitory factor (LIF), granulocyte colony stimulatory factor (G-CSF), macrophage colony stimulating factor (M-CSF), granulocyte-macrophage colony stimulating factor (GM-CSF), interleukins (IL-1, -6, -7, -8, -11, -14, -15), stromal cell-derived factor 1 (SDF-1), Flt-3 ligand, and others.[16] The expression of these factors may be modulated through interactions with other cell types.[17]

Additional interesting and important aspects of MSCs that have come to light include their homing to sites of tissue injury, particularly ischemic regions of heart[18–20] where MSCs may prevent deleterious remodeling.[21] MSCs also have the ability to modify immune responses and engraft in allogeneic recipients, and MSC treatment has been used to clinically treat graft vs. host disease (GVHD).[22] MSCs have surface molecules including HLA-1, VCAM-1, LFA-3, and ICAM-1 that allow intimate interactions with immune cells. MSCs also express galectin-1, a β-galactoside binding protein with growth and immunomodulatory properties. MSCs secrete a number of factors that modulate immune cells including transforming growth factor-β (TGF-β), hepatocyte growth factor (HGF), prostaglandin E2 (PGE2), IL-6 and IL-8. MSCs are also under evaluation for clinical use in children with osteogenesis imperfecta, and glycogen storage diseases.[23,24]

UMBILICAL CORD MSCs

Bone marrow has proven to be a reliable source of MSCs but several other tissues including fetal tissues, amniotic fluid, placenta and umbilical cord have also emerged as sources of MSCs. Fetal tissue can provide MSCs that reflect cells found during early development, but this is a controversial source and access is unreliable.[25,26] Amniotic fluid and the amnion are not easily accessible either, but some researchers have reported successful isolation of MSCs.[27] Placenta is a discarded tissue that may be amenable to large scale processing for MSCs.[28,29] Celgene Cellular Therapy, Inc. is pursuing the isolation of MSCs from placenta for commercial tissue regeneration (http://www.aboutus.org/Celgene.com).

Cord blood has been used for over a decade as a source of hematopoietic stem and progenitor cells for patients requiring a bone marrow transplant as part of their cancer therapy. Cord researchers therefore turned their efforts to isolate MSCs from umbilical cord. When the routine method of term cord blood collection has been used for the isolation of MSC-like cells, results were variable.[30] This is likely due to the very low numbers of MSCs found in free-flowing blood such as peripheral blood, or blood collected from growth factor mobilized donors.[31] Isolations that successfully yielded UCB-MSCs showed that the cells had qualities similar to BM-MSCs in their surface molecules, expression of cytokines and growth factors, and their ability to differentiate. Therefore, it is the ability to expand MSCs in culture that makes cord blood a potentially useful source for MSC isolation.

Another source of umbilical cord-derived MSCs is to use the cord tissue as starting material, rather than cord blood (Fig. 2). The loose connective tissue known as Wharton's jelly surrounding the umbilical arteries and vein have been extruded, minced and placed in culture for the purpose of cell isolation. The cells that attach and proliferate form colonies of spindle-shaped fibroblastic cells, which can differentiate to adipogenic, chondrogenic, and osteogenic lineages.[32] Special interests in Wharton's jelly-derived MSCs has been generated by their potential use in neurological disorders such as Parkinson's or stroke.[33] MSCs derived from Wharton's jelly have also been reported to have the ability to differentiate to muscle[34] and to generate pancreatic islet-like cells for potential use in type I diabetes.[35]

Given the early difficulty isolating MSC-like cells from the cord blood, and knowing that MSCs are usually attached and associated with matrix elements in bone marrow, Smirnov and colleagues took a slightly different approach to isolating UCB cells. The authors discarded the flowing blood and filled the cord vein with medium containing collagenase to break down the extracellular matrix,[36] similar to human umbilical vein endothelial cell (HUVEC) isolation. The dislodged cells were collected and incubated in DMEM rather than the Medium 199 normally used for HUVEC isolation, and serum selected for MSC growth by the manufacturer.

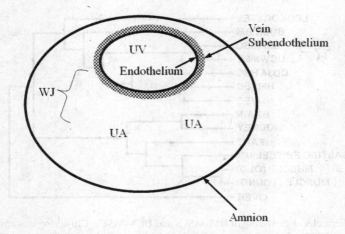

Fig. 2. The umbilical cord, shown in cross-section, provides a variety of domains that have been used to isolate MSC-like cells including the gelatinous connective tissue surrounding the vein and arteries referred to as Wharton's jelly (WJ), the amnion, the vein, and the vein subendothelium. UV: Umbilical vein, UA: umbilical arteries.

Although HUVECs are also isolated by this procedure, most of the resulting cells are fibroblastic MSC-like cells rather than the flat, cuboidal HUVECs, which become less than 1% after 20 days of culture, as shown by antibody staining for the endothelial marker PECAM-1. The remainder of the cells can be considered UCB-MSCs, and were shown to differentiate into adipocytes and osteoblasts.

Serial analysis of gene expression (SAGE) analysis uses molecular hybridization probes for expressed genes and scores the abundance of each RNA within the cell population. This method has been used previously to analyze a limited subset of 2300 genes in BM-MSCs. In order to compare UCB-MSCs and BM-MSCs, Panepucci and colleagues isolated BM-MSCs and then UCB-MSCs from the subendothelial layer of the umbilical cord vein similar to Smirnov and colleagues.[37] The authors then compared the two cell populations for growth characteristics, surface phenotype by flow cytometry, and differentiation ability and found the two populations of cells remarkably similar. They performed SAGE analysis using >100,000 tags representing ~29,000 unique tags that included >18,000 known genes. Not surprisingly, the UCB-MSCs and BM-MSCs

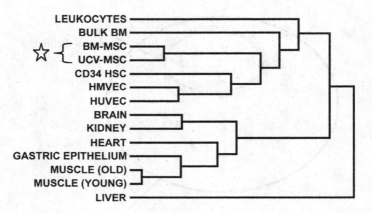

Fig. 3. Hierarchical clustering of BM-MSCs and UCV-MSCs. Clustering was carried out with the first 500 most frequent tags of each of 14 libraries obtained from normal human tissues. BM: Bone marrow, HMVEC: human microvascular endothelial cell, HSC: hematopoietic stem cell, HUVEC: human umbilical vein endothelial cell, MSC: mesenchymal stem cell, UCV: umbilical cord vein. Reprinted with permission from Panepucci *et al.* (2004).[37]

had similar gene expression patterns and strongly resembled each other. Using the most abundant 500 transcripts and cluster analysis, the authors were able to show that the two types of MSCs were closely related to each other, but only distantly related to HUVECs, HSCs, leukocytes, brain, liver, heart and other cell types tested (Fig. 3). Interestingly, UCB-MSCs appeared to express higher levels of the vascular/angiogenesis genes while the BM-MSCs had somewhat higher expression of osteoblast related genes.

Recently, Troyer and Weiss reviewed the literature concerning "primitive stromal cells" derived from Wharton's jelly and the perivascular cells of the cord, while Flynn and colleagues have discussed the MSCs derived from UCB.[38,39] Continued study of UCB-MSCs and comparison with BM-MSCs will yield greater understanding of these important stem cell populations and hopefully lead to significant health benefits in the not-so-distant future. Clearly, MSCs isolated from a variety of tissues represent new, exciting and potentially powerful paradigms for cellular therapy. MSCs from umbilical cord and cord blood, as well as the immunological properties of BM-MSCs and UCB-MSCs are discussed in more detail in the following chapters.

References

1. Caplan AI. Mesenchymal stem cells. *J Orthop Res* 1991 Sep;9(5):641–50.
2. Lazarus HM, Haynesworth SE, Gerson SL, Rosenthal NS, Caplan AI. *Ex vivo* expansion and subsequent infusion of human bone marrow-derived stromal progenitor cells (mesenchymal progenitor cells): implications for therapeutic use. *Bone Marrow Transplant* 1995 Oct;16(4):557–64.
3. Pittenger MF, Mackay AM, Beck SC, Jaiswal RK, Douglas R, Mosca JD, *et al.* Multilineage potential of adult human mesenchymal stem cells. *Science* 1999;284(5411):143–7.
4. Friedenstein AJ, Chailakhjan RK, Lalykina KS. The development of fibroblast colonies in monolayer cultures of guinea-pig bone marrow and spleen cells. *Cell Tissue Kinet* 1970 Oct;3(4):393–403.
5. Friedenstein AJ, Petrakova KV, Kurolesova AI, Frolova GP. Heterotopic of bone marrow. Analysis of precursor cells for osteogenic and hematopoietic tissues. *Transplantation* 1968 Mar;6(2):230–47.
6. Ashton B, Allen T, Howlett C, Eaglesom C, Hattori A, Owen M. Formation of bone and cartilage by marrow stromal cells in diffusion chambers *in vivo*. *Clin Orthop Relat Res* 1980 Sep;151:294–307.
7. Luria E, Owen M, Friedenstein A, Morris J, Kuznetsow S. Bone formation in organ cultures of bone marrow. *Cell Tissue Res* 1987;248(2):449–54.
8. Zuk PA, Zhu M, Ashjian P, De Ugarte DA, Huang JI, Mizuno H, *et al.* Human adipose tissue is a source of multipotent stem cells. *Mol Biol Cell* 2002 Dec;13(12):4279–95.
9. Phinney D, Gene Kopen G, Isaacson R, Prockop D. Plastic adherent stromal cells from the bone marrow of commonly used strains of inbred mice: variations in yield, growth, and differentiation. *J Cell Biochem* 1999;72(4):570–85.
10. Bruder S, Jaiswal N, Ricalton N, Mosca J, Kraus K, Kadiyala S. Mesenchymal stem cells in osteobiology and applied bone regeneration. *Clin Orthop Relat Res* 1998;355(Suppl):S247–56.
11. Wakitani S, Goto T, Pineda S, Young R, Mansour J, Caplan A, *et al.* Mesenchymal cell-based repair of large, full-thickness defects of articular cartilage. *J Bone Joint Surg Am* 1994;76(4):579–92.
12. Kadiyala S, Young R, Thiede M, Bruder S. Culture expanded canine mesenchymal stem cells possess osteochondrogenic potential *in vivo* and *in vitro*. *Cell Transplant* 1997;6(2):125–34.

13. Kraitchman DL, Tatsumi M, Gilson WD, Ishimori T, Kedziorek D, Walczak P, *et al.* Dynamic imaging of allogeneic mesenchymal stem cells trafficking to myocardial infarction. *Circulation* 2005 Sep 6;112(10):1451–61.

14. Murphy J, Fink D, Hunziker E, Barry F. Stem cell therapy in a caprine model of osteoarthritis. *Arthritis Rheum* 2003;48(12):3464–74.

15. Bartholomew A, Patil S, Mackay A, Nelson M, Buyaner D, Hardy W, *et al.* Baboon mesenchymal stem cells can be genetically modified to secrete human erythropoietin *in vivo*. *Hum Gene Ther* 2001;12(12):1527–41.

16. Haynesworth S, Baber M, Caplan A. Cytokine expression by human marrow-derived mesenchymal progenitor cells *in vitro*: effects of dexamethasone and IL-1 alpha. *J Cell Physiol* 1996;166(3):585–92.

17. Aggarwal S, Pittenger MF. Human mesenchymal stem cells modulate allogeneic immune cell responses. *Blood* 2005 Feb 15;105(4):1815–22.

18. Kraitchman DL, Tatsumi M, Gilson WD, Ishimori T, Kedziorek D, Walczak P, *et al.* Dynamic imaging of allogeneic mesenchymal stem cells trafficking to myocardial infarction. *Circulation* 2005 Sep 6;112(10):1451–61.

19. Bittira B, Shum-Tim D, Al-Khaldi A, Chiu RC. Mobilization and homing of bone marrow stromal cells in myocardial infarction. *Eur J Cardiothorac Surg* 2003 Sep;24(3):393–8.

20. Kraitchman DL, Heldman AW, Atalar E, Amado LC, Martin BJ, Pittenger MF, *et al. In vivo* magnetic resonance imaging of mesenchymal stem cells in myocardial infarction. *Circulation* 2003 May 13;107(18):2290–3.

21. Shake J, Gruber P, Baumgartner W, Senechal G, Meyers J, Redmond J, *et al.* Mesenchymal stem cell implantation in a swine myocardial infarct model: engraftment and functional effects. *Ann Thorac Surg* 2002;73(6):1919–25.

22. Le Blanc K, Rasmusson I, Sundberg B, Götherström C, Hassan M, Uzunel M, *et al.* Treatment of severe acute graft-versus-host disease with third party haploidentical mesenchymal stem cells. *Lancet* 2004;363:1439–41.

23. Horwitz EM, Gordon PL, Koo WK, Marx JC, Neel MD, McNall RY, *et al.* Isolated allogeneic bone marrow-derived mesenchymal cells engraft and stimulate growth in children with osteogenesis imperfecta: implications for cell therapy of bone. *Proc Natl Acad Sci USA* 2002 Jun 25;99(13): 8932–7.

24. Koç ON, Day J, Nieder M, Gerson SL, Lazarus HM, Krivit W. Allogeneic mesenchymal stem cell infusion for treatment of metachromatic leukodys-trophy (MLD) and Hurler syndrome (MPS-IH). *Bone Marrow Transplant* 2002 Aug;30(4):215–22.

25. Gotherstrom C, Ringden O, Westgren M, Tammik C, Le Blanc K. Immunomodulatory effects of human foetal liver-derived mesenchymal stem cells. *Bone Marrow Transplant* 2003 Aug;32(3):265–72.

26. In 't Anker P, Noort W, Scherjon S, Kleijburg-van der Keur C, Kruisselbrink A, van Bezooijen R, *et al*. Mesenchymal stem cells in human second-trimester bone marrow, liver, lung, and spleen exhibit a similar immunophenotype but a heterogeneous multilineage differentiation potential. *Haematologica* 2003 Aug 1;88(8):845–52.

27. In 't Anker PS, Scherjon SA, Kleijburg-van der Keur C, Noort WA, Claas FH, Willemze R, *et al*. Amniotic fluid as a novel source of mesenchymal stem cells for therapeutic transplantation. *Blood* 2003 Aug 15;102(4):1548–9.

28. Fukuchi Y, Nakajima H, Sugiyama D, Hirose I, Kitamura T, Tsuji K. Human placenta-derived cells have mesenchymal stem/progenitor cell potential. *Stem Cells* 2004;22(5):649–58.

29. Wulf G, Viereck V, Hemmerlein B, Haase D, Vehmeyer K, Pukrop T, *et al*. Mesengenic progenitor cells derived from human placent. *Tissue Eng* 2004; 10(7–8):1136–47.

30. Mareschi K, Biasin E, Piacibello W, Aglietta M, Madon E, Fagioli F. Isolation of human mesenchymal stem cells: bone marrow versus umbilical cord blood. *Haematologica* 2001 Oct;86(10):1099–100.

31. Tondreau T, Meuleman N, Delforge A, Dejeneffe M, Leroy R, Massy M, *et al*. Mesenchymal stem cells derived from CD133-positive cells in mobilized peripheral blood and cord blood: proliferation, Oct4 expression, and plasticity. *Stem Cells* 2005;23(8):1105–12.

32. Wang H, Hung S, Peng S, Huang C, Wei H, Guo Y, *et al*. Mesenchymal stem cells in the Wharton's jelly of the human umbilical cord. *Stem Cells* 2004;22(7):1330–7.

33. Weiss ML, Medicetty S, Bledsoe AR, Rachakatla RS, Choi M, Merchav S, *et al*. Human umbilical cord matrix stem cells: preliminary characterization and effect of transplantation in a rodent model of Parkinson's disease. *Stem Cells* 2006 Mar;24(3):781–92.

34. Conconi M, Burra P, Di Liddo R, Calore C, Turetta M, Bellini S, *et al*. CD105(+) cells from Wharton's jelly show *in vitro* and *in vivo* myogenic differentiative potential. *Int J Mol Med* 2006;18(6):1089–96.

35. Chao KC, Chao KF, Fu YS, Liu SH. Islet-like clusters derived from mesenchymal stem cells in Wharton's jelly of the human umbilical cord for transplantation to control type 1 diabetes. *PLoS ONE* 2008;3(1):e1451.

36. Romanov YA, Svintsitskaya VA, Smirnov VN. Searching for alternative sources of postnatal human mesenchymal stem cells: candidate MSC-like cells from umbilical cord. *Stem Cells* 2003;21(1):105–10.

37. Panepucci RA, Siufi JL, Silva WA, Jr, Proto-Siquiera R, Neder L, Orellana M, *et al.* Comparison of gene expression of umbilical cord vein and bone marrow-derived mesenchymal stem cells. *Stem Cells* 2004;22(7):1263–78.

38. Flynn A, Barry F, O'Brien T. UC blood-derived mesenchymal stromal cells: an overview. *Cytotherapy* 2007;9(8):717–26.

39. Troyer D, Weiss M. Concise review: Wharton's jelly-derived cells are a primitive stromal cell population. *Stem Cells* 2008;26(3):591–9.

6

Bone Tissue Engineering

Johannes C. Reichert, Travis J. Klein, Kunnika Kuaha,
Bart Rijckaert, Ulrich Nöth and Dietmar W. Hutmacher

ABSTRACT

Currently, well-established clinical therapeutic approaches for bone
reconstruction are restricted to the transplantation of autografts and
allografts, and the implantation of metal devices or ceramic-based
implants to assist bone regeneration. These standard techniques face
significant disadvantages. As a result, research has focused on the
development of alternative therapeutic concepts aiming to design and
engineer unparalleled structural and functional bone grafts. Substantial
academic and commercial interest has been sparked in bone engineering
methods to stimulate, control and eventually replicate key events of
bone regeneration *ex vivo*. Over the years, this interest has further
increased and bone tissue engineering has now become a well-recognized
research discipline in the area of regenerative medicine. The following
chapter gives an overview of bone tissue engineering principles.
It focuses on research related to the combination of scaffolds with
multipotent precursor cells, such as bone marrow-derived mesenchymal
stem cells or human umbilical cord perivascular cells, and the clinical
applications of these tissue engineered bone constructs.

105

INTRODUCTION

Continued advances in the understanding of fundamental physiological and medical processes entail the prospect of increased longevity for mankind. Increased longevity, however, involves susceptibility to disease of the skeletal and other organ systems. Following blood and its components, bone represents the second most transplanted tissue, since current treatment regimes to restore function of damaged bone rely on the transplantation of autologous or allogenic grafts. Autologous bone remains the graft material of choice despite associated donor site morbidity and other complications, as it provides an osteoconductive matrix, stimulating growth factors, and osteogenic cells that are essential for bone repair. Surgeons increasingly face patient groups with limited supply of transplantable autologous bone due to preceding surgeries or poor bone quality. The continuing clinical need to improve existing treatment concepts for bone disorders resulting from congenital deformities, tumors, and trauma has nurtured the interest in novel approaches for skeletal reconstruction.

The most common concept underlying tissue engineering is based on the combination of a scaffold-matrix with living cells, and/or biologically active molecules to promote repair and regeneration of tissues. Bone tissue engineering aims at healing bone using autologous or other cell sources. One approach involves seeding of autologous osteogenic cells on a biodegradable scaffold to create a tissue-engineered construct (TEC) *in vitro*. This requires the isolation of a suitable cell population and its expansion to clinically relevant cell numbers. An appropriate three-dimensional (3D) scaffold may be impregnated with specific growth factors and other chemical cues to facilitate cell colonization, migration, growth and differentiation, as well as tissue ingrowth.[1,2]

Novel approaches to skeletal reconstruction have led to changes in orthopedic, plastic and reconstructive surgery adding bone engineering to the repertoire of applied treatment concepts. The field of regenerative medicine is driven by a clinically oriented perspective and unites aspects of molecular biology, materials science and biomedical engineering. The main objective is directed towards long-term repair and replacement of failing human tissues and organs. The present chapter aims to discuss the advances in the use of adult mesenchymal stem cells and other progenitor cells in scaffold-based bone tissue engineering. The isolation, characterization and *in vitro* differentiation of mesenchymal stem cells (MSCs) and human umbilical cord perivascular cells (HUCPVCs), as well as their application in bone tissue engineering will be discussed.

BONE STRUCTURE AND OSTEOGENESIS

For successful bone tissue engineering, a detailed understanding of bone structure and the osteogenic process is essential. Bone is a highly specialized, dense connective tissue which fulfils a multitude of mechanical, chemical, endocrine and hematological functions. Bone is a dynamic, adaptable tissue, and undergoes continuous remodeling to match its functions and mechanical demands. Bone greatly differs from other connective tissues. Its hardness is attributed to complex mineral deposition within a soft organic matrix mainly consisting of collagen type I. Newly mineralized bone matrix contains a variety of calcium phosphate species that range from relatively soluble complexes to insoluble crystalline hydroxyapatite. As bone matures, the inorganic matrix becomes primarily crystalline hydroxyapatite, although sodium, magnesium, citrate, and fluoride may also be present. The skeletal system protects vital organs. Its 206 bones form a rigid framework to which the softer tissues and organs of the body are attached. Bodily movement is carried out by the interaction of the muscular and skeletal systems. Also, bone tissue containing red marrow serves as a site for hematopoietic activities and acts as a storage area for calcium and phosphorus.

Bone as a composite material is characterized by a variety of mechanical properties determined by its different structural components (Table 1). Collagen, for example, possesses a Young's modulus of 1–2 GPa and an ultimate tensile strength of 50–1000 MPa while hydroxyapatite has a Young's modulus of ~130 GPa and an ultimate tensile strength of ~100 MPa.[3,4]

Macroscopically, cancellous bone, which is found at the ends of long bones and makes up the greater part of cuboidal bones, has a loosely

Table 1. Mechanical properties of human compact and spongy bone.[5]

Property	Compact bone	Spongy bone
Compressive strength (MPa)	100–230	2–12
Flexural, tensile strength (MPa)	50–150	10–20
Strain to failure (%)	1–3	5–7
Fracture toughness (MPam$^{1/2}$)	2–12	—
Young's modulus (GPa)	7–30	0.5–0.05

organized porous matrix where collagen fibrils form concentric lamellae. Compact or cortical bone does not have any spaces or hollows in the bone matrix.

The main cellular components of bone are osteoblasts, osteocytes, bone lining cells, and osteoclasts. Osteoblasts regulate the deposition of mineralized bone matrix. Osteocytes are mature osteoblasts and possess thin processes, which extend from lacunae into canaliculi and enable them to regulate different biologic processes and respond to mechanical stimuli. Bone lining cells are inactive and little is known about their actual function. Osteoclasts are large motile, multinucleated cells located on the bone surface that mediate bone resorption. While osteoclasts originate from fusing mononuclear cells derived from hematopoietic stem cells (HSCs), osteoblasts and osteocytes are derived from mesenchymal stem cells in the bone marrow (BM-MSCs).[6,7] BM-MSCs possess multipotent characteristics and are able to differentiate into different mesenchymal tissues such as muscle, fat, cartilage, tendon, and bone. When stimulated by appropriate growth factors, BM-MSC divide into osteoprogenitor cells that further differentiate to osteoblasts. This process is primarily regulated by the transcription factor long version of Cbfa1 (Cbfa1). Osteogenesis, or the process of bone formation, is initiated by the uncommitted adult multipotent BM-MSC in a regulated sequence of events mediated by a variety of growth factors and cytokines and involves a complex cascade of molecular and morphogenetic processes leading to precisely organized multicellular structures. Hence, BM-MSCs are vital to the repair and remodeling mechanisms in bone.

BONE MARROW STROMAL AND HUMAN UMBILICAL CORD PERIVASCULAR CELLS FOR BONE TISSUE ENGINEERING

Stem and progenitor cells from adult tissues represent an important promise in the therapy of several pathological conditions. Stem cells have the ability to self-replicate over long periods or, in the case of adult stem cells, maintain their differentiation potential throughout the life of the organism. Progenitor cells are derived from stem cells and committed to a single cell lineage. Committed progenitor cells have less capacity for self-replication but are normally in a state of cell division. They are primarily responsible for the generation of mature cells and their mitotic rate reflects the balance between replication, differentiation, and cell death.[8] From a classical view, adult stem cells are restricted to differentiating along the lineage pathways of their own tissue. This concept of lineage restriction has been challenged by experimental evidence over the past few years. Indeed, most stem and progenitor cell types display an amazing plasticity, which is the property of cells to differentiate into phenotypes not restricted to the tissues and, in some cases, to the germ layers from which they are derived.[9]

In 2007, for the first time, methods to reprogram adult human cells to a pluripotent state were described.[10-12] These cells, called induced pluripotent stem (iPS) cells, are genetically modified by the integration of up to four DNA-transcription factors into the adult cell genome. Soon after, it was demonstrated that iPS cells could be generated from a wide variety of adult cells.[13] One of the most valuable aspects of the iPS-cell technology currently, is the ability to perform disease modeling. The properties and potential of iPS cells as well as their safety are currently under investigation, as viruses incorporated in iPS cells could be mutagenic and have the potential to activate oncogenes. Therefore, iPS cells momentarily remain a research tool and not a potential therapeutic agent. But the next step for iPS cells could move them closer to therapeutic applications.

As the definition of the term "stem cell" remains elusive, caution should be exercised when using the terminology in the context of regenerative medicine. The current source of cells used in tissue engineering strategies are better labeled precursor/progenitor cells, although there is experimental evidence demonstrating the plasticity of these cells to differentiate into multiple lineages.

Adult stem/precursor cells were first identified in tissues character-ized by a high rate of cell turnover such as the bone marrow (BM-MSC) and have been well characterized in relation to stem cells originating from other tissues. BM-MSCs reside in close contact with hematopoi-etic progenitors in the bone marrow cavity.[14] Recently, MSCs have also been isolated from the periosteum, trabecular bone, adipose tissue, syn-ovium, and deciduous teeth.[15–19] Adipose tissue has been shown to contain multipotent stem cells which have the capacity to differentiate into cells of connective tissue lineages, including bone, fat, cartilage and muscle when stimulated with lineage specific growth factors.[20] Osteoprogenitor cells have also been isolated from skeletal muscle both in mice and humans. Muscle biopsy and liposuction as minimally inva-sive procedures are hence attractive alternatives in cell-based tissue engineering strategies.

Indeed, MSCs are now considered to be present in almost all body organs, and are mostly found in association with blood vessels.[21] BM-MSCs are now considered of perivascular origin,[14] and the perivascular area has been shown to host mesenchymal progenitor cells in different organs.[22] Therefore, the search for alternative MSC sources focuses more and more on perivascular regions in various tissues.[21,23–29]

One potential alternative source of mesenchymal cells drew interest after reports on cell cultures originated from Wharton's jelly (WJ),[30] the primitive connective tissue of the human umbilical cord (UC). Embryologically, the human UC is derived around day 26 of gestation, and grows to form a 30- to 50-cm-long helical organ at birth. Given this expansion, there must be a mesenchymal precursor cell population within the UC that gives rise to the WJ connective tissue. Sarugaser *et al.* postu-lated that these cells would most likely be located closest to the vasculature, and thus to their source of oxygen and nutrients. The group then described a novel method to extract human umbilical cord perivas-cular cells (HUCPVC).[24] They showed that, due to their low doubling time, high frequency of colony-forming-unit-fibroblast (CFU-F), and a 20% subpopulation that presented neither class I nor class II cell-surface major histocompatibility complexes (MHC$^{-/-}$), HUCPVC represent a possible source of cells suitable for allogenic cell therapies.

Osteogenic Progenitors

The cell-mediated formation of bone marrow and bone was first described when bone marrow-derived fibroblastoid cell populations were transplanted into ectopic sites delineating the osteogenic properties of BM-MSCs.[31] It has been proposed that all highly specialized types of hard tissue, including cortical and trabecular bone, tendons, ligaments and different kinds of cartilage as well as the stromal microenvironment supporting and regulating hematopoiesis, originate from a common type of early mesenchymal progenitor cell.[32–35] The osteogenic precursor cells are presumed to originate from stromal stem cells possessing self-renewing potential; the developmental pathway that mesenchymal stem cells pursue to differentiate into osteoblasts is still under intense investigation.

A sufficient and reproducible supply of cells is of paramount importance for bone engineering. Well-differentiated cells are inadequate due to their limited expansion potential and limited potential for trans-differentiation. Their accessibility and ease to be handled *in vitro* has made BM-MSCs natural candidates for bone tissue engineering. Moreover, as BM-MSCs constitute the stem cell source for osteoprogenitors and osteoblasts in the bone environment *in vivo*, they represent a good candidate for cell-based therapies in bone regeneration. The unambiguous identification of BM-MSCs, however, is presently hampered by the lack of specific cellular markers. Additionally, the frequency of BM-MSCs decreases with age, and thus may not be obtained consistently from older patients. A potential alternative source of MSCs is the umbilical cord. Compared to the remaining tissue of the umbilical cord, cells of the perivascular region of the cord tissue have been shown to have both an increased proliferative capacity and a distinct cell phenotype that is similar to BM-MSCs.[27,36] Furthermore, with an annual global birth rate of 130 million, and a CFU-F frequency in P0 HUCPVC populations of 1:300,[24] the perivascular tissue of the human umbilical cord represents a rich and nearly unlimited alternative supply of MSCs for cell-based therapies.[21]

Phenotype and Isolation

Bone marrow harbors at least two distinct populations of adult stem cells, BM-MSCs and HSCs. The clonogenic stromal progenitor cells, which later came to be described as mesenchymal stem cells, were first described as rapidly adherent, non-phagocytic clonogenic cells capable of extended proliferation *in vitro*.[37,38] Assays of CFU-F from aspirates of human bone marrow yield colony numbers between 1 and 20 per 10^5 mononuclear cells.[39,40] The CFU-F in adult human bone marrow is feeder cell-independent.[41]

Baddoo *et al.* reported a method based on immunodepletion to fractionate fibroblastoid cells from hematopoietic cells within plastic adherent murine marrow cultures. The immunodepleted cells expressed the antigens Sca-1, CD29, CD44, CD81, CD106, and the stem cell marker nucleostemin (NST) but not CD11b, CD31, CD34, CD45, CD48, CD90, CD117, CD135, or the transcription factor Oct-4.[42] The stromal cells obtained were shown to exhibit osteogenic, adipogenic, and chondrogenic differentiation potential *in vitro*.

The Mesenchymal and Tissue Stem Cell Committee of the International Society for Cellular Therapy have proposed three criteria to define MSCs as: "(1) the plastic adherence of the isolated cells in culture; (2) the expression of CD105, CD73, and CD90 in greater than 95% of the culture, and their lack of expression of markers including CD34, CD45, CD14 or CD11b, CD79α or CD19 and HLA-DR in greater than 95% of the culture; and (3) the differentiation of the MSCs into osteoblasts, adipocytes and chondroblasts *in vitro*."

One of the easiest approaches to isolate and enrich stem cells and progenitor populations is by removal of cells expressing antigens that are present on more mature, differentiated cells (negative selection). Examples of lineage-specific antigens are CD19 and CD20 for B-cells, CD3 for T-cells, CD14 for monocytes, and CD66b for granulocytes. Depletion of these and other lineage-positive cells can be achieved by immunomagnetic as well as non-magnetic cell separation approaches. After harvest of bone marrow, adherent cells can be expanded and subcultured to increase the number of pluripotent cells. BM-MSCs have been shown to retain their undifferentiated phenotype through an average of 38 doublings, resulting in over a billion-fold expansion.[43] The use of minimal medium, which is not supplemented by cytokines, facilitates the growth of colony forming units-fibroblasts (CFU-F) that proliferate to form macroscopic fibroblastoid colonies *in vitro*. These can be further propagated by serial passage to become cell strains while still retaining the capacity to differentiate into several mesodermal directions.

Sorting for STRO-1, which does not bind to hematopoietic progenitors, results in a 10- to 20-fold enrichment for CFU-F in fresh aspirates of human bone marrow. In accordance with the previously described properties of the unfractionated STRO-1$^+$ population, STRO-1 bright, VCAM-1$^+$ cells assayed on a clonal level exhibited *in vitro* differentiation into cells with the characteristics of adipose, cartilage and bone cells, and formed human bone tissue following transplantation into SCID mice.[44] Collectively, these data strongly suggest that primitive stromal precursors including putative stromal stem cells with the capacity to differentiate into multiple mesenchymal lineages are restricted to the STRO-1$^+$ fraction in adult human bone marrow.[45,46] BM-MSC enrichment could also be accomplished using several other markers including Thy-1, CD49a, CD10, Muc18/CD146, and, in accord with their response to growth factors, antibodies to receptors for PDGF and EGF.[47,48] Additional antibodies that identify human BM-MSCs have been described but have not yet been verified by other groups.[49]

Other strategies for stem cell isolation use positive selection. Studies have shown that BM-MSCs can be selected directly by virtue of expression of CD49a, the α1-integrin subunit of the very late antigen (VLA)-1, which is a receptor for collagen and laminin.[50] This population is CD49a$^+$/CD45$^{med/low}$ and differentiates into several mesodermal directions. All CFU-F found in human bone marrow are included within this CD49a$^+$/CD45$^{med/low}$ fraction. In contrast to the above, BM-MSCs that are apparently restricted to mesodermal lineages, Jiang *et al.*[51] described the isolation and characterization of cells co-purifying with BM-MSCs, termed mesenchymal adult progenitor cells (MAPCs). These are more plastic than mesenchymal cells and differentiate *in vitro* not only into mesodermal derivatives but also into cells of the visceral mesoderm, neuroectoderm, and endoderm.[52]

The number of progenitor cells and cellularity among tissues varies substantially (Table 2).[53-55] The mean prevalence of alkaline phosphatase positive CFU-Fs after placing bone marrow-derived cells into culture is about 36 cells per one million nucleated cells. Due to contamination with peripheral blood, not more than 2 ml of aspiration volume is recommended

Table 2.   Progenitor cells according to tissue type.

Tissue	Cellularity	Approximate no. of progenitor cells
Bone marrow	4×10^7 cells/ml	2000 cells/ml (1 in 20,000 cells)
Umbilical cord (perivascular)	$2.5–25 \times 10^4$/cm	1 in 10,000 cells
Adipose and skeletal muscle	6×10^6 cells/ml	1 in 4000 cells

from any particular site. Significant patient-to-patient variation in the cellularity of bone marrow aspirates and the prevalence of osteoblast progenitor cells is also observed.

Currently, the search for alternative multipotent cell sources focuses more and more on perivascular organ regions. Studies show that MSC-like populations within bone marrow, dental pulp, periodontal and other tissues share a common perivascular stem cell niche within the microvascular network of their respective tissues.[14,56] These findings are supported by a number of reports demonstrating that MSC-like cells from different postnatal tissues exhibit the characteristics of pericyte-like cells.[14,57] Accordingly, human umbilical cord perivascular cells have been isolated and characterized. HUCPVCs can be isolated using various techniques,[24,25,58] and can easily be characterized by flow cytometric measurements of cellular markers. Several research groups report that HUCPVCs are positive for CD146 (~52%), which is characteristically expressed on endothelial cells and is associated with the actin skeleton, while only ~15% of BM-MSCs are CD146 positive. This makes CD146 a good candidate to distinguish between HUCPVCs and BM-MSCs.[21,27,59,60] Baksh et al.[27] reported that only CD146$^+$ sorted HUCPVCs are capable of multi-lineage differentiation, indicating that sorting HUCPVCs for CD146 is of high importance. The same group also showed that endothelial cells are positive for CD31 (which plays a key role in removing aged neutrophils from the body[61]), while HUCPVCs are not; this makes CD31 a valid target for separating HUCPVCs from a mixed cell population with endothelial cells. Other HUCPVCs surface markers described in the literature include STRO-1$^-$, CD45$^-$,[62] CD105$^+$,[63] and CD90$^+$,[64] of which the latter three indicate mesenchymal phenotype and provide more options of identifying and sorting HUCPVCs from a mixed population of cells.[21,23–25,27,29,60,65,66]

Genetic Modification

Several research groups have focused on genetic engineering to repair or regenerate bone. The trend has been to concentrate on a particular molecular pathway and produce targeted intervention to create enhancements in functionality and/or architecture of the repaired or regenerated bone. However, because almost every known cell-cell and cell-matrix interaction involves multiple genes acting at specific tissue locations at specific times, understanding the entire complex system requires a detailed knowledge of gene functionalities and cellular behaviors as well as the spatial and temporal integration on the tissue level. Osteoinductive factors, such as bone morphogenetic proteins (BMPs), have been successfully applied to augment local bone repair and several formulations are available for clinical application.[67,68] However, the widespread clinical efficacy of these treatments continues to be hampered by inadequate delivery vehicles, release kinetics, dosage, and potency.[68,69] Genetic engineering strategies focusing on osteoinductive factors have emerged as efficient approaches to enhance bone formation and generally consist of two modalities: (a) direct *in vivo* delivery of gene constructs and (b) *ex vivo* transduction and subsequent transplantation of cells expressing the osteoinductive factor. The choice of gene delivery method depends on several factors including the particular gene of interest, indication targeted, desired duration of gene expression, and delivery vector. Gene delivery to musculoskeletal systems has been reviewed elsewhere.[70,71]

In-Vitro Proliferation and Differentiation

Cells are generally cultured in basal medium such as Dulbecco's modified Eagle's medium (DMEM, high glucose) in the presence of 10% fetal bovine serum (FBS). The optimal expansion of MSCs requires the pre-selection of FBS.[34,72] MSCs in culture have a fibroblastic morphology and adhere to the tissue culture substrate. Primary cultures are usually maintained for 12–16 days, during which the non-adherent hematopoietic cell fraction is depleted. The addition of growth factor supplements such as fibroblast growth factor-2 (FGF-2) to primary cultures of human MSCs has been reported to increase expansion while retaining differentiation capacity.[73] In clinical settings, MSCs need to be expanded in large-scale culture. Therapeutic-scale expansion of human MSCs can easily require quantities of more than 1.5×10^8 MSCs (2×10^6/kg $\times$ 75 kg).[74–76] It will be important to identify defined growth media, without or with reduced FBS, to ensure more reproducible culture techniques and enhanced safety. Such media are becoming commercially available (e.g., MesenPRO RS and StemPRO from Invitrogen), yet they must be tested with the specific MSC type, as HUCPVCs have shown less expansion capacity in MesenPRO compared to the standard DMEM with 10% FBS (Fig. 1).

MSCs have the tendency to spontaneously differentiate into multiple lineages when transplanted *in vivo*.[77] Hence, only a small subfraction of the transplanted MSC may differentiate into the tissue type of interest,

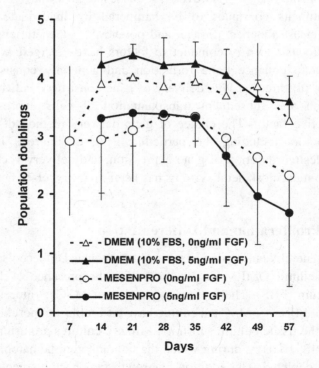

Fig. 1. Proliferation data from HUCPVCs cultured in different media types, with or without 5 ng/ml FGF-2, and passaged weekly — Passage 3 at day 7, to Passage 10 at day 57 (n = 4 donors, mean ± SD).

which in turn could reduce the clinical efficacy of transplantation therapy. Established *in vitro* differentiation protocols are therefore used to direct the osteogenic differentiation of MSCs prior to transplantation. MSCs can either be induced osteogenically while in monolayer culture or after seeding onto 3D porous scaffolds.

The culture environment for stem cell differentiation *in vitro* should be chemically defined, and either be serum-free or utilize synthetic serum replacements with the possible supplementation of specific recombinant cytokines and growth factors.[78,79] However, so far no fully satisfactory serum replacement or serum-free media is available. Therefore, current standard osteogenic media formulations still contain 10% FBS, and are supplemented with 0.1 μM dexamethasone, 50 mM ascorbate-2-phosphate and 10 mM beta-glycerophosphate.[57] Cultures are usually maintained for four weeks. Dexamethasone, a synthetic steroid, has been shown to promote osteogenic differentiation of both embryonic and mesenchymal stem cells.[80,81] Ascorbic acid treatment is accompanied by a five-fold increase in the binding of transcription factor complex Osf2/Cbfa1 (osteoblast-specific transcription factor 2/Runt-related transcription factor 2) to OSE2 (osteoblast-specific *cis*-acting element 2) in MC3T3-E1 cells and osteocalcin and bone sialoprotein promoters are upregulated.[82] Ascorbic acid also promotes relative collagen synthesis by increasing the procollagen stability and secretion. It is required for matrix maturation and subsequent mineralization. Other biologic modulators, which play a role in osteogenic differentiation, include insulin, TGF-β, EGF, LIF, FGF-4, PDGF, calcitropic hormone, and 1,25-dihydroxyvitamin D3. Type I collagen, one of the most commonly occurring extracellular matrix molecules, has been reported to slow down cellular senescence within *in vitro* cultures,[83] and has therefore been proposed as a suitable candidate to reduce the rate of ageing during the *in vitro* expansion of mesenchymal stem cells for tissue engineering applications. Integrin-activated signaling pathways have been implicated in the differentiation of bone marrow stromal precursors into functional osteoblasts.[84] Laminin, an integrin ligand, has been shown to direct osteogenic differentiation *in vitro*.[85] A variety of methods have been developed to incorporate these induction molecules into either culture media or biomimetic scaffolds.

During extracellular matrix deposition, initial collagen matrix accumulation precedes and is essential for sequential expression of differentiation-related proteins e.g. alkaline phosphatase (AP), parathyroid hormone receptor (PTHR), bone sialoprotein (BSP) and osteocalcin (OC). BM-MSCs are highly proliferative and express low levels of bone-specific proteins, while secretory osteoblasts stop dividing and produce large amounts of bone-specific extracelluar matrix (ECM). Osteogenic maturation is then evaluated by measuring AP and OC concentrations in the culture supernatants or directly on the cells using quantitative immunoassays. RT-PCR and Western blot techniques are used to confirm the presence of osteogenic markers on gene and protein level. Alizarin red S and von Kossa staining can be used to visualize mineralization and further allow quantification using extraction techniques.

The developmental sequence of bone has been described to consist of three phases: proliferation with matrix secretion, maturation, and mineralization.[86] Osteopontin (OP) and AP are early osteogenic markers and characterize the matrix maturation phase. AP is a membrane-bound enzyme abundant in early bone formation. Shedding releases the membrane-bound AP into the serum or supernatant where it can be measured by ELISA. Histomorphometric assessment has shown that increasing AP levels correlate with increasing bone formation. Expression of OP, an important protein for matrix-cell interaction, was observed two weeks after induction. For bone-derived cells, AP levels are increased around day 12 after induction. Although these results strongly suggest osteogenesis, these proteins are not considered specific markers of osteogenic differentiation. Osteocalcin is a late and specific osteogenic marker since it determines terminal osteoblast differentiation. Trabercular bone and periosteum-derived cell populations show increasing OC concentrations three weeks after induction.

Osteogenic differentiation induced by mechanical stimulation has been widely documented. The differentiation of human osteoblastic periodontal ligament cells has been reported to be enhanced in response to mechanical stress[87] while a similar result was observed with bone marrow-derived osteogenic progenitors.[88] In the absence of exogenous growth and differentiation factors, mechanical stimuli alone could induce the differentiation of MSCs into the osteogenic lineage.[89] Further investigations showed that there are multiple competing signaling pathways for osteogenic differentiation in response to mechanical stimuli.[90]

Unlike BM-MSCs, the differentiation potential of HUCPVCs is reported to depend on the isolation technique.[24,25,58] Among others, Baksh et al.[27] and Ennis et al.[21] reported that HUCPVCs successfully undergo osteogenic differentiation. Baskh et al. reported higher levels of AP activity in HUCPVCs when compared to BM-MSCs, and von Kossa staining revealed an earlier and greater extent of mineralization in osteogenic cultures of HUCPVCs. The effect of dynamic culture conditions on HUCPVC differentiation has been scarcely researched. However preliminary data from our laboratory suggest that HUCPVCs even undergo osteogenesis when subjected to control media only, and no increased osteogenesis was recorded when HUCPVCs were cultured under dynamic conditions on a rocker plate.

Cell-Host Interactions

The host response to transplanted MSCs is critical for successful long-term outcomes and has received increased attention as application of these cells is considered in a variety of clinical settings. There are several aspects to the transplant cell–host interactions that need to be addressed as we attempt to understand the mechanisms underlying stem cell therapies. These are (a) the host immune response to transplanted cells, (b) the homing mechanisms that guide delivered cells to a site of injury, and (c) differentiation of transplanted cells under the influence of local signals. By virtue of their distinct immunophenotype, associated with the absence of HLA class II expression, as well as low expression of co-stimulatory molecules, BM-MSCs may be non-immunogenic or hypo-immunogenic.[91] HLA Class II expression is also absent from the surface of differentiated BM-MSCs and these cells do not elicit an alloreactive lymphocyte proliferative response, suggesting that BM-MSCs can be transplantable between HLA-incompatible individuals.[92] HUCPVC sub-populations that expresses neither class I nor class II MHC antigens (MHC$^{-/-}$) increase with passage and freeze-thawing, resulting in MHC$^{-/-}$ percentages unattainable in bone marrow-derived cells.[24] Although some authors have found little evidence of an immunogenic response using allogenic[93] and even xenogenic[94] MSC therapy, it has been noted that histocompatibility of the cell source is a significant hurdle to be addressed for safe and effective application of cell-based therapies.[95] In this context, an MHC$^{-/-}$ cell population may represent a promising avenue to surmount the potential hazards of a host immune response leading to graft versus host disease.[24] For further detail, see Chapter 8 by Pittenger and LeBlanc.

BONE TISSUE ENGINEERING

From a clinical perspective, the main therapeutic goal is tissue reconstitution to a minimal level required to re-establish systemic homeostasis. More ambitious projects aim for *restitutio ad integrum*. Debatable, though not necessarily out of reach, are approaches to enhance general fitness to a "supranatural" level (*enhancement*). Currently, our explicit aim to develop "reconstructive therapies" implies that our clinical targets exclude pure enhancement. The challenge now facing regenerative medicine is one of its own making — namely, how to address the problems associated with the progressive increase in patients' lifespan. The tendency of people to live longer has two effects on the use of implants and/or bone grafts: more patients require treatment, and the treatments themselves are required to have longer lasting effects. A shift in emphasis is therefore needed from current methods to replace tissues to more biological approaches aiming to regenerate tissues.

A wide variety of matrices has been used as carriers to deliver MSCs including ceramics, collagen sponges and gels, and biodegradable polymers. A comprehensive report on scaffold design and fabrication for bone engineering is beyond the scope of this chapter and has been reviewed elsewhere.[96,97] There is ample evidence that the nature and properties of a scaffold play an important role in bone engineering (Table 3). It is important to emphasize that the field is still young and many different approaches are under investigation. Thus, it is by no means clear what defines an ideal scaffold-cell or scaffold-neo-tissue construct, even for a specific tissue type. Indeed, since some tissues perform multiple functional roles, it is unlikely that a single scaffold would serve as a universal foundation for the regeneration of even a single tissue.

For each envisioned application, a successful TEC will have to meet minimum requirements regarding biochemical and physical properties. Scaffold architecture has to allow for initial cell attachment and subsequent migration into and through the matrix, mass transfer of nutrients and metabolites, and provision of sufficient space for development and remodeling of organized tissue. In addition to these essentials of mechanics and geometry, a construct will need to possess surface properties

Table 3. Minimal scaffold requirements for bone tissue engineering.

	Minimal scaffold requirements
Biocompatibility	The material should not elicit an immunological or chemically detectable primary or secondary foreign body reaction.
Biodegradability	Degradation should be at a controlled rate and degradation products should be non-toxic and easily excreted by metabolic pathways.
Mechanical strength	Strength should be sufficient to maintain structural integrity during culture, as well as to support and transfer loads after implantation; mechanical properties should adequately address short-term functions and should not interfere with long-term functions.
Porosity	Porous architecture should be controlled and adequate (i.e., 100% interconnectivity, average pore size of bigger than 200 μm, sufficient surface to volume ratio) to allow cell attachment, cell growth, tissue regeneration, vascularization and clearance of waste.
Osteoconductivity	Vascular invasion, cell infiltration and attachment as well as appositional bone formation should be facilitated
Drug delivery	Should allow for release or attachment of active compounds.
Ability to integrate	Should be able to integrate with the host tissue following *in vivo* implantation.
Availability	Should be available to surgeons on short notice; have adequate shelf-life and handling properties.
Sterilization	Should be able to be sterilized without major loss of significant properties.
Remodeling	Should allow *in vitro* grown constructs to continuously remodel *in vivo*.

facilitating attachment and migration of cell types of interest. External size and shape of the construct must also be considered; especially if a construct is customized for an individual patient. Cell and tissue remodeling is important to achieve stable biomechanical conditions and vasuclarization at host sites. Hence, a 3D scaffold/tissue construct should maintain sufficient structural integrity during the *in vitro* and/or *in vivo* growth and remodeling process. The growth of neo-tissue is dependent on the immediate environment the constructs are subjected to, including temperature, culture medium, chemical factors, and mechanical factors. Therefore, adaptive remodeling responses to simple stimuli of force, pressure, and chemical gradients are desirable. Most remarkable is the plasticity of the biological remodeling processes. Considering the requirements for ideal tissue engineering constructs, it is unnecessary to attempt to provide a too detailed guide to regenerating tissue.'Instead, it might be more worthwhile to concentrate on the general conditions and principal stimuli which guide the natural remodeling processes to produce a functional, mature tissue.

BIOREACTORS

In all *in vitro* and *in vivo* settings, access to substrate molecules (oxygen, glucose, and amino acids) and clearance of metabolic products (CO_2, lactate, and urea) are critical to cell survival.[98] Transfer of nutrients and waste-products in three-dimensional constructs *in vitro* may rely on passive diffusion or more active delivery mechanisms such as convection. Numerous long-term tissue engineering studies have shown the limited capacity of static conditions to promote high-density three-dimensional *in vitro* growth of human tissues *ex vivo*, where they are deprived of their normal vascular supply of nutrients and gas exchange.[99–101]

The culture dish/flask can be defined as the simplest and most widely used bioreactor today. This product provides an environment that is sterile, easy to use, simple, and economical to manufacture. Cells in monolayer culture are generally not nutrient-limited since passive diffusion is more than adequate to supply a tissue of 10 µm thickness. However, the fact that the culture dish offers only a static 2D environment and requires individual manual handling for medium exchange and cell seeding ultimately limits its usefulness when large cell numbers are required.

Scale-up of the culture dish only goes so far as, for example, having multi-well plates or larger surfaces.

In contrast to monolayer cultures, the supply of oxygen and soluble nutrients becomes critically limiting for tissue cultures exceeding a thickness of 100–200 μm. *In vivo*, cells benefit from close proximity of blood capillaries to satisfy mass-transfer requirements, as in most tissues cells are no more than 100 μm from these capillaries.[102] The diffusion limitation *in vitro* can partly be alleviated by stirring the culture medium.

First-generation reactors were simple Petri dishes with mechanical stimulation provided by an orbital shaker or rocker plate. Constructs grown in orbitally mixed Petri dishes were of larger diameter and contained higher amounts of ECM than constructs cultured under static conditions.[103,104] For bone, the thickest tissues generated in an agitated dish, were of 0.5 mm in diameter.[105] However, if transplantable constructs are to be generated, larger sized tissues are required.

External mass-transfer limitations can also be reduced by culturing constructs in a stirred flask. As one of the most basic bioreactors, the stirred flask induces mixing of oxygen and nutrients throughout the medium and reduces the concentration boundary layer at the construct surface.[106] Although cultured cells on a 3D scaffold showed increased synthesis of ECM in a stirred flask, formation of fibrous capsules was observed. This was probably caused by turbulent eddies generated within the flask, which increased the rate of ECM components released into the culture medium and were associated with cell dedifferentiation.[106] Furthermore, tissue damage was observed with stirred flasks.[107]

A dynamic laminar flow generated by a rotating fluid environment is an alternative and efficient way to reduce diffusional limitations of nutrients and wastes while producing low levels of shear. The efficacy of rotating wall vessel (RWV) bioreactors for the generation of tissue equivalents has been demonstrated using chondrocytes,[108] cardiac cells,[109] tumor cells,[110] and pre-osteoblasts.[111] In each case, cultured constructs had biochemical and biomechanical properties, cellular architecture, and composition superior to those of static or stirred-flask cultures and it was proposed that RWV bioreactors support the engineering of tissues and organoids as *in vitro* model systems of tissue development and function.[112]

Bioreactors that perfuse medium either through or around semi-permeable hollow fibers have been used successfully to maintain the function of highly metabolic cells. This concept has been extended by perfusing culture medium directly through pores of cell-seeded 3D scaffolds, thereby reducing mass transfer limitations both at the construct periphery and within its internal pores. In the context of bone tissue engineering, direct perfusion bioreactors have been shown to enhance growth, differentiation, and mineralized matrix deposition by bone cells.[113,114] Direct perfusion can therefore be used as a valuable tool to enhance cell survival, growth, and function. However, the effects of direct perfusion depend on medium flow-rate and the maturation stage of the constructs.[115] Therefore, perfusion bioreactors must provide a careful balance between mass transfer of nutrients and waste products, the retention of newly synthesized ECM, and fluid-induced shear stresses.

Mechanical forces are important modulators of cell physiology. Increasing evidence suggests that these forces might increase biosynthetic activity of cells in bioartificial matrices and improve structural and functional properties of engineered tissue *in vitro*.[116] Various studies have demonstrated the validity of this principle, particularly in the context of musculoskeletal tissue engineering.[117–120] However, little is known about the optimal mechanical forces or regimes of force application (i.e., magnitude, frequency, continuous or intermittent, duty cycle). In this context, bioreactors might play an important role, providing a controlled environment for reproducible and accurate application of mechanical forces to 3D constructs.

CLINICAL BONE ENGINEERING

The first generation of clinically applied tissue engineering concepts in the area of skin, cartilage, and bone regeneration was based on the isolation, expansion, and implantation of cells from patient's own tissue. Although successful in selected treatments, bone tissue engineering needs to overcome major challenges to allow widespread clinical application with predictable outcomes. This includes the isolation and expansion of cells with the highest potential to form bone-like tissue in a 3D environment. Another major challenge is to transplant the cells in a matrix that allows cell survival despite contraction forces and mechanical loading.

Although many characteristics regarding the behavior of MSCs are conserved across the phylogenetic tree, great care must be exercised when extrapolating results from studies on animal MSCs to human applications.

Oral and Maxillofacial Reconstruction

The paucity of techniques in cranial reconstructive surgery emphasizes the need for alternative bone formation strategies. The repair of large cranial defects is often unsuccessful due to the limited amount of autologous bone available. Calvarial bone is hard and brittle, which makes the contouring of a graft extremely difficult. In instances where corticocancellous bone from the ilium was used, grafts appeared to be more prone to resorption than bone of membranous origin. Calvarial reconstruction strategies have hence employed the use of adult MSCs for bone regeneration. Adipose-derived stem cells were recently used in the repair of a calvarial defect of the size of 120 cm^2 in a seven-year-old girl who sustained multi-fragment fractures following head injury.[121] Autologous fibrin glue was used to attach the cells onto milled cancellous bone, which was used as an osteoconductive scaffold. CT-scans showed new bone formation and near complete calvarial continuity three months after reconstruction.

Studies by Schantz *et al.*[122] to investigate the differentiation potential of BM-MSCs and to compare their potential in bone regeneration against trabecular osteoblasts in a rabbit calvarial defect model showed promising results of site-specific differentiation of unconditioned BM-MSCs. The transplanted BM-MSCs mostly committed to the osteogenic phenotype, formed islands of bone tissue and the amount of mineralized tissue increased and filled out the graft sites almost entirely. In a clinical pilot study by the same group,[123] a burr hole of 14 mm diameter resulting from trepanation was closed using a biodegradable polymer coated with autologous marrow cells from the calvarium in five patients who underwent evacuation of subdural hematomas. The follow-up CT scans showed good implant integration into the surrounding calvarial bone, with new bone filling the porous space of the scaffold.

Warnke *et al.*[124] reported the fabrication of a mandibular transplant for a patient suffering from a large defect after resection of mandibular bone.

The transplant consisted of a titanium mesh cage filled with bone mineral blocks, which were infiltrated with a combination of autologous bone marrow from the iliac crest and rhBMP-7. The transplant was then implanted into the right latissimus dorsi muscle and left there for a period of seven weeks. Skeletal scintigraphy showed bone remodeling and mineralization within the mandibular transplant both before and after transplantation. Computed tomography provided evidence of new bone formation. Seven weeks post-transplantation, the transplant was excised with an adjoining part of the latissimus dorsi muscle containing the thoracodorsal artery and vein ensuring blood supply for the entire transplant. It was then transplanted into the mandibular defect for reconstruction. As a result, the patient showed an improved degree of mastication and a satisfactory aesthetic outcome.

In yet another pilot study by Sittinger and colleagues,[125] augmentation of the posterior maxilla was carried out in two patients using a TEC derived from mandibular periosteal cells on a polymer fleece. The cell polymer transplants were cultured for one week prior to implantation with autologous serum, ascorbic acid, dexamethasone, and beta-glycerophosphate. Bone biopsies from both patients revealed mineralized trabecular bone with remnants of the biomaterial and osteocytes were apparent within bone lacunae.

From a clinical point of view, little has been reported on the application of tissue engineering for regeneration of periodontal tissues. Yamada and colleagues[126] applied a technique based on tissue engineering principles to periodontology. BM-MSCs were cultured and expanded, and platelet-rich plasma (PRP) from the same patient was isolated from peripheral blood. Full-thickness periodontal flaps were elevated and the root surfaces were scaled and planed. Using the expanded BM-MSCs and PRP, a gel was prepared and applied to the root surface and adjacent defect space. Follow-up examinations revealed that the application of BM-MSCs in combination with PRP at periodontal sites with angular defects resulted in a 4-mm reduction in probing depths and a 4-mm clinical attachment gain, while bleeding and tooth mobility disappeared. Radiographic assessments showed reduction of the bone defect in depth. Moreover, it could be observed that interdental papillae showed signs of regeneration. Kawagushi *et al.* demonstrated that transplantations of *ex vivo* expanded autologous MSCs can regenerate new cementum, alveolar bone, and periodontal ligament in class III periodontal defects in dogs. Morphometric analysis revealed a 20% increase in new cementum length and bone area in animals treated with MSCs. In a subsequent study, the same group reported a similar approach in humans[127] when they transplanted 2×10^7 cells/ml of autologous expanded BM-MSCs mixed with Atelocollagen into periodontal osseous defects. All patients showed significant improvement.

Orthopedic Surgery

Outstanding progress has been made in the area of skeletal reconstruction and the treatment of osseous defects, but non-union fractures remain critical to numerous orthopedic and craniofacial treatment concepts. Although commonplace in orthopedic surgery, current approaches like autografting and allografting cancellous bone, applying vascularized grafts of the fibula and iliac crest, and other bone transport methods have a number of limitations. The lack of techniques and limited approaches in reconstructive surgery emphasize the potential benefit of tissue-engineered bone in clinical orthopedic applications.

Vacanti *et al.* reported the replacement of an avulsed phalanx with tissue-engineered bone, which resulted in the functional restoration of a biomechanically stable thumb of normal length.[128] Periosteal cells, which have been shown to shed osteoblastic cells, were obtained from the distal radius and seeded onto a coral scaffold. A stable calcium alginate hydrogel encapsulating the cells saturated the coral implant. MRI examination showed evidence of vascular perfusion and biopsy revealed new bone formation of lamellar architecture.

Cancedda *et al.* have reported the use of cell-based tissue engineering approaches to treat large bone defects.[129] Osteoprogenitor cells were isolated from bone marrow, expanded, and placed on macroporous scaffolds. Follow-up radiographs showed abundant callus formation along the implants and good integration at the interfaces with the host bones. All patients reported recovery earlier than expected with a traditional bone graft approach. The group had earlier reported the use of BM-MSC-ceramic composites in the treatment of full-thickness gaps in sheep tibiae.[130] *In vitro* expanded BM-MSCs were loaded onto highly porous ceramic cylinders and implanted in critical sized stable segmental defects. Similar results in animal models were also reported where cultured marrow cells were implanted with a coral scaffold in large segmental defects in sheep metatarsals.[131] Bruder *et al.* have successfully treated experimentally induced non-union defects in adult dog femurs with autologous cells loaded onto a hydroxyapatite-beta tri-calcium phosphate scaffold.[132]

Bajada *et al.*[133] reported the successful healing of a nine-year-old's tibial mid-shaft non-union following a high-speed road traffic accident that was resistant to various surgical procedures including the application of a monolateral external fixation, functional bracing, and two programs of ring circular external fixation with autologous bone grafting. Using BM-MSCs expanded *in vitro* to 5×10^6 cells within a period of three weeks combined with calcium sulfate ($CaSO_4$) in pellet form, the defect was reconstructed observing clinical and radiological convalescence two months after implantation.

The use of OP-1 (BMP-7) in the treatment of tibial non-union was studied by Friedlaender *et al.* in a randomized controlled, prospective clinical trial.[134] Clinical and radiographic results were compared to assess the efficacy of OP-1 versus autograft in the treatment of tibial non-unions that had persisted for at least nine months. One hundred and twenty-four tibial non-unions in 122 patients were randomized to either intramedullary nail and autograft or intramedullary nail and implantation of OP-1 at the non-union site. Nine months after surgery, 81% of the OP-1 group and 85% of the autograft group had achieved clinical union. Radiographic analysis indicated that 75% of the fractures treated with OP-1 and 84% of the autograft-treated group had healed. There was no statistically significant difference between groups clinically or radiographically. More than 20% of the bone graft group complained of persistent donor site pain. The authors concluded that OP-1 was a safe and effective alternative to bone grafting in the treatment of tibial non-unions. This study led to multiple regulatory approvals worldwide. Numerous studies in the literature suggest that OP-1 is a safe and effective treatment option for fractures and atrophic non-unions not only of the lower but also of the upper extremities.

The overall goal of treatments for adult femoral head necrosis is to preserve the femoral head and therefore avoid total hip replacement

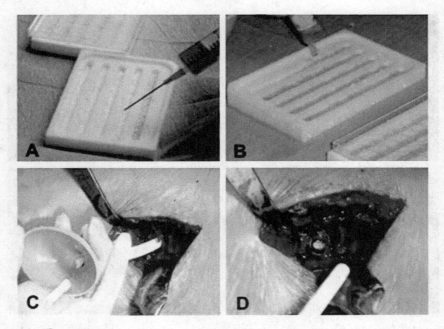

Fig. 2. Preparation and application of stem cell-TCP matrix. (**A**) The stem cell suspension is added to the β-TCP granules with a syringe. (**B**) Addition of serum isolated from patient blood via centrifugation. (**C**) The stem cell-TCP matrix is installed through the boring duct using a special funnel-formed applicator. (**D**) Completely filled boring duct.[135] Reprinted from Ref. 135, with kind permission of Springer Science+Business Media.

surgery. Core decompression has been shown to decrease intraosseous pressure, while additionally providing the opportunity to deliver bioactive materials and/or progenitor cells to enhance healing. Noth *et al.*[135] presented a therapeutic approach for patients suffering from femoral head necrosis stage ARCO II using BM-MSCs in combination with a beta-TCP matrix (Figs. 2 and 3). Kawate *et al.*[136] reported on three cases of steroid-induced femoral head osteonecrosis stage Steinberg 4A (one patient) and C (two patients) treated with MSCs cultured with beta-TCP ceramics and with a free vascularized fibula. All hips showed preoperative collapse and radiographic progression was observed in two hips postoperatively, although osteonecrosis did not progress any further within the time frame reported. Recently, Hernigou *et al.*[137] demonstrated that autologous bone marrow transplantations combined with core decompression

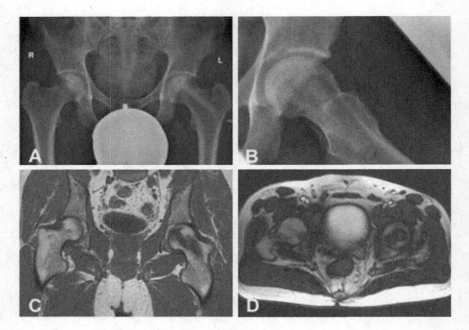

Fig. 3. Imaging diagnostic six weeks post application of the stem cell-TCP matrix. **(A)** X-ray of the pelvis. **(B)** The axial projection (Lauenstein) demonstrates the anterior and posterior boring duct. **(C)** The MRI (T1) shows the anterior boring duct directly reaching the necrotic area in the coronar plain. **(D)** In the horizontal plain (T2) the anterior and posterior boring ducts are presented adjacently. Reprinted from Ref. 135, with kind permission of Springer Science + Business Media.

before collapse of the femoral head made hip replacement surgery necessary in only nine of 145 cases, compared to 25 of 44 cases when operated after manifest collapse. As well, Gangji et al.[138] reported on successful treatment of 18 patients treated with bone marrow cells harvested from the iliac crest, suggesting that the application of cell-based treatment concepts in case of femoral head necrosis might play a decisive role in future therapeutics.

CONCLUSION

Bone formation within the body, as part of a development, healing and/or repair process, is a complex event in which cell populations in combination with extracellular matrix self-assemble into structural and functional units. There is intense academic and commercial interest in finding methods to stimulate and control these events and eventually to replicate these events outside the body as closely as possible. This interest has accelerated and resulted in bone tissue engineering becoming a well-recognized research area in the arena of regenerative medicine. The increasing number of papers reporting on formerly unknown function, structure, and plasticity of MSCs has spawned a major switch in the perception of their nature, and ramifications of their potential surgical application in tissue engineering approaches are not only envisioned but already implemented. The ability to isolate MSCs with the most extensive replication and differentiation potential is therefore of utmost importance for the regeneration of mesenchymal tissues via bone engineering strategies. Several tissues including bone marrow, adipose tissue, and umbilical cord contain MSCs that can be isolated with relatively simple techniques, expanded to a large extent, and differentiated into bone-forming cells. It still remains to be determined which of these tissues provide the optimal source of cells for bone tissue engineering.

Although early pre-clinical and clinical data demonstrate the safety and effectiveness of TEC-based bone engineering, there are still many questions to be answered such as when, why and how such concepts work. Additional information is required concerning the therapeutic efficacy of transplanted TEC and the mechanisms of engraftment, homing and *in vivo* differentiation. There is also a need to carry out appropriately designed toxicology studies to demonstrate the long-term safety of these therapies. The widespread use of bone engineering will also depend upon the availability of validated methods for large-scale culture, storage and distribution. As these areas are addressed, new applications will be developed leading to novel therapeutic opportunities. Much has been learned about bone engineering therapy in the past few years, and much remains to be learned.

References

1. Hutmacher DW, Schantz JT, Lam CX, Tan KC, Lim TC. State of the art and future directions of scaffold-based bone engineering from a biomaterials perspective. *J Tissue Eng Regen Med* 2007 Jul–Aug;1(4):245–60.

2.   Drosse I, Volkmer E, Capanna R, De Biase P, Mutschler W, Schieker M. Tissue engineering for bone defect healing: an update on a multi-component approach. *Injury* 2008 Sep;39(Suppl 2):S9–20.

3.   Park J, Lakes R (eds.). *Biomaterials: An Introduction,* 2nd edn. New York: Plenum Press, 1992.

4.   Martin R, Burr D. *The Structure, Function and Adaption of Compact Bone.* New York: Raven Press, 1989.

5.   Hench L, Wilson J. *Introduction to Bioceramics.* Singapore: World Scientific, 1993.

6.   Owen M. Lineage of osetogenic cells and their relationship to the stromal system. In: Peck W (ed.) *Bone and Mineral Research.* Amsterdam: Elsevier, 1985, pp. 1–24.

7.   Owen M, Friedenstein AJ. Stromal stem cells: marrow-derived osteogenic precursors. *Ciba Found Symp* 1988;136:42–60.

8.   Robinson SH. Physiology of hematopoiesis. In: Robinson SH, Reich PR (eds.) *Hematology Pathological Basis for Clinical Practice*, 3rd edn. Boston: Little, Brown and Company, 1993, pp. 1–13.

9.   Lakshmipathy U, Verfaillie C. Stem cell plasticity. *Blood Rev* 2005 Jan;19(1):29–38.

10. Nakagawa M, Koyanagi M, Tanabe K, Takahashi K, Ichisaka T, Aoi T, *et al.* Generation of induced pluripotent stem cells without Myc from mouse and human fibroblasts. *Nat Biotechnol* 2008 Jan;26(1):101–6.

11. Okita K, Ichisaka T, Yamanaka S. Generation of germline-competent induced pluripotent stem cells. *Nature* 2007 Jul 19;448(7151):313–7.

12. Wernig M, Meissner A, Foreman R, Brambrink T, Ku M, Hochedlinger K, *et al. In vitro* reprogramming of fibroblasts into a pluripotent ES-cell-like state. *Nature* 2007 Jul 19;448(7151):318–24.

13. Mitalipov S, Wolf D. Totipotency, pluripotency and nuclear reprogramming. *Adv Biochem Eng Biotechnol* 2009 Nov 1;335(1):179–87.

14. Shi S, Gronthos S. Perivascular niche of postnatal mesenchymal stem cells in human bone marrow and dental pulp. *J Bone Miner Res* 2003 Apr;18(4): 696–704.

15. Ringe J, Leinhase I, Stich S, Loch A, Neumann K, Haisch A, *et al.* Human mastoid periosteum-derived stem cells: promising candidates for skeletal tissue engineering. *J Tissue Eng Regen Med* 2008 Mar–Apr;2(2–3): 136–46.

16. Nöth U, Osyczka AM, Tuli R, Hickok NJ, Danielson KG, Tuan RS. Multilineage mesenchymal differentiation potential of human trabecular bone-derived cells. *J Orthop Res* 2002 Sep;20(5):1060–9.

17. Rider DA, Dombrowski C, Sawyer AA, Ng GH, Leong D, Hutmacher DW, *et al.* Autocrine fibroblast growth factor 2 increases the multipotentiality of human adipose-derived mesenchymal stem cells. *Stem Cells* 2008 Jun;26(6):1598–608.

18. Fan J, Varshney RR, Ren L, Cai D, Wang DA. Synovium-derived mesenchymal stem cells: a new cell source for musculoskeletal regeneration. *Tissue Eng Part B Rev* 2009 Mar;15(1):75–86.

19. Perry BC, Zhou D, Wu X, Yang FC, Byers MA, Chu TM, *et al.* Collection, cryopreservation, and characterization of human dental pulp-derived mesenchymal stem cells for banking and clinical use. *Tissue Eng Part C Methods* 2008 Jun;14(2):149–56.

20. Zuk PA, Zhu M, Ashjian P, De Ugarte DA, Huang JI, Mizuno H, *et al.* Human adipose tissue is a source of multipotent stem cells. *Mol Biol Cell* 2002 Dec;13(12):4279–95.

21. Ennis J, Sarugaser R, Gomez A, Baksh D, Davies JE. Isolation, characterization, and differentiation of human umbilical cord perivascular cells (HUCPVCs). *Methods Cell Biol* 2008;86:121–36.

22. da Silva Meirelles L, Chagastelles PC, Nardi NB. Mesenchymal stem cells reside in virtually all post-natal organs and tissues. *J Cell Sci* 2006 Jun 1;119(Pt 11):2204–13.

23. Panepucci RA, Siufi JL, Silva WA, Jr, Proto-Siquiera R, Neder L, Orellana M, *et al.* Comparison of gene expression of umbilical cord vein and bone marrow-derived mesenchymal stem cells. *Stem Cells* 2004;22(7):1263–78.

24. Sarugaser R, Lickorish D, Baksh D, Hosseini MM, Davies JE. Human umbilical cord perivascular (HUCPV) cells: a source of mesenchymal progenitors. *Stem Cells* 2005 Feb;23(2):220–9.

25. Lu LL, Liu YJ, Yang SG, Zhao QJ, Wang X, Gong W, *et al.* Isolation and characterization of human umbilical cord mesenchymal stem cells with hematopoiesis-supportive function and other potentials. *Haematologica* 2006 Aug;91(8):1017–26.

26. Song L, Webb NE, Song Y, Tuan RS. Identification and functional analysis of candidate genes regulating mesenchymal stem cell self-renewal and multipotency. *Stem Cells* 2006 Jul;24(7):1707–18.

27. Baksh D, Yao R, Tuan RS. Comparison of proliferative and multilineage differentiation potential of human mesenchymal stem cells derived from umbilical cord and bone marrow. *Stem Cells* 2007 Jun;25(6):1384–92.
28. Liu TM, Martina M, Hutmacher DW, Hui JH, Lee EH, Lim B. Identification of common pathways mediating differentiation of bone marrow- and adipose tissue-derived human mesenchymal stem cells into three mesenchymal lineages. *Stem Cells* 2007 Mar;25(3):750–60.
29. Kermani AJ, Fathi F, Mowla SJ. Characterization and genetic manipulation of human umbilical cord vein mesenchymal stem cells: potential application in cell-based gene therapy. *Rejuvenation Res* 2008 Apr;11(2):379–86.
30. McElreavey KD, Irvine AI, Ennis KT, McLean WH. Isolation, culture and characterisation of fibroblast-like cells derived from the Wharton's jelly portion of human umbilical cord. *Biochem Soc Trans* 1991 Feb;19(1):29S.
31. Friedenstein AJ, Chailakhyan RK, Latsinik NV, Panasyuk AF, Keiliss-Borok IV. Stromal cells responsible for transferring the microenvironment of the hemopoietic tissues. Cloning *in vitro* and retransplantation *in vivo*. *Transplantation* 1974 Apr;17(4):331–40.
32. Caplan AI. Mesenchymal stem cells. *J Orthop Res* 1991 Sep;9(5):641–50.
33. Prockop DJ. Marrow stromal cells as stem cells for nonhematopoietic-tissues. *Science* 1997 Apr 4;276(5309):71–4.
34. Pittenger MF, Mackay AM, Beck SC, Jaiswal RK, Douglas R, Mosca JD, *et al.* Multilineage potential of adult human mesenchymal stem cells. *Science* 1999 Apr 2;284(5411):143–7.
35. Krause DS, Theise ND, Collector MI, Henegariu O, Hwang S, Gardner R, *et al.* Multi-organ, multi-lineage engraftment by a single bone marrow-derived stem cell. *Cell* 2001 May 4;105(3):369–77.
36. Friedman R, Betancur M, Boissel L, Tuncer H, Cetrulo C, Klingemann H. Umbilical cord mesenchymal stem cells: adjuvants for human cell transplantation. *Biol Blood Marrow Transplant* 2007 Dec;13(12):1477–86.
37. Friedenstein AJ, Petrakova KV, Kurolesova AI, Frolova GP. Heterotopic of bone marrow. Analysis of precursor cells for osteogenic and hematopoietic tissues. *Transplantation* 1968 Mar;6(2):230–47.
38. Friedenstein AJ, Chailakhjan RK, Lalykina KS. The development of fibroblast colonies in monolayer cultures of guinea-pig bone marrow and spleen cells. *Cell Tissue Kinet* 1970 Oct;3(4):393–403.

39. Simmons PJ, Torok-Storb B. Identification of stromal cell precursors in human bone marrow by a novel monoclonal antibody, STRO-1. *Blood* 1991 Jul 1;78(1):55–62.

40. Waller EK, Olweus J, Lund-Johansen F, Huang S, Nguyen M, Guo GR, *et al.* The "common stem cell" hypothesis re-evaluated: human fetal bone marrow contains separate populations of hematopoietic and stromal progenitors. *Blood* 1995 May 1;85(9):2422–35.

41. Kuznetsov SA, Friedenstein AJ, Robey PG. Factors required for bone marrow stromal fibroblast colony formation *in vitro*. *Br J Haematol* 1997 Jun;97(3):561–70.

42. Baddoo M, Hill K, Wilkinson R, Gaupp D, Hughes C, Kopen GC, *et al.* Characterization of mesenchymal stem cells isolated from murine bone marrow by negative selection. *J Cell Biochem* 2003 Aug 15;89(6):1235–49.

43. Bruder SP, Jaiswal N, Haynesworth SE. Growth kinetics, self-renewal, and the osteogenic potential of purified human mesenchymal stem cells during extensive subcultivation and following cryopreservation. *J Cell Biochem* 1997 Feb;64(2):278–94.

44. Gronthos S, Graves SE, Ohta S, Simmons PJ. The STRO-1+ fraction of adult human bone marrow contains the osteogenic precursors. *Blood* 1994 Dec 15;84(12):4164–73.

45. Gronthos S, Zannettino AC, Hay SJ, Shi S, Graves SE, Kortesidis A, *et al.* Molecular and cellular characterisation of highly purified stromal stem cells derived from human bone marrow. *J Cell Sci* 2003 May 1;116(Pt 9):1827–35.

46. Dennis JE, Carbillet JP, Caplan AI, Charbord P. The STRO-1+ marrow cell population is multipotential. *Cells Tissues Organs* 2002;170(2–3):73–82.

47. Simmons PJ, Gronthos S, Zannettino A, Ohta S, Graves S. Isolation, characterization and functional activity of human marrow stromal progenitors in hemopoiesis. *Prog Clin Biol Res* 1994;389:271–80.

48. Filshie RJ, Zannettino AC, Makrynikola V, Gronthos S, Henniker AJ, Bendall LJ, *et al.* MUC18, a member of the immunoglobulin superfamily, is expressed on bone marrow fibroblasts and a subset of hematological malignancies. *Leukemia* 1998 Mar;12(3):414–21.

49. Haynesworth SE, Baber MA, Caplan AI. Cell surface antigens on human marrow-derived mesenchymal cells are detected by monoclonal antibodies. *Bone* 1992;13(1):69–80.

50. Deschaseaux F, Gindraux F, Saadi R, Obert L, Chalmers D, Herve P. Direct selection of human bone marrow mesenchymal stem cells using an anti-CD49a antibody reveals their CD45med, low phenotype. *Br J Haematol* 2003 Aug;122(3):506–17.

51. Jiang Y, Vaessen B, Lenvik T, Blackstad M, Reyes M, Verfaillie CM. Multipotent progenitor cells can be isolated from postnatal murine bone marrow, muscle, and brain. *Exp Hematol* 2002 Aug;30(8):896–904.

52. Jiang Y, Jahagirdar BN, Reinhardt RL, Schwartz RE, Keene CD, Ortiz-Gonzalez XR, *et al.* Pluripotency of mesenchymal stem cells derived from adult marrow. *Nature* 2002 Jul 4;418(6893):41–9.

53. Muschler GF, Boehm C, Easley K. Aspiration to obtain osteoblast progenitor cells from human bone marrow: the influence of aspiration volume. *J Bone Joint Surg Am* 1997 Nov;79(11):1699–709.

54. Majors AK, Boehm CA, Nitto H, Midura RJ, Muschler GF. Characterization of human bone marrow stromal cells with respect to osteoblastic differentiation. *J Orthop Res* 1997 Jul;15(4):546–57.

55. Muschler GF, Nitto H, Boehm CA, Easley KA. Age- and gender-related changes in the cellularity of human bone marrow and the prevalence of osteoblastic progenitors. *J Orthop Res* 2001 Jan;19(1):117–25.

56. Zannettino AC, Harrison K, Joyner CJ, Triffitt JT, Simmons PJ. Molecular cloning of the cell surface antigen identified by the osteoprogenitor-specific monoclonal antibody, HOP-26. *J Cell Biochem* 2003 May 1;89(1):56–66.

57. Zuk PA, Zhu M, Mizuno H, Huang J, Futrell JW, Katz AJ, *et al.* Multilineage cells from human adipose tissue: implications for cell-based therapies. *Tissue Eng* 2001 Apr;7(2):211–28.

58. Mitchell KE, Weiss ML, Mitchell BM, Martin P, Davis D, Morales L, *et al.* Matrix cells from Wharton's jelly form neurons and glia. *Stem Cells* 2003;21(1):50–60.

59. Ennis J, Gotherstrom C, Le Blanc K, Davies JE. *In vitro* immunologic properties of human umbilical cord perivascular cells. *Cytotherapy* 2008; 10(2):174–81.

60. Crisan M, Huard J, Zheng B, Sun B, Yap S, Logar A, *et al.* Purification and culture of human blood vessel-associated progenitor cells. *Curr Protoc Stem Cell Biol* 2008 Mar; Chapter 2:Unit 2B 1–2B 13.

61. Newman PJ, Newman DK. Signal transduction pathways mediated by PECAM-1: new roles for an old molecule in platelet and vascular cell biology. *Arterioscler Thromb Vasc Biol* 2003 Jun 1;23(6):953–64.

62. Ishikawa H, Tsuyama N, Abroun S, Liu S, Li FJ, Taniguchi O, *et al.* Requirements of src family kinase activity associated with CD45 for myeloma cell proliferation by interleukin-6. *Blood* 2002 Mar 15;99(6):2172–8.

63. Lopez-Casillas F, Cheifetz S, Doody J, Andres JL, Lane WS, Massague J. Structure and expression of the membrane proteoglycan betaglycan, a component of the TGF-beta receptor system. *Cell* 1991 Nov 15;67(4):785–95.

64. Ades EW, Zwerner RK, Acton RT, Balch CM. Isolation and partial characterization of the human homologue of Thy-1. *J Exp Med* 1980 Feb 1;151(2):400–6.

65. Covas DT, Siufi JL, Silva AR, Orellana MD. Isolation and culture of umbilical vein mesenchymal stem cells. *Braz J Med Biol Res* 2003 Sep;36(9):1179–83.

66. Zhang ZY, Teoh SH, Chong MS, Schantz JT, Fisk NM, Choolani MA, *et al.* Superior osteogenic capacity for bone tissue engineering of fetal compared to perinatal and adult mesenchymal stem cells. *Stem Cells* 2009 Jan;27(1):126–37.

67. Turgeman G, Aslan H, Gazit Z, Gazit D. Cell-mediated gene therapy for bone formation and regeneration. *Curr Opin Mol Ther* 2002 Aug;4(4):390–4.

68. Yoon ST, Boden SD. Osteoinductive molecules in orthopaedics: basic science and preclinical studies. *Clin Orthop Relat Res* 2002 Feb;395:33–43.

69. Winn SR, Uludag H, Hollinger JO. Carrier systems for bone morphogenetic proteins. *Clin Orthop Relat Res* 1999 Oct; 367(Suppl):S95–106.

70. Oligino TJ, Yao Q, Ghivizzani SC, Robbins P. Vector systems for gene transfer to joints. *Clin Orthop Relat Res* 2000 Oct;379(Suppl):S17–30.

71. Hutmacher DW, Garcia AJ. Scaffold-based bone engineering by using genetically modified cells. *Gene* 2005 Feb 28;347(1):1–10.

72. Digirolamo CM, Stokes D, Colter D, Phinney DG, Class R, Prockop DJ. Propagation and senescence of human marrow stromal cells in culture: a simple colony-forming assay identifies samples with the greatest potential to propagate and differentiate. *Br J Haematol* 1999 Nov;107(2):275–81.

73. Martin I, Muraglia A, Campanile G, Cancedda R, Quarto R. Fibroblast growth factor-2 supports *ex vivo* expansion and maintenance of osteogenic precursors from human bone marrow. *Endocrinology* 1997 Oct;138(10):4456–62.

74. Studeny M, Marini FC, Champlin RE, Zompetta C, Fidler IJ, Andreeff M. Bone marrow-derived mesenchymal stem cells as vehicles for interferon-beta delivery into tumors. *Cancer Res* 2002 Jul 1;62(13):3603–8.

75. Ringden O, Uzunel M, Rasmusson I, Remberger M, Sundberg B, Lonnies H, *et al.* Mesenchymal stem cells for treatment of therapy-resistant graft-versus-host disease. *Transplantation* 2006 May 27;81(10):1390–7.
76. Le Blanc K, Rasmusson I, Sundberg B, Gotherstrom C, Hassan M, Uzunel M, *et al.* Treatment of severe acute graft-versus-host disease with third party haploidentical mesenchymal stem cells. *Lancet* 2004 May 1;363(9419):1439–41.
77. Mackenzie TC, Flake AW. Multilineage differentiation of human MSC after *in utero* transplantation. *Cytotherapy* 2001;3(5):403–5.
78. Wong M, Tuan RS. Nuserum, a synthetic serum replacement, supports chondrogenesis of embryonic chick limb bud mesenchymal cells in micromass culture. *In Vitro Cell Dev Biol Anim* 1993 Dec;29A(12):917–22.
79. Goldsborough M, Tilkins M, Price P, Lobo-Alfonso J, Morrison J, Stevens M. Serum-free culture of murine embryonic stem (ES) cells. *Focus* 1998;20:8–12.
80. Buttery LD, Bourne S, Xynos JD, Wood H, Hughes FJ, Hughes SP, *et al.* Differentiation of osteoblasts and *in vitro* bone formation from murine embryonic stem cells. *Tissue Eng* 2001 Feb;7(1):89–99.
81. Rogers JJ, Young HE, Adkison LR, Lucas PA, Black AC, Jr. Differentiation factors induce expression of muscle, fat, cartilage, and bone in a clone of mouse pluripotent mesenchymal stem cells. *Am Surg* 1995 Mar;61(3):231–6.
82. Xiao G, Cui Y, Ducy P, Karsenty G, Franceschi RT. Ascorbic acid-dependent activation of the osteocalcin promoter in MC3T3-E1 preosteoblasts: requirement for collagen matrix synthesis and the presence of an intact OSE2 sequence. *Mol Endocrinol* 1997 Jul;11(8):1103–13.
83. Volloch V, Kaplan D. Matrix-mediated cellular rejuvenation. *Matrix Biol* 2002 Oct;21(6):533–43.
84. Gronthos S, Simmons PJ, Graves SE, Robey PG. Integrin-mediated interactions between human bone marrow stromal precursor cells and the extracellular matrix. *Bone* 2001 Feb;28(2):174–81.
85. Roche P, Goldberg HA, Delmas PD, Malaval L. Selective attachment of osteoprogenitors to laminin. *Bone* 1999 Apr;24(4):329–36.
86. Owen TA, Aronow M, Shalhoub V, Barone LM, Wilming L, Tassinari MS, *et al.* Progressive development of the rat osteoblast phenotype *in vitro*: reciprocal relationships in expression of genes associated with osteoblast proliferation and differentiation during formation of the bone extracellular matrix. *J Cell Physiol* 1990 Jun;143(3):420–30.

87. Matsuda N, Morita N, Matsuda K, Watanabe M. Proliferation and differentiation of human osteoblastic cells associated with differential activation of MAP kinases in response to epidermal growth factor, hypoxia, and mechanical stress *in vitro. Biochem Biophys Res Commun* 1998 Aug 19;249(2):350–4.

88. Yoshikawa Y, Hirayama F, Kanai M, Nakajo S, Ohkawara J, Fujihara M, *et al.* Stromal cell-independent differentiation of human cord blood CD34+CD38- lymphohematopoietic progenitors toward B cell lineage. *Leukemia* 2000 Apr;14(4):727–34.

89. Altman GH, Horan RL, Martin I, Farhadi J, Stark PR, Volloch V, *et al.* Cell differentiation by mechanical stress. *FASEB J* 2002 Feb;16(2):270–2.

90. Kapur S, Baylink DJ, Lau KH. Fluid flow shear stress stimulates human osteoblast proliferation and differentiation through multiple interacting and competing signal transduction pathways. *Bone* 2003 Mar;32(3):241–51.

91. Majumdar MK, Keane-Moore M, Buyaner D, Hardy WB, Moorman MA, McIntosh KR, *et al.* Characterization and functionality of cell surface molecules on human mesenchymal stem cells. *J Biomed Sci* 2003 Mar–Apr; 10(2):228–41.

92. Le Blanc K, Tammik C, Rosendahl K, Zetterberg E, Ringden O. HLA expression and immunologic properties of differentiated and undifferentiated mesenchymal stem cells. *Exp Hematol* 2003 Oct;31(10):890–6.

93. Koc ON, Day J, Nieder M, Gerson SL, Lazarus HM, Krivit W. Allogeneic mesenchymal stem cell infusion for treatment of metachromatic leukodystrophy (MLD) and Hurler syndrome (MPS-IH). *Bone Marrow Transplant* 2002 Aug;30(4):215–22.

94. Saito T, Kuang JQ, Bittira B, Al-Khaldi A, Chiu RC. Xenotransplant cardiac chimera: immune tolerance of adult stem cells. *Ann Thorac Surg* 2002 Jul;74(1):19–24; discussion.

95. Donovan PJ, Gearhart J. The end of the beginning for pluripotent stem cells. *Nature* 2001 Nov 1;414(6859):92–7.

96. Hutmacher DW. Scaffolds in tissue engineering bone and cartilage. *Biomaterials* 2000 Dec;21(24):2529–43.

97. Hutmacher DW, Sittinger M. Periosteal cells in bone tissue engineering. *Tissue Eng* 2003;9(Suppl 1):S45–64.

98. Freed LE, Vunjak-Novakovic G. Spaceflight bioreactor studies of cells and tissues. *Adv Space Biol Med* 2002;8:177–95.

99. Sutherland RM, Sordat B, Bamat J, Gabbert H, Bourrat B, Mueller-Klieser W. Oxygenation and differentiation in multicellular spheroids of human colon carcinoma. *Cancer Res* 1986 Oct;46(10):5320–9.

100. Ishaug SL, Crane GM, Miller MJ, Yasko AW, Yaszemski MJ, Mikos AG. Bone formation by three-dimensional stromal osteoblast culture in biodegradable polymer scaffolds. *J Biomed Mater Res* 1997 Jul;36(1):17–28.

101. Martin I, Obradovic B, Freed LE, Vunjak-Novakovic G. Method for quantitative analysis of glycosaminoglycan distribution in cultured natural and engineered cartilage. *Ann Biomed Eng* 1999 Sep–Oct;27(5):656–62.

102. Vander AJ, Sherman JH, Luciano DS. *Vander's Human Physiology: The Mechanisms of Body Function*, Widmaier EP, Raff H, Strang H (eds.). New York: McGraw-Hill, 1985.

103. Gooch KJ, Kwon JH, Blunk T, Langer R, Freed LE, Vunjak-Novakovic G. Effects of mixing intensity on tissue-engineered cartilage. *Biotechnol Bioeng* 2001 Feb 20;72(4):402–7.

104. Freed LE, Marquis JC, Langer R, Vunjak-Novakovic G, Emmanual J. Composition of cell-polymer cartilage implants. *Biotechnol Bioeng* 1994 Mar 25;43(7):605–14.

105. Martin I, Padera RF, Vunjak-Novakovic G, Freed LE. *In vitro* differentiation of chick embryo bone marrow stromal cells into cartilaginous and bone-like tissues. *J Orthop Res* 1998 Mar;16(2):181–9.

106. Wendt D, Jakob M, Martin I. Bioreactor-based engineering of osteochondral grafts: from model systems to tissue manufacturing. *J Biosci Bioeng* 2005 Nov;100(5):489–94.

107. Martin Y, Vermette P. Bioreactors for tissue mass culture: design, characterization, and recent advances. *Biomaterials* 2005 Dec;26(35):7481–503.

108. Vunjak-Novakovic G, Martin I, Obradovic B, Treppo S, Grodzinsky AJ, Langer R, *et al.* Bioreactor cultivation conditions modulate the composition and mechanical properties of tissue-engineered cartilage. *J Orthop Res* 1999 Jan;17(1):130–8.

109. Carrier RL, Papadaki M, Rupnick M, Schoen FJ, Bursac N, Langer R, *et al.* Cardiac tissue engineering: cell seeding, cultivation parameters, and tissue construct characterization. *Biotechnol Bioeng* 1999 Sep 5;64(5):580–9.

110. Rhee HW, Zhau HE, Pathak S, Multani AS, Pennanen S, Visakorpi T, *et al.* Permanent phenotypic and genotypic changes of prostate cancer cells

cultured in a three-dimensional rotating-wall vessel. *In Vitro Cell Dev Biol Anim* 2001 Mar;37(3):127–40.

111. Klement BJ, Young QM, George BJ, Nokkaew M. Skeletal tissue growth, differentiation and mineralization in the NASA rotating wall vessel. *Bone* 2004 Mar;34(3):487–98.

112. Unsworth BR, Lelkes PI. Growing tissues in microgravity. *Nat Med* 1998 Aug;4(8):901–7.

113. Goldstein AS, Juarez TM, Helmke CD, Gustin MC, Mikos AG. Effect of convection on osteoblastic cell growth and function in biodegradable polymer foam scaffolds. *Biomaterials* 2001 Jun;22(11):1279–88.

114. Bancroft GN, Sikavitsas VI, van den Dolder J, Sheffield TL, Ambrose CG, Jansen JA, *et al.* Fluid flow increases mineralized matrix deposition in 3D perfusion culture of marrow stromal osteoblasts in a dose-dependent manner. *Proc Natl Acad Sci USA* 2002 Oct 1;99(20):12600–5.

115. Davisson T, Sah RL, Ratcliffe A. Perfusion increases cell content and matrix synthesis in chondrocyte three-dimensional cultures. *Tissue Eng* 2002 Oct;8(5):807–16.

116. Butler DL, Goldstein SA, Guilak F. Functional tissue engineering: the role of biomechanics. *J Biomech Eng* 2000 Dec;122(6):570–5.

117. Klein-Nulend J, van der Plas A, Semeins CM, Ajubi NE, Frangos JA, Nijweide PJ, *et al.* Sensitivity of osteocytes to biomechanical stress *in vitro. FASEB J* 1995 Mar;9(5):441–5.

118. Owan I, Burr DB, Turner CH, Qiu J, Tu Y, Onyia JE, *et al.* Mechanotransduction in bone: osteoblasts are more responsive to fluid forces than mechanical strain. *Am J Physiol* 1997 Sep;273(3 Pt 1):C810–5.

119. Sikavitsas VI, Bancroft GN, Holtorf HL, Jansen JA, Mikos AG. Mineralized matrix deposition by marrow stromal osteoblasts in 3D perfusion culture increases with increasing fluid shear forces. *Proc Natl Acad Sci USA* 2003 Dec 9;100(25):14683–8.

120. Simmons CA, Matlis S, Thornton AJ, Chen S, Wang CY, Mooney DJ. Cyclic strain enhances matrix mineralization by adult human mesenchymal stem cells via the extracellular signal-regulated kinase (ERK1/2) signaling pathway. *J Biomech* 2003 Aug;36(8):1087–96.

121. Lendeckel S, Jodicke A, Christophis P, Heidinger K, Wolff J, Fraser JK, *et al.* Autologous stem cells (adipose) and fibrin glue used to treat

widespread traumatic calvarial defects: case report. *J Craniomaxillofac Surg* 2004 Dec;32(6):370–3.

122. Schantz JT, Hutmacher DW, Lam CX, Brinkmann M, Wong KM, Lim TC, *et al.* Repair of calvarial defects with customised tissue-engineered bone grafts II. Evaluation of cellular efficiency and efficacy *in vivo. Tissue Eng* 2003;9(Suppl 1):S127–39.

123. Schantz JT, Lim TC, Ning C, Teoh SH, Tan KC, Wang SC, *et al.* Cranioplasty after trephination using a novel biodegradable burr hole cover: technical case report. *Neurosurgery* 2006 Feb;58(1 Suppl):ONS-E176; discussion ONS-E.

124. Warnke PH, Springer IN, Wiltfang J, Acil Y, Eufinger H, Wehmöller M, *et al.* Growth and transplantation of a custom vascularised bone graft in a man. *Lancet* 2004 Aug 28–Sep 3;364(9436):766–70.

125. Schmelzeisen R, Schimming R, Sittinger M. Making bone: implant insertion into tissue-engineered bone for maxillary sinus floor augmentation — a preliminary report. *J Craniomaxillofac Surg* 2003 Feb;31(1):34–9.

126. Yamada Y, Ueda M, Hibi H, Baba S. A novel approach to periodontal tissue regeneration with mesenchymal stem cells and platelet-rich plasma using tissue engineering technology: a clinical case report. *Int J Periodontics Restorative Dent* 2006 Aug;26(4):363–9.

127. Kawaguchi H, Hayashi H, Mizuno N, Fujita T, Hasegawa N, Shiba H, *et al.* Cell transplantation for periodontal diseases. A novel periodontal tissue regenerative therapy using bone marrow mesenchymal stem cells. *Clin Calcium* 2005 Jul;15(7):99–104.

128. Vacanti CA, Bonassar LJ, Vacanti MP, Shufflebarger J. Replacement of an avulsed phalanx with tissue-engineered bone. *N Engl J Med* 2001 May 17; 344(20):1511–4.

129. Quarto R, Mastrogiacomo M, Cancedda R, Kutepov SM, Mukhachev V, Lavroukov A, *et al.* Repair of large bone defects with the use of autologous bone marrow stromal cells. *N Engl J Med* 2001 Feb 1;344(5):385–6.

130. Cancedda R, Mastrogiacomo M, Bianchi G, Derubeis A, Muraglia A, Quarto R. Bone marrow stromal cells and their use in regenerating bone. *Novartis Found Symp* 2003;249:133–43.

131. Petite H, Viateau V, Bensaid W, Meunier A, de Pollak C, Bourguignon M, *et al.* Tissue-engineered bone regeneration. *Nat Biotechnol* 2000 Sep;18(9): 959–63.

132. Bruder SP, Kraus KH, Goldberg VM, Kadiyala S. The effect of implants loaded with autologous mesenchymal stem cells on the healing of canine segmental bone defects. *J Bone Joint Surg Am* 1998 Jul;80(7):985–96.

133. Bajada S, Harrison PE, Ashton BA, Cassar-Pullicino VN, Ashammakhi N, Richardson JB. Successful treatment of refractory tibial nonunion using calcium sulphate and bone marrow stromal cell implantation. *J Bone Joint Surg Br* 2007 Oct;89(10):1382–6.

134. Friedlaender GE, Perry CR, Cole JD, Cook SD, Cierny G, Muschler GF, *et al.* Osteogenic protein-1 (bone morphogenetic protein-7) in the treatment of tibial nonunions. *J Bone Joint Surg Am* 2001;83-A Suppl 1(Pt 2):S151–8.

135. Nöth U, Reichert J, Reppenhagen S, Steinert A, Rackwitz L, Eulert J, *et al.* Cell based therapy for the treatment of femoral head necrosis. *Orthopade* 2007 May;36(5):466–71.

136. Kawate K, Yajima H, Ohgushi H, Kotobuki N, Sugimoto K, Ohmura T, *et al.* Tissue-engineered approach for the treatment of steroid-induced osteonecrosis of the femoral head: transplantation of autologous mes-enchymal stem cells cultured with beta-tricalcium phosphate ceramics and free vascularized fibula. *Artif Organs* 2006 Dec;30(12):960–2.

137. Hernigou P, Beaujean F. Treatment of osteonecrosis with autologous bone marrow grafting. *Clin Orthop Relat Res* 2002 Dec; 405:14–23.

138. Gangji V, Hauzeur JP, Matos C, De Maertelaer V, Toungouz M, Lambermont M. Treatment of osteonecrosis of the femoral head with implantation of autologous bone-marrow cells. A pilot study. *J Bone Joint Surg Am* 2004 Jun;86-A(6):1153–60.

7

Neural Differentiation of Umbilical Cord Blood Derived Cells and Applications in Neurological Disorders

Gerald Udolph

ABSTRACT

Stem cell and progenitor cells derived from the bone marrow and the cord blood have been shown to possess multilineage differentiation potential such as hematopoietic, mesenchymal as well as neural differentiation capabilities. This review summarizes recent advances in the field of neural differentiation *in vitro* and applications of cord blood derived cells in neurological preclinical disease models as well as in clinical applications in neurological diseases.

INTRODUCTION

Bone marrow as well as cord blood derived stem cells are widely used as clinical tools for treatment of different diseases in humans. Cord blood derived hematopoietic stem cells have gained increasing recognition in the treatment of blood disorders such as lymphoma and leukemia.[1] Besides hematopoietic stem cells, cord blood also harbors an endothelial as well as a mesenchymal-like stem cell population.[2–4] Common to all of these stem cells is their presence in the mononuclear cell (MNC) fraction of the cord blood which is abundantly available after birth and can be easily collected after normal deliveries without any risk to either mother or child. The mesenchymal-like compartment of the cord blood has been shown to be multipotent as such cells can give rise to multiple cell lineages of mesodermal nature as well as other cells types of other germ layers such as the ectoderm and the endoderm. This review will focus on cells derived from the cord blood and their neurogenic potential *in vitro*, their potential role in preclinical models of brain related diseases *in vivo* as well as their clinical applications in human patients with central nervous system disorders.

HOW IT ALL BEGAN: *IN VITRO* DIFFERENTIATION INTO THE NEURAL LINEAGE OF BONE MARROW MESENCHYMAL CELLS

Mesenchymal stem cells (MSCs) have been first identified by Friedenstein *et al.* from guinea-pig bone marrow.[5] With the report of Pittenger *et al.* it became obvious that human mesenchymal stem cells from the bone marrow possessed multilineage differentiation potential along the mesenchymal lineages including adipocytes, chondrocytes and osteocytes.[6] Subsequently, two reports described *in vitro* differentiation of MSCs into cells expressing neural marker genes.[7,8] Neural differentiation was achieved by either co-culturing MSCs with primary brain tissue or by applying neurogenic cytokine cocktails to the cells. Essentially, both studies reported that treatment of MSCs resulted in morphological changes such as neurite-like outgrowth and upregulation of neuronal as well as glial marker gene expression. Furthermore, differentiation of MSCs to neural progenitors (NPCs) was achieved in the rat.[9] By applying cytokine supplemented media to rat MSCs the authors were able to convert adherently growing MSCs into sphere-shaped structures which expressed markers typical for neural progenitor cells such as nestin and neurogenin1. Upon further differentiation, cells developed into neurons and glia as judged by the expression of specific marker genes indicative for neuronal and glial cell fate. A similar report showed that MSC-derived NPCs expressed neuroectodermal marker genes such as Otx1, neuroD1, neurogenin2, as well as nestin.[10] These authors were able to demonstrate clonality of these spheres, suggesting self-renewal properties of at least some cells within these spheres. Furthermore, cells in the spheres, upon applying specific differentiation protocols, differentiated into neurons and glial cells. Interestingly, MSCs, once converted into NPCs did not differentiate into mesenchymal lineages anymore, indicating that they had lost differentiation potential along the mesengenic lineage. Therefore, the conversion from a mesenchymal to an ectodermal fate appeared irreversible. The 3D environment of spheres seemed to be beneficial for the neural differentiation of MSCs, possibly promoting a stepwise differentiation similar to what is observed in a context of a developing organism. This suggests that the 3D environment might be supportive for a switch of MSCs into neural lineage, although no further mechanistic explanation for this conversion was provided.[11]

Two earlier papers reported the contribution of bone marrow derived cells to brain tissue after transplanting bone marrow cells into either lethally irradiated or immunodeficient mice.[12,13] The integrated cells expressed neuronal antigens, albeit at a low frequency. It was concluded that cells gained access to the brain and assumed characteristics of typical CNS neurons, and that bone marrow derived cells might serve as an additional source for new neurons besides the brain's resident neural stem cells. Therefore, there is some evidence that *in vitro* MSCs from bone marrow can be directed to express genes typically found in neurons and as such those cells are capable of acquiring some characteristics of the neuronal lineage.

IN VITRO DIFFERENTIATION OF HUMAN CORD BLOOD DERIVED CELL TYPES INTO THE NEURAL LINEAGE

With the knowledge derived from the bone marrow field, researchers started to explore the potential of neural differentiation of other sources of stem cells, including umbilical cord blood cells. In this context, Sanchez-Ramos *et al.* demonstrated that adherently growing mononuclear cells (MNCs) from cord blood expressed neural specific antigens after exposure to a neural-inducing cocktail which included nerve growth factor (NGF) and retinoic acid (RA).[14] However, these initial cells were not a well-defined cell population, and most likely constituted a heterogeneous population. Cultures under these conditions are likely to have included cells of the hematopoietic lineage such as macrophages, dendritic cells as well as hematopoietic stem and progenitor cells which are able to adhere to cell culture plastic. Following the cytokine cocktail treatment, cells were observed to upregulate markers typically involved in early neurogenesis and early neuronal differentiation including the expression of the neuronal marker β3-tubulin, as well as the glia specific marker, GFAP. It is worth noting that most of the markers studied in this report were already expressed in the undifferentiated cells, although at lower levels. A microarray analysis also revealed that a number of genes involved in neural differentiation were upregulated, and, conversely, hematopoietic markers were down-regulated upon differentiation. This switch of marker gene expression might indicate a shift from a mesodermal hematopoietic to an ectodermal neural fate of at least some of cells within the differentiating population. Although this study did not provide any functional aspects of the neural differentiation process, it was still the first study showing that some cells in the cord blood mononuclear cell population potentially have neurogenic differentiation capabilities. In the same year, Ha *et al.*, reported that cultured human cord blood monocytes gave rise to cellular spheres. Upon replating of such spheres onto laminin-coated surfaces, cells with neural morphologies appeared and upregulation of neural specific marker genes was observed.[15] It was concluded that some cells in the cord blood possessed neural differentiation potential. These two studies were confirmed by a study using isolated plastic-adherent and proliferative cells devoid of any hematopoietic markers from cord blood.[16] Although no further characterization was done for this adherent and expanding cell population, the authors isolated a different and better defined population of cells which showed some similarities to MSCs derived from bone marrow. These cells were termed umbilical cord multipotent stem cells (UC-MCs). UC-MCs, upon growth factor exposure, expressed a neuronal (β3-tubulin) as well as glial markers (GFAP, Gal-C). It was also noted that some cells co-expressed both neuronal as well as glial markers, indicating that these co-expressing cells are most likely not finally differentiated, as cells expressing markers of both glial and neuronal cells are also found during normal neurogenesis.

Integration and differentiation of umbilical cord blood MNCs into the brain of experimental animals was also studied.[17] Cells were either cultured under growth or under neural differentiation conditions before implantation into the developing rat brain which still provides developmental cues to implanted cells. It was shown that MNCs were found in several areas as well as around blood vessels in the brain. Some of them expressed the glial marker GFAP (~2%) and to a lesser extend cells with neuronal marker gene expression were found as well (~0.2%). No difference in the cell numbers expressing neural markers was observed for undifferentiated or predifferentiated cells. This study at least provided some evidence that MNCs from cord blood survived in such a xenotransplant assay and furthermore some of these cells expressed markers typical for the neural lineage. Therefore, it appears that MNCs from cord blood also possess *in vivo* neural differentiation potential.

Nestin expressing progenitor cells could also be derived from the non-hematopoietic CD34 negative and CD45 negative fraction of MNCs from cord blood by treating adherent cells with epidermal growth factor (EGF).[18] These nestin positive cells then could be differentiated into cells expressing marker genes of the neuronal as well as glial lineages by

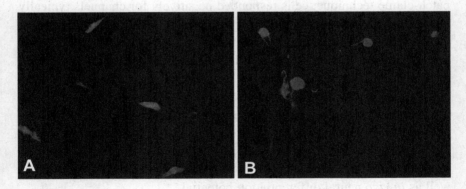

Fig. 1. Cord blood derived MSC-like cells differentiated into neuronal cells expressing the neuronal markers neurofilament (**A**) and the neurotransmitter γ-aminobutyric acid (GABA) (**B**), as shown by immunofluorescence in red. Blue represents nuclear staining. Magnification 400x.

either adding retinoic acid or by co-culturing these cells on rat primary cortical cultures. Two further studies by the same group[19,20] confirmed the neural differentiation potential of cord blood derived cells by using varying protocols to generate NSC-like stem cells from non-adherent cord blood cells. Such cells readily differentiated into the neural lineage. Strikingly, expression of the pluripotent embryonic stem cell markers such as Oct-4 and Sox-2 at the mRNA level was also observed in these cells, although no attempts were made to examine expression on the protein level or to monitor down regulation of these markers following differentiation.

In 2004 Koegler *et al.* reported the isolation of a novel stem cell type from cord blood which they named unrestricted somatic stem cell (USSC).[21] USSCs were isolated and grew similarly to bone marrow derived MSCs in a plastic adherent fashion with a largely similar surface marker expression profile. The further characterization and description of the multipotent differentiation potential of this cell type is described elsewhere in this book in more detail (see Chapter 10). The neural differentiation potential of these cells was tested *in vitro* and it was shown that USSCs, upon exposure to a neurogenic cytokine cocktail, expressed a number of neuronal as well as glial marker genes as detected by immunocytochemistry. Interestingly, a substantial fraction of the differentiated cells expressed tyrosine hydroxylase (TH), an enzyme typically found in dopaminergic neurons. Although cells were tested functionally with the patch-clamp technique, cells responded negatively for channels typically found on mature neurons. It was concluded that USSCs might differentiate into precursor cells but not into mature and fully functional neurons. Furthermore, it was demonstrated that after transplantation into cortical areas of rats numerous USSCs could be found surviving and some of these cells also expressed the human isoform of the neuronal marker *tau*. Further quantification of the frequency of integrated or *tau* expressing cells was not provided in this study. However, if implanted into a preimmune sheep model, transplanted cells contributed significantly to the brain. In summary, these authors showed that USSCs exhibited *in vitro* as well as *in vivo* neural differentiation potential.

A further study showed that cord blood-derived USSCs expressed genes normally associated with dopaminergic neuron differentiation or maintenance.[22] This was particularly the case when cells were cultured on laminin-coated surfaces. Strikingly, this study showed that neuronal genes were already expressed in the undifferentiated conditions. Further support for dopaminergic differentiation of USSCs came from a report by Greschat *et al.* (2008).[23] Incubation of USSCs in a defined cytokine medium followed by a dopamine containing medium led to close to 80% of cells being positive for the dopaminergic marker tyrosine hydroxylase. HPLC analysis also revealed that the neurotransmitters dopamine and serotonin were secreted into the medium upon stimulation, suggesting that cells were equipped with the functional machinery for neurotransmitter synthesis as well as release. Furthermore, it was demonstrated that differentiated cells expressed voltage-gated sodium channels. Accordingly, functional voltage-gated sodium channels, which could be reversibly blocked, were measured in a subset of the cells. Moreover, while potassium currents were recorded, no fast-activating and inactivating sodium currents could be detected.

A different study compared two different populations of cells derived from umbilical cord blood for their neural differentiation potential i.e., an adherent fraction of cells carrying the hematopoietic marker CD45 and an initially floating but thereafter replated fraction of cells.[24] It was shown that the adherent fraction in the absence of any differentiation protocol already expressed neural markers, although the expression of mature neuronal markers was limited. Some of the cells also co-expressed the hematopoietic marker CD45. In contrast, the non-adherent fraction expressed neural markers with a high frequency upon re-adherence. The authors hypothesized that the presence of such neuronal markers raises the possibility of transdifferentiation of hematopoietic cells into neuronal cells. However, due to the lack of functional data, alternative explanations for their findings were discussed such as non-specificity of markers used as well as cellular toxicity leading to neural marker gene expression.

Microarray analyses of differentiating cells revealed that a number of neurotransmitter receptors and ion channels were expressed and were upregulated in the differentiated conditions.[25] Furthermore, the expression of some receptor subtypes were studied on the level of immunohistochemistry and it was found that cells mainly expressed the kainate GluR2 receptor (88%), the GABA-AR (93%), GlyR (20%), nicotinic AchR

(39%), 5HT1CR (90%), and DA receptor D2 (85%). The presence of these receptors suggested that cells are equipped with the neurotransmitter receptor machinery typical for functional neurons. As a consequence, it was tested whether such cells were electrophysiologically active. In electrophysiological assays such as whole cell patch-clamp analysis it was found that the electrophysiological properties recorded were similar to properties normally seen in immature neurons or glia. However, the hallmark of mature functional neurons, voltage-gated sodium channel activity could not be detected. The authors concluded that the process of functional maturation is at least incomplete under the experimental conditions used.[25]

Cryopreserved cord blood mononuclear cells have also been directed into the neural lineage using various differentiation protocols.[26] This finding might have important implications for any future applications, as the long-term storage of any cell types without losing the differentiation potential is an important issue for off-the-shelf supply of therapeutic cells.

Mechanisms of Neural Differentiation of Cord Blood Derived Cells

Although the amount of literature on umbilical cord blood stem cell differentiation along the neuronal lineage strongly suggests that cord blood stem cells can differentiate into neurons, reports addressing the underlying molecular and biochemical mechanisms of differentiation of cord blood cells are still scarce. As a consequence, the molecular mechanisms of neural differentiation from cord blood derived cells are only poorly understood to date. Two papers have been published so far which tried to address mechanistical aspects. In 2006, Jurga et al. reported the involvement of ID1 as a negative regulator of neuronal differentiation of cord blood derived stem cells. ID1 was expressed in undifferentiated cord blood stem cells where it was found in the cytoplasm and is translocated into the nucleus in partially committed cells. Translocation correlated with the upregulation of a key neurogenic gene, NeuroD1.[20] In contrast, fully differentiated cells strongly down-regulated ID1 expression.

In cord blood derived MSC-like cells, the PKA signaling pathway, but not MAPK signaling, plays a role in neural differentiation of cord blood derived MSC-like cells.[27] Activation of the PKA signaling pathway through increasing the intracellular cAMP levels led to phosphorylation of CREB, a direct target of PKA, and concomitant upregulation of neural specific marker genes including neurofilament, GFAP and Nurr1. In this context, the activation of the PKA pathway was both necessary as well as sufficient for the observed cellular and molecular changes since specific inhibition of PKA abolished both CREB phosphorylation and upregulation of the neural marker genes studied.

Attempts were also made to dissect a complex medium used for neural induction in cord blood cells, in order to derive a better understanding about the contribution of each of the factors in the medium.[28] For example, components raising intracellular cAMP were found critical for the induction of marker gene expression associated with general neuronal differentiation, including expression of dopaminergic markers. Furthermore, tyrosine hydroxylase was upregulated during differentiation and was also phosphorylated which is an important step in activation of the TH enzyme. Also, it was shown that different components of the differentiation medium had defined and reproducible effects on the cellular response of the cells such as neurite outgrowth, marker gene expression, and phosphorylation of TH showing that umbilical cord blood MSC-like cells could respond differentially to components of the medium.[28]

IS NEURAL DIFFERENTIATION OF CORD BLOOD CELLS AN ARTIFACT?

Recently, a number of reports raised concerns that *in vitro* neural differentiation of MSCs from bone marrow might be due to stress response of cells to the cell culture conditions, rather than cells entering a true neuronal differentiation program.[29–31] However, based on the available published data discussed above it is evident that cord blood derived cells can be triggered to express markers specific for neural cells. At this point it is unclear whether the conclusions drawn for bone marrow MSCs also apply to any of the cell types from the cord blood with inducible neuronal differentiation potential. Moreover, in differentiating cells morphological changes were always accompanied by complex changes in marker gene expression known to be active during neural differentiation programs and during differentiation into functional neuronal and glial cells. This indicates that complex and specific molecular programs are activated in these cells. There is furthermore indication of post translational modifications of enzymes involved in the neurotransmitter synthesis suggesting that defined biochemical programs are activated in those differentiating cells.[28] In addition, inducible release of neurotransmitters and electrophysiological properties in cells differentiated from cord blood cells are similar to electrophysiological properties of immature primary neurons.[23,25] In summary, it has been demonstrated that cells from the cord blood can be induced to differentiate into neurons using criteria such as cell morphology, cell type specific marker gene expression and functional criteria such as neurotransmitter release and electrophysiological parameters. Whether differentiation programmes leading to MSC neural differentiation share common mechanisms with the differentiation of primary neurons during development or differentiation of neural stem cells *in vitro* remains to be determined. Further investigations are required to understand underlying molecular and cellular mechanisms of this differentiation process.

CORD BLOOD CELLS IN PRECLINICAL STUDIES OF NEUROLOGICAL DISEASES

As discussed above, cord blood cells are capable of neuronal differentiation including the acquisition of some functional aspects of neurons *in vitro*. Therefore, it seemed tempting to determine whether such cells, undifferentiated or differentiated, are of any therapeutic value in the treatment of brain-related disorders. Next, the evidence for successful application of cord blood derived cells in a number of different animal model systems of neurological diseases will be discussed in more detail.

Stroke and brain injury models

Stroke

The best studied neurological disorder using cord blood (stem) cells as a potential treatment modality is ischemic stroke. Although there is an excellent review of umbilical cord blood cells for stroke therapy,[32] a brief summary of studies and achievements is summarized below. Several studies showed that systemic infusion of cord blood nucleated cells into animal stroke models resulted in behavioral recovery of treated animals.[33,34] Histological and immunohistochemical analyses revealed that human cord blood derived cells could be found in the brain, mainly in the ischemic boundary zone. Furthermore, a small but substantial amount of cells expressed markers typically found in neurons such as NeuN and MAP2 (2% and 3%, respectively). A better overall improvement in recovery was observed when cells were injected peripherally into the bloodstream rather than directly into the brain. Moreover, rescue was found to be cell dose dependent.[35] Four weeks after infusing cells intravenously, significant recovery of behavioral performance were observed when more than 1×10^6 UCB cells were delivered. This behavioral recovery was accompanied by a reduction of infarct volume that was significant when more than 1×10^7 cells were injected. Furthermore, human cells could be detected in several tissues including the brain, although cells seemed to be more closely associated with blood vessels than with the brain parenchyma. Cells were only found in animals of the higher cell dose groups, suggesting that cell incorporation into organs was a rare event although detailed quantification of such incorporation were not provided. In the same year, a study demonstrated that peripherally infused UCB did reduce the infarct size and also reduced behavioral deficits.[36] Moreover, infused cells could not always be found in the brains of animals with improved behavioral deficits, suggesting that the observed recovery may be independent of cell integration into the brain.[36]

Another study noted that even though cells were found at the site of injury they did not engraft into the peri-infarct region. Nevertheless, neural tissue was protected from progressive damage normally seen in untreated animals.[37] Thus, cord blood cells seem to exert a regenerative effect through mechanisms other than cell replacement. Instead, transplanted cord blood cells appeared to induce recovery by decreasing inflammatory processes by reducing immune infiltration of cells such as CD45/CD11b- and CD45/B220-positive cells in the infarct area. This reduction in lymphocyte numbers was accompanied by a decrease of pro-inflammatory cytokines and a reduction in nuclear factor κ-B (NF-κB). Besides the immune modulatory activity of cord blood cells it was also shown that neuroprotective mechanisms play a role in the recovery process. It was suggested that cord blood cells are unlikely operating via a cell replacement mechanism.[37]

It was known that in stroke models umbilical cord blood cells migrated to the site of lesion upon systemic injection of cells into the vascular system, and it could be show that two factors, MCP-1 and MIP-1α, produced by the injury site were responsible for the migration of transplanted cells to the injury site.[38] This migration could be prevented by blocking antibodies. Furthermore, chemokine receptors were expressed on the transplanted cells, suggesting that cord blood cells were chemotactically attracted to the injury site, and thus this study highlighted a cytokine based active migration of cells to the injured tissue.

Hemorrhagic brain injuries

In hemorrhagic brain injury it was shown that UCB cells were capable of reducing neurological deficits in a rat model of hemorrhagic brain injury.[39] MNCs derived from umbilical cord blood were injected into animals which had a collagenase induced hemorrhagic injury in the brain. Behavioral test such as the stepping test, elevated body-swing test, and the limb-placement test showed statistically significant improved outcomes in animals which received cord blood cell injections. In addition, it was demonstrated that some UCB cells entered into the brain and some of them expressed the neuron specific marker, NeuN. However, only few cells entered the brain, raising the question whether recovery was due to the cells found in the brain or whether other mechanisms such as cytokine secretion by umbilical cord blood cells contributed to the amelioration of the brain injury.

Traumatic brain injuries

Following traumatic injury of the rat brain it was shown that intravenous administration of UCB cells reduced neurological as well as motor deficits.[40] Furthermore, injected cells entered the brain and preferentially migrated to the injured brain hemispheres. In addition, a few of the implanted cells were found at the boundary of the injury site and expressed neuronal and astrocytic markers, suggesting that UCB cells might be useful for treating traumatic brain injuries.

Amyotrophic lateral sclerosis (ALS)

The first report using umbilical cord blood cells in animal models of nervous system degeneration demonstrated that infusing MNCs from cord blood into a genetic model of amyotrophic lateral sclerosis led to a significant increase in the life span of treated versus untreated animals.[41] Human cells were present in some organs of the transplanted mice for up to one year. In a follow-up study it was highlighted that survival of animals was dependent on the cell dose, with increasing cell doses resulting in increase of the overall survival of animals compared to the untreated control groups.[42] Similar observations were made by Garbuzova-Davis *et al.* showing that infusion of MNCs from cord blood let to extended survival and furthermore delayed disease progression.[43] Infusion of cells also resulted in a decrease of pro-inflammatory cytokine levels. The authors concluded that cord blood cells may have a therapeutic potential in the treatment of ALS, possibly by providing protective function to motor neurons by inhibiting various inflammatory cytokines normally contributing to ALS disease progression and motor neuron death.

Parkinson's and Alzheimer's disease

In a Parkinson disease model, MNCs from cord blood extended the onset of symptoms as well as survival significantly, as compared to control,[44] suggesting that cells derived from the umbilical cord blood had potential regenerative functions. In a genetic model of Alzheimer's disease, Nikolic *et al.* demonstrated that cord blood derived MNCs, which were administered systemically, were able to reduce parenchymal as well as vascular β-amyloid deposits, although no evidence was found for any of the transplanted cells entering the brain. Further experiments revealed that the systemic injection of cells suppressed the CD40-CD40L axis thereby supporting the clearance of parenchymal and vascular β-amyloid into the serum.[45] As such, mechanistically, MNCs from cord blood appear to achieve their regenerative potential via influencing the host immune response, rather than by a cellular replacement.

San Fillipo B disease

San Fillipo B disease is caused by an enzyme deficiency leading to the accumulation of heparin sulfate which can affect neuron functionality and viability. Transplantations of cord blood derived cells into a murine genetic model of San Fillipo B showed that cord blood cells engrafted long term and about 1%–2% of the transplanted cells could still be detected up to seven months after administration of cells.[46] Some intraparenchymal donor cells were positive for typical neuronal, and, more importantly, for astrocytic markers. Furthermore, transplanted mice appeared to have more surviving endogenous neurons as compared to non-treated mice, suggesting that cord blood cells may differentiate into functional astrocytes and may provide trophic support to degenerating neurons.

Regeneration of the optical system

Recently it was reported that MSCs derived from umbilical cord blood could support the recovery of experimental lesions of the optical nerve by promoting regeneration of axotomized neurons.[47] The study showed that transplanted MSCs survived up to two weeks without any obvious signs of differentiation. These cells contributed to the survival of retinal ganglion cells (RGC) and their connectivity to the existing circuitry as shown by anterograde and retrograde labeling experiments post transplantation. These experiments also demonstrated that endogenous RGC were able to regrow axons which connected to the normal target areas. Interestingly, non-stem cells such as fibroblasts had a similar although not identical regenerative effect on endogenous RGCs. It was also highlighted that MSCs as well as fibroblasts secreted neuroprotective factors *in vitro,* although the relevance of those factors for the observed *in vivo* repair was not further examined. In summary, this study showed that stem cells such as MSCs or non-stem cells such as fibroblasts infused into axotomized areas of the optic system resulted in significant regeneration of the injury in the absence of any signs of differentiation of the infused cells, suggesting that implanted cells exerted their effect via a paracrine effect by possibly secreting regenerative molecules which enhanced endogenous repair.

Other animal studies

Long-term cultured cord blood cells, which *in vitro* showed signs of neural differentiation, were transplanted into the striatum of immune compromised NOD/SCID mice.[48] In this model, cord blood derived cells could be detected up to five days after implantation, with cells expressing neural-specific markers and displaying a neuronal morphology. However, after one month, human cells were undetectable, raising the possibility that long-term survival of the cells was compromised. This showed that in the short term there seemed to be some degree of cell replacement, while long-term engraftment could not be maintained even under conditions of partial immune incompetence of the NOD/SCID mice.

Interestingly, a recent report showed that peripheral infusion of cord blood cells resulted in a statistical significant increase in endogenous neurogenesis in the hippocampus of aged animals as determined by BrdU incorporation and expression of doublecortin, a marker for newly born neurons.[49] It was also shown that dividing stem cells and new neurons were not derived from the implanted cord cells but were rather produced by endogenous cells. The increase of cell proliferation and neurons expressing early differentiation markers was maintained for up to 15 days after implantation. Furthermore, the increase of proliferative cells correlated with a decrease in microglial activation which is consistent with reports that demonstrated a decrease of inflammatory response after infusion of cord blood cells.[35,37,50]

Negative studies

In stroke models, several studies have however demonstrated a lack of regenerative potential of umbilical cord blood cells. The administration of CD34+ cord blood cells did not result in major regenerative outcomes in experimental versus control animals[51] when parameters such as sensory-motor functions and infarct volume were analyzed. There was a trend of improved water-maze performance by the animals which had received CD34+ cells and also some improvement in impaired forelimbs test was observed. Human derived cells could not be detected in the brain of experimental stroke animals between three to four weeks after implantation. In an almost identical study using whole MNCs from cord blood, as well as hematopoietic lineage negative (Lin-) selected cells there was little donor cell induced recovery. The authors concluded that UCB cells did not improve functional recovery or histological outcome in their stroke model.[52] Taken together, these two studies showed that CD34[+] as well as the entire MNCs or the Lin- fraction of cells had little impact on regeneration. This was also supported by an additional study which noted the lack of sufficient numbers of cells integrating into the brain parenchyma of experimental stroke animals after transplantation, paired with the observation of no behavioral recovery or reduction in infarct volume in animals receiving cells as compared to the non-transplanted control animals.[53]

In the case of ALS, Habisch *et al.* reported that intrathecal application of cells (either the entire undifferentiated or neural differentiated MNC fraction) did not result in any regenerative effect in a genetic model of ALS, although transplanted cells could still be found when animals were sacrificed.[54] Another study reported that when USSCs were injected into the developing and healthy murine brain, cells did not integrate into the brain parenchyma and also showed very little signs of differentiation into brain specific cell types,[55] suggesting that the neurogenic environment of the developing brain does not promote the neural phenotype of UCB-derived cells, at least not in adherently grown USSCs.

NEUROPROTECTIVE EFFECTS OF CORD BLOOD STEM CELLS *IN VITRO*

In *in vitro* systems it was shown that cord blood cells protected cortical neurons against glutamate induced apoptosis.[56] Cord blood cells exerted these protective effects against glutamate excitotoxicity via decreasing activity of apoptotic mediators such as caspase-3 and -7, down regulation of NMDR receptors normally mediating glutamate toxicity, and activation of the anti-apoptotic AKT pathway, ultimately leading to neuroprotection. Thus, cord blood cells are capable to exert protective functions on distressed cells. As currently there are not many *in vitro* studies, the development of *in vitro* disease model systems would be considered an extremely helpful tool for the field.

CLINICAL STUDIES USING CORD BLOOD CELLS IN NEUROLOGICAL DISEASES

Krabbe Disease

One study highlighted the feasibility of using umbilical cord blood MNCs for treating Infantile Krabbe Disease, a demyelinating disorder of the central nervous system in children. Krabbe disease leads to early disabilities and childhood death due to an autosomal recessive disorder of the lysosomal enzyme galactocerebrosid which, if missing, leads to defects in myelination of the central and peripheral nervous system.[57] Although Krabbe disease is primarily a metabolic disease it can be regarded a neurological disorder as myelinating glia of the peripheral and central nervous system are affected by the loss of the enzymatic activity of galactocerebrosid. The study by Escolar *et al.* showed that transplantations of mononuclear cells derived from sibling cord blood favorably altered the natural history of the disease and children receiving cells at an asymptomatic stage of the disease showed greatly improved neurological function and survival. In contrast, children treated at a symptomatic stage did not show any major improvements suggesting that cord blood cells are mostly effective in disease prevention/propagation but not disease reversal. Thus, it seemed that transplanted MNCs were able to rescue the enzymatic deficiency particular if myelination defects had not progressed to a stage where clinical symptoms could be detected. These data suggested that umbilical cord blood cells possessed long-term regenerative capabilities able to prevent the onset and/or the progression of the disease in asymptomatic children. However, the mechanisms of how cord blood derived cells might have contributed to the beneficial effects were not studied and as such the underlying mechanisms of regeneration remain to be examined in greater detail. Nevertheless, this study constitutes an important proof-of-principle for the usefulness of these cells in regenerative approaches of human neurological disorders.

Spinal Cord Injury

The use of cord blood cells for preclinical treatment of spinal cord injuries are not discussed in detail in this review although there is some evidence for beneficial roles of cord blood cells in the preclinical setting.[58,59] In a clinical study Kang *et al.* reported a case[60] in which a spinal cord injured patient was treated with MSC-like cells from cord blood. It was observed that the patient showed signs of improved sensory perception and mobility, both functionally and morphologically. It was speculated that this regenerative effect could be due to an endocrine effect of MSCs or due to reconstitution of the injured spinal cord. No further evidence for any of these two mechanisms was provided and this study has remained a single-case study until now. Further clinical trials using UCB cells are in the planning and are in progress such as for cerebral palsy, pediatric traumatic brain injuries and spinal cord injuries.[61]

CONCLUDING REMARKS AND FURTHER QUESTIONS

Although there is a convincing body of literature showing that umbilical cord blood derived cells can be differentiated into cells of the neural lineage and are potentially useful in preclinical models of neurological disorders, there are certain areas of concern which need to be addressed. Progress into translational aspects could be accelerated if some obstacles could be removed as suggested below.

(1) As it is evident from this review, multiple and different cells types derived from cord blood have been explored for both the *in vitro* as well as *in vivo* studies ranging from the entire population of mononuclear (blood) cells, marker-selected cell types mostly of hematopoietic origin, plastic adherence selected mesenchymal-like stem cells and USSCs. There is a need to firstly characterize some of these cell types further. This would enable different labs and researchers to focus their activities and experiments around specified and maybe standardized cell types. This is of particular importance if a translational direction towards the clinic is the major objective.

(2) It is also evident that multiple isolation, growth and expansion as well as differentiation protocols are currently used in different laboratories. For example, there are multiple different protocols being applied for inducing the neural lineage which range from converting MSCs into neural progenitor cells first, or differentiating them directly into neural lineage. It would be helpful if protocols could be standardized so that experimental outcomes between labs can be directly compared. Characterizing individual cell cultures or "lines" and standardizing them as done in the embryonic stem cell (ESC) field, is probably not feasible as most cells derived from cord blood are not immortal and therefore any standardization would be only of short term value. However, standardizing protocols for differentiation, read-out assays and functional tests in animals would greatly speed-up the translation of such cells into clinical applications.

(3) It is important to standardize experimental models as well as experimental parameters used such as route of delivery of cells, cell dosage-outcome relationship, and follow-up time points. Such standardization will ultimately lead to comparable information about cell types generated in different labs.

(4) Most importantly, more mechanistic studies are needed in order to gain insight into cellular, biochemical, molecular and genetic mechanism of neural differentiation of cord blood derived cells *in vitro*. As demonstrated by the ESC field, better understanding of mechanisms of differentiation directly translated into improved differentiation protocols and outcomes of neural differentiation. Unfortunately, most studies of cord blood neural differentiation still center on the proof-of-principle of neural potential/capabilities and end-point analysis, rather than addressing the underlying mechanisms in order to gain further insight into the mechanisms converting a cell in the cord blood into a neuron. Knowledge of the mechanisms could lead to novel strategies for improved directed manipulations of the molecular pathways involved in differentiation process.

Do Cord Blood Derived Cells Replace Diseased/Lost Cells? Are Cells Required for the Regenerative Effects?

Current knowledge suggests that it is unlikely that with the culture and differentiation protocols used so far, cellular replacement is the predominant mechanism for observed regenerative effect. Although multiple studies showed beneficial effects on clinical parameters, most studies found either no or too little cells to account for any cell replacement based recovery. This suggests that at least the long-term presence of cells is not required for the observed regenerative effects, and this implies that the mode of recovery after cord blood derived cell transplantation is possibly through support of endogenous repair mechanisms by delivery of molecules beneficial to this process. This view is supported by *in vitro* work suggesting that cord blood cells, particularly MSCs and USSCs, release several therapeutic cytokines under standard cell culture conditions which can be supportive for other cell types such as the expansion of hematopoietic stem cells.[62] Other studies highlighted that cells in culture produce cytokines with implications for brain repair[63] and suppression of immune rejection.[64] The latter is reviewed in detail elsewhere in this book (Chapter 8). There is very little evidence that cytokines secreted *in vitro* are also playing a role in the regenerative processes *in vivo*. Additional studies are required to investigate such mechanisms in further detail. Current knowledge about MSCs from bone marrow clearly suggests that MSCs have a supportive function, probably through release of cytokines that support clinically relevant recovery.[65–68] It is likely that the different cell types from cord blood are capable of providing regenerative support, and efforts need to be undertaken to identify the cell type with the highest impact in each disease setting.

There is furthermore still room for studies attempting to improve the *in vitro* differentiation potential of cells, probably supported by procedures for selecting favorable cell types during the differentiation process, to derive more defined, well-characterized differentiated cells and populations of high purity for transplantation in the context of cell replacement therapies. It is tempting to speculate that such defined and enriched cell populations might result in clinically relevant cell replacement.

Does It Have to be a Stem Cell?

It is apparent that currently the field between *in vitro* differentiation and *in vivo* regenerative studies is drifting apart in the sense that *in vitro* studies tend to focus on the potential of neural differentiation whereas the preclinical studies seem to move towards supporting endogenous recovery. At this stage it seems that the classical capability of stem cells to differentiate into multiple cell lineages and cell types does not seem to be of essential importance for the *in vivo* recovery effect. As a consequence, the question arises whether for clinical therapies the favored cell type to transplant necessarily has to be a stem or progenitor cell with multilineage differentiation capability or the *in vitro* generated cell types which had degenerated in the disease context. It could be argued that any cell type capable of producing and secreting molecules supportive of regeneration would be a feasible alternative to stem cells.

Is the Regenerative Effect Long-Lasting?

It has not been shown with sufficient evidence that secretion of factors by cord blood cells or stem cells in the transplantation experiments was causative for the regenerative effect. Additional efforts are required to study this in more detail as a means to possibly identify the molecules required for beneficial effects. This could ultimately lead to strategies trying to replace cells with the relevant molecules. Furthermore, the time points studied in the current animal models have mostly been short term. Long-term studies are required to address the question whether cellular therapy produces long-lasting effects and if so, for how long. If cells are not required for the effect it could be argued that therapeutic effects are possibly highest and most long-lasting when disease progression is still in the early stages. As such the issue of timing of therapeutic intervention and the window of opportunity for such strategies becomes an essential issue. Furthermore, the application of multiple injections over different time intervals needs to be considered.

Lastly, although stem cells or cells derived from stem cells as tools for cellular replacement in disease contexts might not be the way forward, *in vitro* differentiated cell types from stem cells including umbilical cord blood (stem) cells with the potential to support endogenous recovery processes independent of any bona-fide cell replacement could be a feasible alternative for clinical applications.

References

1. Rocha V, Gluckman E. Clinical use of umbilical cord blood hematopoietic stem cells. *Biol Blood Marrow Transplant* 2006 Jan;12(1 Suppl 1):34–41.
2. Broxmeyer HE, Douglas GW, Hangoc G, Cooper S, Bard J, English D, *et al.* Human umbilical cord blood as a potential source of transplantable hematopoietic stem/progenitor cells. *Proc Natl Acad Sci USA* 1989 May;86(10):3828–32.
3. Zhang L, Yang R, Han ZC. Transplantation of umbilical cord blood-derived endothelial progenitor cells: a promising method of therapeutic revascularisation. *Eur J Haematol* 2006 Jan;76(1):1–8.
4. Erices A, Conget P, Minguell JJ. Mesenchymal progenitor cells in human umbilical cord blood. *Br J Haematol* 2000;109(1):235–42.
5. Friedenstein AJ, Chailakhjan RK, Lalykina KS. The development of fibroblast colonies in monolayer cultures of guinea-pig bone marrow and spleen cells. *Cell Tissue Kinet* 1970 Oct;3(4):393–403.
6. Pittenger MF, Mackay AM, Beck SC, Jaiswal RK, Douglas R, Mosca JD, *et al.* Multilineage potential of adult human mesenchymal stem cells. *Science* 1999 Apr 2;284(5411):143–7.
7. Sanchez-Ramos J, Song S, Cardozo-Pelaez F, Hazzi C, Stedeford T, Willing A, *et al.* Adult bone marrow stromal cells differentiate into neural cells *in vitro*. *Exp Neurol* 2000;164(2):247–56.
8. Woodbury D, Schwarz EJ, Prockop DJ, Black IB. Adult rat and human bone marrow stromal cells differentiate into neurons. *J Neurosci Res* 2000 Aug 15;61(4):364–70.
9. Kabos P, Ehtesham M, Kabosova A, Black KL, Yu JS. Generation of neural progenitor cells from whole adult bone marrow. *Exp Neurol* 2002 Dec;178(2):288–93.
10. Hermann A, Gastl R, Liebau S, Popa MO, Fiedler J, Boehm BO, *et al.* Efficient generation of neural stem cell-like cells from adult human bone marrow stromal cells. *J Cell Sci* 2004 Sep 1;117(Pt 19):4411–22.
11. Hermann A, Liebau S, Gastl R, Fickert S, Habisch HJ, Fiedler J, *et al.* Comparative analysis of neuroectodermal differentiation capacity of human

bone marrow stromal cells using various conversion protocols. *J Neurosci Res* 2006 Jun;83(8):1502–14.

12. Brazelton TR, Rossi FM, Keshet GI, Blau HM. From marrow to brain: expression of neuronal phenotypes in adult mice. *Science* 2000 Dec 1;290(5497):1775–9.

13. Mezey E, Chandross KJ, Harta G, Maki RA, McKercher SR. Turning blood into brain: cells bearing neuronal antigens generated *in vivo* from bone marrow. *Science* 2000 Dec 1;290(5497):1779–82.

14. Sanchez-Ramos JR, Song S, Kamath SG, Zigova T, Willing A, Cardozo-Pelaez F, *et al.* Expression of neural markers in human umbilical cord blood. *Exp Neurol* 2001 Sep;171(1):109–15.

15. Ha Y, Choi JU, Yoon DH, Yeon DS, Lee JJ, Kim HO, *et al.* Neural phenotype expression of cultured human cord blood cells *in vitro*. *Neuroreport* 2001;12(16):3523–7.

16. Bicknese AR, Goodwin HS, Quinn CO, Henderson VC, Chien SN, Wall DA. Human umbilical cord blood cells can be induced to express markers for neurons and glia. *Cell Transplant* 2002;11(3):261–4.

17. Zigova T, Song S, Willing AE, Hudson JE, Newman MB, Saporta S, *et al.* Human umbilical cord blood cells express neural antigens after transplantation into the developing rat brain. *Cell Transplant* 2002;11(3):265–74.

18. Buzanska L, Machaj EK, Zablocka B, Pojda Z, Domanska-Janik K. Human cord blood-derived cells attain neuronal and glial features *in vitro*. *J Cell Sci* 2002 May 15;115(Pt 10):2131–8.

19. Buzanska L, Jurga M, Stachowiak EK, Stachowiak MK, Domanska-Janik K. Neural stem-like cell line derived from a nonhematopoietic population of human umbilical cord blood. *Stem Cells Dev* 2006 Jun;15(3):391–406.

20. Jurga M, Buzanska L, Malecki M, Habich A, Domanska-Janik K. Function of ID1 protein in human cord blood-derived neural stem-like cells. *J Neurosci Res* 2006 Oct;84(5):993–1002.

21. Kogler G, Sensken S, Airey JA, Trapp T, Muschen M, Feldhahn N, *et al.* A new human somatic stem cell from placental cord blood with intrinsic pluripotent differentiation potential. *J Exp Med* 2004 Jul 19;200(2):123–35.

22. Fallahi-Sichani M, Soleimani M, Najafi SM, Kiani J, Arefian E, Atashi A. *In vitro* differentiation of cord blood unrestricted somatic stem cells expressing dopamine-associated genes into neuron-like cells. *Cell Biol Int* 2007 Mar;31(3):299–303.

23. Greschat S, Schira J, Kury P, Rosenbaum C, de Souza Silva MA, Kogler G, *et al.* Unrestricted somatic stem cells from human umbilical cord blood can be differentiated into neurons with a dopaminergic phenotype. *Stem Cells Dev* 2008 Apr;17(2):221–32.

24. Chen N, Hudson JE, Walczak P, Misiuta I, Garbuzova-Davis S, Jiang L, *et al.* Human umbilical cord blood progenitors: the potential of these hematopoietic cells to become neural. *Stem Cells* 2005 Nov;23(10):1560–70.

25. Sun W, Buzanska L, Domanska-Janik K, Salvi RJ, Stachowiak MK. Voltage-sensitive and ligand-gated channels in differentiating neural stem-like cells derived from the nonhematopoietic fraction of human umbilical cord blood. *Stem Cells* 2005 Aug;23(7):931–45.

26. Lee MW, Moon YJ, Yang MS, Kim SK, Jang IK, Eom YW, *et al.* Neural differentiation of novel multipotent progenitor cells from cryopreserved human umbilical cord blood. *Biochem Biophys Res Commun* 2007 Jun 29;358(2): 637–43.

27. Wang TT, Tio M, Lee W, Beerheide W, Udolph G. Neural differentiation of mesenchymal-like stem cells from cord blood is mediated by PKA. *Biochem Biophys Res Commun* 2007 Jun 15;357(4):1021–7.

28. Tio M, Tan KM, Lee W, Wang TT, Udolph G. Roles of db-cAMP, IBMX and RA in aspects of neural differentiation of cord blood derived mesenchymal-like stem cells. *PLoS One* 2010 Feb;5(2):e9398.

29. Croft AP, Przyborski SA. Generation of neuroprogenitor-like cells from adult mammalian bone marrow stromal cells *in vitro*. *Stem Cells Dev* 2004 Aug; 13(4):409–20.

30. Lu P, Blesch A, Tuszynski MH. Induction of bone marrow stromal cells to neurons: differentiation, transdifferentiation, or artifact? *J Neurosci Res* 2004 Jul 15;77(2):174–91.

31. Neuhuber B, Gallo G, Howard L, Kostura L, Mackay A, Fischer I. Reevaluation of *in vitro* differentiation protocols for bone marrow stromal cells: disruption of actin cytoskeleton induces rapid morphological changes and mimics neuronal phenotype. *J Neurosci Res* 2004 Jul 15;77(2):192–204.

32. Yu G, Borlongan CV, Stahl CE, Hess DC, Ou Y, Kaneko Y, *et al.* Systemic delivery of umbilical cord blood cells for stroke therapy: a review. *Restor Neurol Neurosci* 2009;27(1):41–54.

33. Chen J, Sanberg PR, Li Y, Wang L, Lu M, Willing AE, *et al.* Intravenous administration of human umbilical cord blood reduces behavioral deficits after stroke in rats. *Stroke* 2001 Nov;32(11):2682–8.

34. Willing AE, Lixian J, Milliken M, Poulos S, Zigova T, Song S, *et al.* Intravenous versus intrastriatal cord blood administration in a rodent model of stroke. *J Neurosci Res* 2003 Aug 1;73(3):296–307.

35. Vendrame M, Cassady J, Newcomb J, Butler T, Pennypacker KR, Zigova T, *et al.* Infusion of human umbilical cord blood cells in a rat model of stroke dose-dependently rescues behavioral deficits and reduces infarct volume. *Stroke* 2004 Oct;35(10):2390–5.

36. Borlongan CV, Hadman M, Sanberg CD, Sanberg PR. Central nervous system entry of peripherally injected umbilical cord blood cells is not required for neuroprotection in stroke. *Stroke* 2004 Oct;35(10):2385–9.

37. Vendrame M, Gemma C, Mesquita DD, Collier L, Bickford PC, Sanberg CD, *et al.* Anti-inflammatory effects of human cord blood cells in a rat model of stroke. *Stem Cells Dev* 2005 Oct;14(5):595–604.

38. Jiang L, Newman M, Saporta S, Chen N, Sanberg C, Sanberg PR, *et al.* MIP-1alpha and MCP-1 induce migration of human umbilical cord blood cells in models of stroke. *Curr Neurovasc Res* 2008 May;5(2):118–24.

39. Nan Z, Grande A, Sanberg CD, Sanberg PR, Low WC. Infusion of human umbilical cord blood ameliorates neurologic deficits in rats with hemorrhagic brain injury. *Ann NY Acad Sci* 2005 May;1049:84–96.

40. Lu D, Sanberg PR, Mahmood A, Li Y, Wang L, Sanchez-Ramos J, *et al.* Intravenous administration of human umbilical cord blood reduces neurological deficit in the rat after traumatic brain injury. *Cell Transplant* 2002;11(3):275–81.

41. Ende N, Weinstein F, Chen R, Ende M. Human umbilical cord blood effect on sod mice (amyotrophic lateral sclerosis). *Life Sci* 2000 May 26;67(1):53–9.

42. Chen R, Ende N. The potential for the use of mononuclear cells from human umbilical cord blood in the treatment of amyotrophic lateral sclerosis in SOD1 mice. *J Med* 2000;31(1–2):21–30.

43. Garbuzova-Davis S, Sanberg CD, Kuzmin-Nichols N, Willing AE, Gemma C, Bickford PC, *et al.* Human umbilical cord blood treatment in a mouse model of ALS: optimization of cell dose. *PLoS ONE* 2008;3(6):e2494.

44. Ende N, Chen R. Parkinson's disease mice and human umbilical cord blood. *J Med* 2002;33(1–4):173–80.

45. Nikolic WV, Hou H, Town T, Zhu Y, Giunta B, Sanberg CD, *et al.* Peripherally administered human umbilical cord blood cells reduce parenchymal and vascular beta-amyloid deposits in alzheimer mice. *Stem Cells Dev* 2008 Jun;17(3):423–40.

46. Garbuzova-Davis S, Willing AE, Desjarlais T, Davis Sanberg C, Sanberg PR. Transplantation of human umbilical cord blood cells benefits an animal model of Sanfilippo syndrome type B. *Stem Cells Dev* 2005 Aug;14(4):384–94.

47. Zwart I, Hill AJ, Al-Allaf F, Shah M, Girdlestone J, Sanusi AB, *et al.* Umbilical cord blood mesenchymal stromal cells are neuroprotective and promote regeneration in a rat optic tract model. *Exp Neurol* 2009 Apr;216(2):439–48.

48. Walczak P, Chen N, Eve D, Hudson J, Zigova T, Sanchez-Ramos J, *et al.* Long-term cultured human umbilical cord neural-like cells transplanted into the striatum of NOD SCID mice. *Brain Res Bull* 2007 Sep 14;74(1–3):155–63.

49. Taguchi A, Soma T, Tanaka H, Kanda T, Nishimura H, Yoshikawa H, *et al.* Administration of CD34$^+$ cells after stroke enhances neurogenesis via angiogenesis in a mouse model. *J Clin Invest* 2004 Aug;114(3):330–8.

50. Vendrame M, Gemma C, Pennypacker KR, Bickford PC, Davis Sanberg C, Sanberg PR, *et al.* Cord blood rescues stroke-induced changes in splenocyte phenotype and function. *Exp Neurol* 2006 May;199(1):191–200.

51. Nystedt J, Makinen S, Laine J, Jolkkonen J. Human cord blood CD34+ cells and behavioral recovery following focal cerebral ischemia in rats. *Acta Neurobiol Exp (Wars)* 2006;66(4):293–300.

52. Makinen S, Kekarainen T, Nystedt J, Liimatainen T, Huhtala T, Narvanen A, *et al.* Human umbilical cord blood cells do not improve sensorimotor or cognitive outcome following transient middle cerebral artery occlusion in rats. *Brain Res* 2006 Dec 6;1123(1):207–15.

53. Zawadzka M, Lukasiuk K, Machaj EK, Pojda Z, Kaminska B. Lack of migration and neurological benefits after infusion of umbilical cord blood cells in ischemic brain injury. *Acta Neurobiol Exp (Wars)* 2009;69(1):46–51.

54. Habisch HJ, Janowski M, Binder D, Kuzma-Kozakiewicz M, Widmann A, Habich A, *et al.* Intrathecal application of neuroectodermally converted stem cells into a mouse model of ALS: limited intraparenchymal migration and survival narrows therapeutic effects. *J Neural Transm* 2007 Nov; 114(11):1395–406.

55. Coenen M, Kogler G, Wernet P, Brustle O. Transplantation of human umbilical cord blood-derived adherent progenitors into the developing rodent brain. *J Neuropathol Exp Neurol* 2005 Aug;64(8):681–8.

56. Dasari VR, Veeravalli KK, Saving KL, Gujrati M, Fassett D, Klopfenstein JD, *et al.* Neuroprotection by cord blood stem cells against glutamate-induced apoptosis is mediated by Akt pathway. *Neurobiol Dis* 2008 Dec;32(3):486–98.

57. Escolar ML, Poe MD, Provenzale JM, Richards KC, Allison J, Wood S, *et al.* Transplantation of umbilical-cord blood in babies with infantile Krabbe's disease. *N Engl J Med* 2005 May 19;352(20):2069–81.

58. Saporta S, Kim JJ, Willing AE, Fu ES, Davis CD, Sanberg PR. Human umbilical cord blood stem cells infusion in spinal cord injury: engraftment and beneficial influence on behavior. *J Hematother Stem Cell Res* 2003 Jun;12(3):271–8.

59. Kuh SU, Cho YE, Yoon DH, Kim KN, Ha Y. Functional recovery after human umbilical cord blood cells transplantation with brain-derived neutrophic factor into the spinal cord injured rat. *Acta Neurochir (Wien)* 2005 Sep;147(9):985–92; discussion 92.

60. Kang KS, Kim SW, Oh YH, Yu JW, Kim KY, Park HK, *et al.* A 37-year-old spinal cord-injured female patient, transplanted of multipotent stem cells from human UC blood, with improved sensory perception and mobility, both functionally and morphologically: a case study. *Cytotherapy* 2005;7(4):368–73.

61. Harris DT. Cord blood stem cells: a review of potential neurological applications. *Stem Cell Rev* 2008 Dec;4(4):269–74.

62. Kogler G, Radke TF, Lefort A, Sensken S, Fischer J, Sorg RV, *et al.* Cytokine production and hematopoiesis supporting activity of cord blood-derived unrestricted somatic stem cells. *Exp Hematol* 2005 May;33(5):573–83.

63. Newman MB, Willing AE, Manresa JJ, Sanberg CD, Sanberg PR. Cytokines produced by cultured human umbilical cord blood (HUCB) cells: implications for brain repair. *Exp Neurol* 2006 May;199(1):201–8.

64. Winter M, Wang XN, Daubener W, Eyking A, Rae M, Dickinson AM, *et al.* Suppression of cellular immunity by cord blood-derived unrestricted somatic stem cells is cytokine dependent. *J Cell Mol Med* 2009 Aug;13(8B):2465–75.

65. Chopp M, Li Y. Treatment of neural injury with marrow stromal cells. *Lancet Neurol* 2002 Jun;1(2):92–100.

66. Li Y, Chopp M. Marrow stromal cell transplantation in stroke and traumatic brain injury. *Neurosci Lett* 2009 Jun 12;456(3):120–3.

67. Chopp M, Li Y, Zhang ZG. Mechanisms underlying improved recovery of neurological function after stroke in the rodent after treatment with neurorestorative cell-based therapies. *Stroke* 2009 Mar;40(3 Suppl):S143–5.

68. Zhang ZG, Chopp M. Neurorestorative therapies for stroke: underlying mechanisms and translation to the clinic. *Lancet Neurol* 2009 May;8(5):491–500.

8

Immunological Properties of Mesenchymal Stem Cells Isolated from Bone Marrow and Umbilical Cord

Mark F. Pittenger and Katarina LeBlanc

ABSTRACT

Bone marrow-derived mesenchymal stem cells (BM-MSCs) have generated great interest due to their ease of isolation, propagation and their ability to differentiate. Umbilical cord blood has been useful as a source of hematopoietic stem and progenitor cells, and investigators have sought to isolate umbilical cord blood (UCB-) or cord tissue-derived MSCs (UC-MSCs). These appear to be similar to BM-MSCs in their propagation, ability to differentiate and their immunological properties, suggesting they will be useful for tissue repair and transplantation. Subtle differences may exist that could be useful in tissue regeneration strategies. Initial efforts to isolate MSCs from bone marrow anticipated they would have immunological properties similar to hematopoietic stem cells, stimulate T-cell proliferation and be rejected by allogeneic T-cells. However, when tested in the mixed lymphocyte reaction

173

(lymphocytes to MSC ratio of ~10:1) MSCs failed to stimulate proliferation of T-cells. If MSCs were genetically engineered to express the missing co-stimulatory surface molecules, then they became immunogenic and stimulated the allo-T-cell response, but still not as actively as most cell types. Many researchers demonstrated that MSCs could actively suppress immune cell responses. MSCs express a number of molecules including hepatocyte growth factor (HGF), transforming growth factor β (TGF-β), prostaglandin E2 (PGE2), indoleamine 2,3-dioxygenase (IDO), nitric oxide, interleukin-10 (IL-10), IL-1 receptor antagonist (IL-1Ra), galectin-1 (Gal-1) and HLA-G that modulate the activity of the adaptive immune cells including inflammatory $T_{Helper}1$ (Th1), dendritic cells 1 (DC1), responding B-cells, as well as anti-inflammatory regulatory T-cells (T_{Reg}), $T_{Helper}2$ (Th2) and dendritic cells 2 (DC2). MSCs also have demonstrable interactions with the Natural Killer (NK) cells of the innate immune response. Animal studies support the notion that MSCs are immune privileged, although engraftment of transplanted MSCs remains low and leaves certain questions unanswered as to their universal transplantability or their best preparation for transplantation. Nevertheless, promising clinical trials have begun with BM-MSCs, and are in the planning stages with UC- or UCB-MSCs.

INTRODUCTION

The bone marrow stroma, a heterogeneous mixture of marrow cells often used to aid culture of hematopoietic stem cells, has been known since the 1980s to regulate proliferation and maturation of myeloid and lymphoid lineages. Mixed populations of stromal cells and clonal cell lines have been found to influence B- and T-cell maturation *in vitro*.[1-4] Mesenchymal stem cells (MSCs) are very rare but can be found in many tissues including bone marrow, adipose tissue, and umbilical cord blood. Though rare, MSCs can be culture expanded easily to produce a very homogeneous cell population with the ability to differentiate to multiple lineages.[5] The discovery of the immune modulation properties of MSCs had to wait for the isolation of these more homogeneous populations and subsequent testing of their interactions with allogeneic mixed lymphocyte populations. The *in vitro* study of the interaction of various preparations of MSCs with mixed lymphocyte cell populations or isolated immune cells, as well as the *in vivo* interactions with damaged tissue, and the allogeneic transplantation response, is adding to our understanding of immune modulation. This chapter will review some of the *in vitro* experiments leading to our current understanding of the immune modulating effects of MSCs, and the current state of clinical studies.

LACK OF IMMUNE RESPONSE TO ALLOGENEIC MSCs

Initial experiments performed by Kevin McIntosh, Elana Klyushnenkova and Joseph Mosca at Osiris Therapeutics, Inc. in the late 90s were designed to evaluate the antigen presenting properties of MSCs.[a] These experiments gave the surprising result that MSCs did not stimulate an immune response in the mixed lymphocyte assay using peripheral blood mononuclear cells.[6] These researchers demonstrated that MSCs expressed surface MHC class I (human lymphocyte antigen, HLA-A, B), ICAM, VCAM and VLA-3 proteins, and therefore would expectedly interact with cognate receptors on T-cells. However, the MSCs did not have surface expression of MHC class II (HLA-DR), nor did they express co-stimulatory molecules CD40, CD80, or CD86, and the MSCs failed to activate CD25 or CD134 expression on T-cells. In the mixed lymphocyte reaction, MSCs did not stimulate T-cell proliferation and could subdue an on-going proliferating T-cell response (Fig. 1). These studies, presented at international meetings, sparked great interest from immunologists and transplantation researchers, although the results were not published for several years.[6] The studies coincided with, and were used to support, an on-going Phase I clinical trial of using MSCs to support hematopoietic stem cell engraftment in blood cancer patients transplanted with bone marrow or peripheral blood stem cells (PBSCs).[7,8] The clinical experience at the time, when matched allogeneic BM cells or PBSCs were transplanted in conjunction with the MSCs, showed surprisingly limited graft versus host disease (GVHD). Encouraged by such *in vitro* and *in vivo* early reports, many investigators abandoned lead efforts to develop autologous MSC clinical therapies and switched to testing allogeneic MSCs

---

[a] This original work was supported by DARPA Agreement #MDA972-96-3-0018.

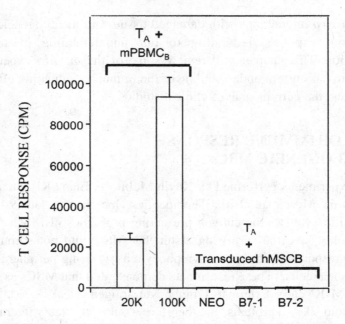

Fig. 1. MSCs do not stimulate T-cell proliferation in the mixed lymphocyte reaction. Responding T-cells from donor A were mixed with peripheral blood mononuclear cells (PBMCs) from donor B and incorporate up to 90–100,000 CPM of ^{3}H. However, the T-cells do not proliferate when mixed with a similar number (100K) of MSCs from donor B, even when the MSCs have been transduced to express the co-stimulatory molecules B7-1 or B7-2 and further treated with interferon γ. Reprinted from McIntosh, Mosca and Klyushnenkova, Figure 1, US Patent 6,368,636, at USPTO.gov.

where large scale production potentially could yield an off-the-shelf cellular therapy to be used in multiple patients and/or for multiple ailments.

MSCs FROM CORD BLOOD OR UMBILICAL CORD TISSUE

Subsequent to the successful isolation of hematopoietic stem cells from umbilical cord blood (UCB), investigators have attempted to also isolate MSCs from UCB. UCB might provide a convenient source of MSCs given its important role in early development, and the fact that the tissue is considered medical waste. While MSCs have proven to be almost non-existent in free-flowing blood under normal (healthy) conditions, it was thought they might be much more abundant in UCB. Results however, have been mixed and researchers have found MSCs in UCB roughly 60% of the time. However, it is the ability to propagate MSCs *in vitro* that leads to sufficient numbers for study.[9] Another interesting finding was the isolation of MSCs from umbilical cord tissue rather than from the extruded blood. Romanov *et al.* took advantage of reports that MSCs may be found associated with the microvasculature such as mural cells or microvascular pericytes. Brief enzymatic digestion of the cord tissue, followed by cell culturing led to a robust population of umbilical cord-derived MSCs (UC-MSCs).[10] The isolated cells have properties similar to MSCs isolated from bone marrow or other tissues. A similar approach by Davies and colleagues yielded human umbilical cord perivascular (HUCPCs) cells.[11,12]

Beyond the unique availability of post-birth UC and its nature as a neo-natal tissue containing a variety of progenitor cells, there have not been many special characteristics demonstrated for UC-MSCs. Exceptions might be that UC- or UCB-MSCs do not appear to respond well to the adipogenic differentiation conditions established for BM-MSCs,[13–15] and that UC-MSCs may express more mRNAs related to angiogenesis and vasculogenesis. Placenta-derived MSCs have quite similar properties as BM-derived MSCs and do undergo adipogenic differentiation, but otherwise also resemble UC- or UCB-MSCs, including the lack of eliciting an allogeneic T-cell response.[16,17] Otherwise, the culture expanded UCB- or UC-MSCs have very similar properties as culture expanded BM-MSCs, and additional studies are needed to evaluate and contrast differences among these MSC populations further.

IMMUNE MODULATORY ACTIVITY OF MSCs

Prompted by the first reports of "immune privileged" MSCs, there was much discussion at scientific meetings and in clinical circles about the immune modulating properties of BM-MSCs. The first peer-reviewed publications describing the BM-MSCs' ability to modulate the immune response were published in 2002 including reports by Di Nicola *et al.*, Le Blanc *et al.*, and Bartholomew *et al.*[18–20] Since that time more than

100 published studies have investigated the effects of MSCs on various populations of mixed lymphocytes, peripheral blood mononuclear cells or isolated immune cells. It is clear that in addition to stem cell properties of proliferation and differentiation to mesenchymal lineages (at least for osteogenesis, chondrogenesis and adipogenesis), MSCs from BM, UC and UCB can play an important role in the regulation of immune and inflammatory responses.

The knowledge of the immune modulating activity of MSCs pointed to their potential therapeutic benefits for deleterious immune-mediated reactions occurring during GVHD, solid organ transplantation, and autoimmune diseases. A first case report describing successful prospective treatment of a boy with severe refractory acute GVHD was published in 2004 by Le Blanc and colleagues.[21] This work is especially significant because the patient relapsed with GVHD and was successfully retreated with BM-MSCs a second time providing a clear treatment-response relationship. This and several more recent reports support the potential use of MSCs for the treatment of conditions characterized by an excessive immune response.[11,22–24] The accumulating data indicate that MSCs can modulate both the innate and the adaptive immune responses. The effects of the interactions of immune cells and MSCs using isolated populations of T-cells, B-cells, dendritic cells and NK-cells are described in detail below.

MSCs AND T-LYMPHOCYTES

Suppression of T-lymphocyte proliferation *in vitro* was one of the first described immune regulatory effects of MSCs. Co-culture of responding T-cells and MSCs showed that the MSCs inhibit about equally the proliferation of $CD4^+$ or $CD8^+$ subsets of T-lymphocytes induced to proliferate by a variety of T-cell stimuli including (1) allogeneic cells such as in the *in vitro* mixed lymphocyte reaction (MLR), (2) known T-cell mitogens (phytohemaglutinin (PHA) or conconavalin A), (3) antibodies that stimulate T-cell proliferation (anti-CD3/CD28 antibodies), and (4) other antigens.[6,11,16–20,25] The effect is not HLA-restricted, may be reversible and depends on the ratio between MSCs and immune cells. At a low ratio

(1:100, or lower, of MSCs to lymphocytes) MSCs stimulate rather than inhibit immune cell proliferation, likely due to the production of growth factors and cytokines by MSCs.[20]

It has been shown that, at least, for human cells, immunosuppressive effects of MSCs are mediated mostly by soluble factors.[6,16–20,25] However, the direct contact with lymphocytes seems to be critical for mouse MSCs to suppress the immune response.[26,27] At least nine molecules that are important for inhibition of T-cell functions have been identified as produced by MSCs or induced in MSC-lymphocyte co-cultures, including hepatocyte growth factor (HGF) and transforming growth factor β (TGF-β),[19] prostaglandin E2 (PGE2),[28] IL-10,[28,29] nitric oxide,[30] IL-1Ra,[31] IDO enzyme,[32] galectin-1[33] and HLA-G (see Fig. 2).[34] The accumulating data indicate that the MSCs may produce different sets of these inhibitory factors depending on the type of immune cells in co-cultures and the type of stimuli used in the experiments.[28,35] Also, some investigators have isolated subpopulations of MSCs producing a specific subset of cytokines from the larger pool of cultured MSCs.[31] Therefore, the responses seen *in vitro*, and presumably may operate *in vivo*, between MSCs and T-cells are specific to the concentration of cells, timing of the cell interactions, level of stimulation, and extracellular milieu present. Put another way, two cell types, MSCs and T-cells, each designed and equipped to respond to the multiple features of their immediate environment, will interact to produce a complex outcome that varies over time.

MSCs may constitutively produce low levels of immunosuppressive factors, but the production of suppressive factors is up-regulated by pro-inflammatory cytokines such as TNF-α, IL-1 and IFN-γ secreted by immune cells in response to stimulation.[6,28,36] The findings that secretion of cytokines from MSCs may be regulated by factors and cells in the local microenvironment suggest that some of the serious side effects due to systemic immunosuppression caused by modern immunosuppressive drugs may be avoided with the use of MSC-based therapies.

Further investigations of molecular mechanisms underlying MSC effects on T-lymphocyte functions revealed that MSCs can arrest T-cells in the G0/G1 phase of the cell cycle rather than induce T-cell apoptosis.[37] MSCs support survival of unstimulated T-cells,[38] however, MSC-induced

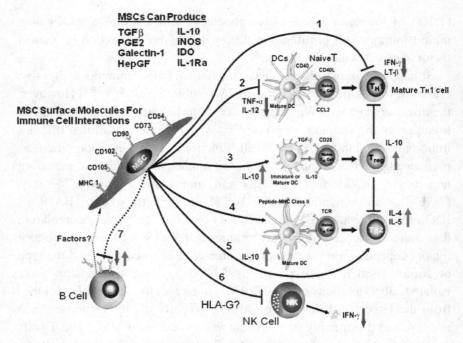

Fig. 2. MSCs interact with immune cells in a complex manner. In an ongoing inflammatory response, MSC interaction with T cells generally results in an anti-inflammatory response due to the inhibition of expression of IFN-γ from mature Th1 cells (pathway 1), along with a decrease in the expression of TNF-α and IL-12 caused by interaction with mature DC1 cells (pathway 2), as well as an increase in anti-inflammatory IL-10 expression (pathway 4). Early in an inflammatory response, MSCs may interact with immature DCs (pathway 3) to cause release of IL-10 and promote T$_{Reg}$ induction from immature T-cells, and release of additional IL-10. Additionally, MSCs can interact with Th2 cells present to release anti-inflammatory IL-4 and IL-5 (pathway 5). MSCs co-cultured with NK cells caused a decrease in IFN-γ expression, inhibited NK proliferation (possibly due to HLA-G) and did not cause lysis of MSCs (pathway 6). However further studies with IL-15 activated NK cells in contact with MSC did cause MSC lysis. MSCs inhibit B-cell proliferation and maturation resulting in less secretion (pathway 7), but can also increase B-cell maturation responses in other conditions. There seem to be unknown factors from B cells acting on MSC as well. This research was originally published in *Blood* © the American Society of Hematology. Aggarwal and Pittenger (2005).[28]

apoptosis of activated T-cells also has been reported.[26,39] Thus, depending on the T-cell activity state, MSCs may protect quiescent T-cells from death, arrest T-cell division at the early steps of activation and promote apoptosis of already activated T-cells.

Both CD4 (helper) and CD8 (cytotoxic) T-cell functions are regulated by MSCs. The addition of MSCs to immune cell cultures decreases secretion of the pro-inflammatory cytokines TNF-α and IFN-γ while simultaneously increasing the production of anti-inflammatory cytokines IL-10 and IL-4 indicating a shift from inflammatory Th1 toward anti-inflammatory Th2 responses.[28] MSCs also inhibit proliferation and maturation of cytotoxic T-lymphocytes (CTL), however MSCs do not block lysis of cell targets by active, matured CTL[35] suggesting that MSCs inhibit T-cells mostly at the early activation steps rather than at the later effector stages.

Most of the published studies investigated MSC effects on naïve T-cells. Data regarding how MSCs regulate memory T-cell activity are limited and contradictory. It has been shown that human MSCs do not change T-cell responses *in vitro* to recall antigens such as *Candida albicans*, *Bordetella pertusis* and tetanus toxoid, suggesting that MSCs do not affect memory cell functions.[6,28] In contrast, inhibition of HY-specific memory T-cells by MSCs was observed in a mouse model where the MSCs inhibited naïve and memory cell proliferation, cytotoxicity and secretion of IFN-γ.[40] Thus, mouse systems may differ from higher vertebrates and MSC effects on memory T-cells require further investigation.

The mechanisms of MSC-mediated T-cell suppression can be direct and/or indirect depending on the other cell types present in the MSC/T-cell co-cultures. Indirect MSC effects can be mediated by upregulation of inhibitory antigen presenting cells or T_{Reg} cells, numbers and activity of which are elevated by MSCs.[28,41–43] The underlying mechanisms of MSC-mediated T-cell suppression suggest the possibility of immune tolerance induction that is extremely important for successful treatment of autoimmune diseases. Indeed, in the mouse model of experimental autoimmune encephalomyelitis (EAE) the injection of MSCs ameliorated symptoms of the disease, and antigen-specific T-cell anergy was induced.[44,45] However, the induction of tolerance in this model was mediated neither by DC nor by T_{Reg} cells.[45] Collectively, accumulated data show that MSC effects on T-lymphocyte functions and underlying mechanisms are driven by factors and cells in the local microenvironment. The MSC immune modulating effect(s) will depend also on the ratio between the MSCs and the immune cells and the state and the stage of immune cell activation or maturation.

MSCs AND B-LYMPHOCYTES

The ability of MSCs to suppress T-lymphocyte functions suggests that MSCs will affect B-lymphocytes by at least indirect mechanisms involving inhibition of Th1 cells that are required for the majority of B-cell mediated immune responses.[28] However, it has been shown recently that MSCs regulate B-cell functions by both direct and indirect mechanisms.[46,47] Human MSCs inhibited proliferation and maturation of B-cells purified from peripheral blood resulting in a low number of Ig-secreting B-cells and low levels of Ig M, G and A in culture supernatants. Although it is clear that MSCs can inhibit the early steps of B-cell activation, the study design did not allow answering whether the decrease in Ig-secreting B-cells is due to a direct MSC inhibition of B-cell differentiation, or whether it is mediated by an indirect mechanism following inhibition of B-cell activation and proliferation.

Corcione and colleagues[46] demonstrated that MSCs present in B-cell cultures affect the B-cell phenotype. MSCs decreased B-cell expression of CXCR4, CXCR5 and CCR7 chemokine receptors without significant changes in expression of HLA-DR and accessory molecules CD80, CD86 and CD40. These findings indicate that MSCs may affect chemokine-mediated B-cell homing to secondary lymphoid organs, but not antigen-presenting functions of B-cells. Similar to the mechanisms of T-cell suppression, MSCs inhibit B-cell functions via soluble factors, secretion of which is induced by signaling molecules from activated B-cells. These MSC-derived soluble factors arrest B-cells in G0/G1 phase of cell cycle without triggering apoptosis.[46]

Rasmussen *et al.* showed that MSCs can inhibit and stimulate B-cell activity depending on the source and purity of B-cells and nature and dose of stimuli.[47] The MSCs increased the number of IgG-producing B-cells in co-cultures without externally added stimuli. For unpurified B-cells, this MSC effect was mediated by soluble factors whereas for purified B-cells direct MSC-B-cell contact was required. In the case of bacterial and viral stimuli such as LPS, CMV and VSV, the MSC effect was stimulating and MSC-dose dependent: in the presence of sub-optimal concentrations of stimuli, the MSCs increased the level of B-cell activation, while in the presence of high concentrations of stimuli MSCs inhibit B-cell functions.[47]

These data demonstrate that MSCs are able to regulate B-cells in both positive and negative ways depending on the cell type and concentration of factors in local microenvironment. Although the same MSC soluble factors that inhibit T-cell responses may play a role in suppression of B-cells, the nature of these factors as well as the nature of B-cell-derived signaling molecules triggering secretion of suppressive factors by MSCs remains to be further investigated.

MSCs AND DENDRITIC CELLS (DCs)

DCs are the most potent antigen-presenting cells (APCs) and have the highest potential to stimulate naïve T-cells. The presence of MSCs in DC cultures inhibited maturation as demonstrated by reduced expression of the CD83 maturation marker, HLA-DR and DC1α molecules, which are important for antigen presentation.[48] MSCs also inhibited expression of accessory molecules such as CD40, CD80, and CD86, which provide co-stimulatory signals essential for activation of T-cells. These changes in DC phenotype are associated with decreased DC capacity to activate naïve T-cells.

Investigation of DC effector functions has revealed that MSCs decreased secretion of the inflammatory cytokine TNF-α by DC1 (induced by Th1 cells) and increased secretion of anti-inflammatory cytokine IL-10 by DC2 (induced by Th2 cells).[28] These results support the hypothesis that MSCs can induce an immunological tolerance via inhibition of DC1 functions. The latest studies show that MSCs affect phenotype, cytokine secretion and immunostimulatory activity of both immature and mature DCs.[49–51] The initial steps of DC differentiation from monocytes include down-regulation of CD14 expression and up-regulation of CD1α, CD83 and CD80. In the presence of MSCs, no decrease of CD14 and no increase in CD1α, CD83 and CD80 were observed, indicating blockage of DC maturation by MSCs. One potential mechanism of blocking DC maturation from monocytes by MSCs may be due to blocking monocytes from entering the G1 phase of the cells cycle.[51] Experiments using a trans-well system indicate that suppression of DC maturation is mediated by MSC-derived soluble factors. While IL-6 and M-CSF are responsible for maintenance of CD14 expression, these factors are not involved in suppression of CD1α expression.[49,50]

Human MSCs have been known for a long time to express IL-6, and similar results, that inhibition of DC maturation by MSCs is partially mediated by IL-6, were observed in a mouse model.[52] The cytokine(s) counteracting DC-specific GM-CSF/IL-4-induced CD1α expression remains unknown. IL-12 is a cytokine critical for the maturation and functions of DCs. By interfering with IL-12 production by DCs, the MSCs affect maturation and the DC ability to stimulate T-cell proliferation. Effects of MSC co-culture on mature DCs include a significant decrease in expression of HLA-DR, CD1α, CD80 and CD86 without changes in CD14. Thus, MSCs can reverse the mature DCs to an immature state and such changes in DC immunophenotype would correlate with the impaired DC immune stimulatory activity.[49]

MSCs AND NATURAL KILLER (NK) CELLS

The first MSC-NK cell study reported that MSCs did not affect the ability of freshly isolated NK cells to lyse the K562 tumor cells *in vitro*, and also that allogeneic MSCs were resistant to NK-mediated lysis.[35] However, MSCs did affect generation and functions of lymphokine-activated NK cells.[28,53] The co-culture of MSCs with NK cells in the presence of IL-2 resulted in a significant decrease of IFN-γ production by the NK cells.[28] In addition to inhibition of cytokine secretion, MSCs inhibited IL-15-induced NK proliferation, and altered their phenotype and cytotoxicity against HLA class I positive targets.[53] The effect of MSCs on IL-15-induced NK cell proliferation is likely mediated by the soluble factors TGF-β and PGE2, arresting NK cells in G0/G1 phase of cell cycle. However, inhibition of activating receptors 2B4 and NKG2D in NK cells and NK-mediated cellular cytotoxicity against HLA class I positive targets required MSC-NK cellular contacts. Therefore, although MSCs are resistant to freshly isolated NK-mediated lysis,[35] they may be lysed by lymphokine-activated NK cells.[53] Taken together, these data show that MSCs are regulators of the innate immunity in addition to adaptive responses.

NITRIC OXIDE PRODUCED BY MSCs

Nitric oxide (NO) is a powerful signaling molecule in many cell-cell interactions and may work over short or long distances, although its half-life is very short. MSCs have been shown to express NO but its effect in immune modulation was unclear. Recently, Ozawa and colleagues demonstrated that interaction of MSCs with CD4[+] and CD8[+] T-cells induced expression of nitric oxide synthase in MSCs. The results showed that STAT5 phosphorylation, associated with proliferation of T-cells, was inhibited by NO produced when MSCs were present and that inhibitors of prostaglandin synthesis or NO resulted in an increase in T-cell proliferation.[30]

GALECTIN-1 EXPRESSION BY MSCs

Galectin-1 (Gal-1) is a β-galactoside binding protein found as a homo-dimer on the surface of cells and it can modulate several cellular activities. It has no transmembrane domain but interacts strongly with galactose-bearing proteins and may enhance binding and signaling of cell surface counter-receptors on interacting cells. Gal-1 is expressed abundantly by MSCs.[33] Studies with T-cells demonstrates that ~10 μM Gal-1 causes apoptosis of activated T-cells if the exposure lasts 30 minutes or more.[54] The expression of Gal-1 in some cell types is inducible and this may explain some conflicting data on the interaction of MSCs and T-cells, where apoptosis of T-cells was detected in one set of experiments but not another. Membrane-bound Gal-1 may also represent the cell-cell contact factor that establishes the MSC-immune cell interactions that has been lacking in studies of soluble immune modulators expressed by MSCs.

HLA-G EXPRESSION BY MSCs

MSCs from bone marrow and umbilical cord have been shown to exhibit surface expression of HLA-A, B. Another related HLA surface molecule is HLA-G, a class I surface molecule with powerful immune inhibitory functions that likely plays a role in the mother's body accepting the fetus during gestation. Nasef *et al.* analyzed the expression of HLA-G on MSCs in the mixed lymphocyte reaction.[34] HLA-G expression was detectable on MSCs by flow cytometry and immunofluorescence, and soluble molecules could be detected in the culture supernatant. The HLA-G expression was not altered by co-culture with allogeneic lymphocytes. Significantly, a blocking antibody to HLA-G partially restored the lymphocyte proliferation in the co-culture experiments.

MSCs AS UNIVERSAL DONOR CELLS

Accumulated data support the concept of MSCs as universally tolerated stem cells, allowing the use of allogeneic MSCs without donor-recipient HLA matching. This opens the unique opportunity to have off-the-shelf MSCs ready for treatment of patients in acute settings. The universal MSC concept is supported by the low immunogenic phenotype of MSCs and by the absence of immune response against allogeneic MSCs demonstrated in a number of *in vitro* and *in vivo* models.

MSCs express low levels of HLA class I molecules and do not express HLA class II molecules on the cell surface. In the presence of inflammatory cytokines such as IFN-γ, the surface expression of HLA class I is increased and surface expression of HLA class II is induced. However, even after exposure to IFN-γ, MSCs do not express the accessory molecules CD40, CD80, and CD86 which are essential for activation of a T-cell immune response.[17,55] It has been shown that allogeneic MSCs do not induce lymphocyte proliferation when used as a target in a mixed lymphocyte reaction (MLR) *in vitro*.[6,8,9,17] Long-term allogeneic MSC engraftment in multiple tissues *in vivo* suggests that MSCs are not rejected in rats, dogs, sheep, pigs or goat. Several human clinical trials are underway and support the use allogeneic MSCs either by systemic or local injection for treatment of severe refractory acute GVHD, inflammatory bowel disease, support of HSC transplantation, genetic disorders, or cardiac disease. Collectively, these data support the future clinical development of allogeneic MSC use, particularly in the acute setting.

However, there are reports suggesting that allogeneic MSCs are not "universal" and that transplantation of these cells without donor-recipient matching has limited lasting effects. In contrast to the data previously discussed, it has been shown that mouse allogeneic MSCs are immunogenic and will be rejected if used in the non-myeloablative setting.[56] High immunogenicity can be partially explained by differences between the purity of mouse MSCs and MSCs derived from other species, as well as potential cell surface differences. For example, in contrast to human cells, mouse MSCs may express the accessory molecule CD80.[56] The presence of these accessory molecules is a characteristic of highly immunogenic cells, and thus, may play a critical role in rejection of allogeneic mouse MSCs.

On the other hand, some high MSC immunogenicity may be linked to MSC isolation, growth and preparation methods. For example, the use of MSCs expanded *ex vivo* in the presence of fetal bovine serum (FBS) may lead to detectable anti-FBS antibodies in animals[57] and in humans. This may then correlate with limited cell engraftment,[58] but another study did not see an effect on patients.[59] Thus, MSC culture and purification methods may be found to play a critical role in the success or failure of MSC cellular therapy and each step must be evaluated properly. Moreover, accumulating evidence suggests that MSCs are functionally defective in patients with autoimmune diseases and therefore allogeneic MSCs might form the basis for a more effective treatment for these patients.[60,61]

PRECLINICAL STUDIES USING ALLOGENEIC UC-, UCB- AND BM-MSCs

The *in vitro* data provides compelling evidence that MSCs do not stimulate allogeneic immune cells. However, it is still necessary to demonstrate that allo-MSCs are not rejected in the *in vivo* setting, and many studies have been presented with BM-MSCs, while studies utilizing UC-MSCs, UCB-MSCs and placental-MSCs are still in the early stages. A few examples exist where human MSCs have been placed in non-human animal models to test rejection and early adaptation of the MSCs to the tissue. Although interesting in the context of immune response to xenografted MSCs, the results of xenograft experiments may not be predictive of same-species transplantation and results of xenografted human MSCs into a rat, sheep, etc., are not enough to qualify human clinical studies. Rather, same species allo-transplant study results are preferred.

Orthopedic Studies Using Allo-MSCs

Many of the earliest studies with MSCs recognized their bone forming ability and a number of studies in small and large animals using autologous

MSCs have been published. With the recognition that allogeneic MSCs may heal orthopedic injuries, researchers have tested several pertinent animal models. Allogeneic canine BM-MSCs were implanted on a ceramic carrier to rebuild a 2 cm segment of surgically removed femur. Allogeneic BM-MSCs repaired the bone completely over eight to 16 weeks without rejection and there were no responding immune cells detected against the allo-MSCs.[62] Another study evaluated the repair of an alveolar ridge defect in the canine jaw using allogeneic canine MSCs. New bone formation and substantial repair was evident after nine weeks.[63] These studies are significant as inflammation and a rejection response caused by an immune response to the allogeneic cells would have prevented new bone formation around the implants.

Treating Cardiac Ischemia with Allo-MSCs

Heart failure is the leading cause of death in the developed world and demands our attention as even slight improvements in cardiac care have the potential to aid thousands of patients. The cardiac field has embraced MSCs as a potential repair/regeneration method to treat damaged tissue following myocardial infarction. Results of animal studies have indicated functional improvement of damaged hearts, likely by their expression of growth factors and cytokines that limit deleterious remodeling that would otherwise result in heart failure. But there are also indications of MSC differentiation as they express a number of cardiac specific proteins. Toma *et al.* injected lacZ-labeled human MSCs into the left ventricle of a healthy mouse and analyzed the cardiac tissue for lacZ positive cells over six weeks. There were no lacZ positive cells expressing cardiac proteins at four days. Over several weeks, the number of human lacZ MSCs diminished but at four and six weeks, each lacZ positive MSC was also positive for cardiac specific proteins (β-myosin heavy chain, α-actinin, SERCA2, troponin and phospholamban).[64] That is, each xenografted MSC appeared to assimilate to the *in vivo* conditions and acquired morphology and gene expression characteristic of cardiomyocytes. However, in these healthy hearts, it was not possible to test for a functional improvement as too few cells were engrafted. That all human MSCs detected after three weeks expressed cardiac proteins argued against the possibility that they were the result of the very rare cell-cell fusion that has been seen in some embryonic stem cell engraftment experiments.

Several recent studies utilizing UCB-MSCs have been published that evaluated their cardiac differentiation potential.[65–67] Umezawa and colleagues co-cultured human UCB-MSCs with fetal murine cardio-myocytes. The UCB-MSCs were labeled with the gene for green fluorescence protein (GFP) to distinguish them from mouse cells. After five days of co-culture, up to 50% of the UCB-MSCs appeared to contract rhythmically with the mouse cardiomyocytes, and showed expression of connexin-43 and striations when stained for troponin I. Electrical meas-urements indicated a cardiac-like action potential. Interestingly, the UCB-MSCs also began to differentiate to cardiomyocyte-like cells when separated from cardiomyocytes by a porous membrane, suggesting that soluble factors may induce cardiac differentiation and ruling out fusion between UCB-MSCs and cardiomyocytes. In another study, the UCB-MSCs isolated by Prat-Vidal *et al.* expressed proteins for α-actinin, the mem-brane calcium pump SERCA2 and the gap junction protein connexin-43, but the cells did not express the cardiac transcription factors GATA4, NKX2.5 and Tbx5, nor cardiac β-myosin heavy chain or troponin I. *In vivo* studies will be necessary to see if the *in vivo* environment can stim-ulate greater cardiac differentiation.[67]

Homing of MSCs to Sites of Injury

Many experiments have been performed to place MSCs at the site of tis-sue damage and following the repair process. It is not clear how MSCs might circulate in the body. If they are present in bone marrow and other locations, how do they respond to reach the injury site? Many investiga-tors have searched for MSCs in free flowing blood but the number of MSCs present is usually much, much less than in bone marrow, where they are already rare. This then raises the question of whether endogenous MSCs circulate or are recruited from local sites. The second important question is whether endogenous MSCs are present in large enough numbers, even if they proliferate *in situ*, to be effective in the healing process. This second question remains open but experimental evidence for trafficking of MSCs to sites of injury is known, and a few examples are described below. For this, cultured MSCs were labeled either with gene tags (lacZ or membrane dyes such as carboxymethyl-di I or other) to allow their identification and their movements or location were determined. More recent experiments have labeled MSCs with superparamagnetic iron-oxide particles that allow imaging by magnetic resonance imaging (MRI) (see below).

Chiu and colleagues studied mouse MSCs, labeled with the LacZ gene for later identification, and given systemically to a healthy rat. One week later, the rat received a surgical procedure to perform an experimental myocardial infarction. At several time points out to 12 weeks, the rats were sacrificed and the xenogeneic mouse MSCs were detected in abundance within the infracted regions of the heart while animals without infarcts only revealed LacZ-MSCs in their bone marrow.[68] In another study, allogeneic rat MSCs migrated to the site of myocardial infarction when given intravenously. A greater number of allo-MSCs migrated to the cardiac injury when given immediately after the infarction as opposed to 14 days later, suggesting inflammatory factors are involved in MSC migration to the injury site (Fig. 3).[69] These studies demonstrate that MSCs can be systemically delivered by intravenous injection and that the cells migrate to sites of tissue injury. This migration to inflammation sites is another MSC attribute that is not fully understood, but which is being utilized in clinical studies today. *In vivo* imaging studies have been performed to study the migration of allogeneic MSCs to cardiac infarcts in canine models. For this, the MSCs were labeled with superparamagnetic ferric oxide particles that can be visualized by magnetic resonance imaging (Fig. 4).[70,71]

Hare and colleagues used a catheter to deliver allogeneic porcine MSCs to pigs subjected to experimental infarcts and were able to demonstrate robust migration and engraftment in the injured tissue, with beneficial physiologic outcomes.[72] These experiments led to a phase I clinical trial described later.

Finally, an important *in vivo* experiment utilizing parabiosis that was recently reported further illustrates the migration of MSCs to the injury site. Parabiotic mouse pairs were created by suturing the skin of two mice, side by side, together. These mice appear to co-exist without complaint and develop a shared circulatory system. Kaneda and colleagues used a wild type mouse which was gamma-irradiated to

Allo-MSCs "Homing" to Myocardial Infarction

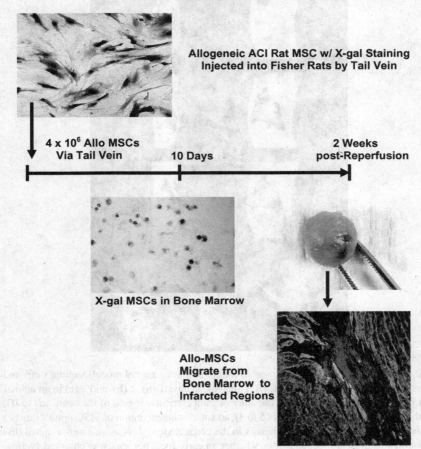

Allogeneic ACI Rat MSC w/ X-gal Staining Injected into Fisher Rats by Tail Vein

4×10^6 Allo MSCs Via Tail Vein

10 Days

2 Weeks post-Reperfusion

X-gal MSCs in Bone Marrow

Allo-MSCs Migrate from Bone Marrow to Infarcted Regions

Fig. 3. Allogeneic MSCs will migrate to sites of injury in the heart. Allogeneic rat MSCs from the ACI strain were transduced with a vector carrying the lacZ gene coding for the expression of β-galactosidase and injected by the tail vein into Fisher rats. After ten days, the lacZ-MSCs could be found in the bone marrow. The Fisher rat then received an experimental infarct by tying off the left anterior descending artery. After two weeks the animal was sacrificed and the hearts analyzed for β-gal cells (see blue staining in heart) and - histology sections were tested with antibody to β-gal (red) and the surrounding tissue stained for cardiac α-actinin. The MSCs were only found in the damaged infarct tissue. Reprinted from Pittenger and Martin (2004).[69]

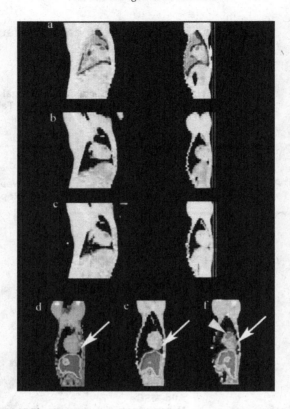

Fig. 4. Infused MSCs migrate to site of infarct in a large animal model. Sagittal (left) and coronal (right) view of fused SPECT/CT images on days 1 (**a**), 2 (**b**), and 7 (**c**) in an animal that demonstrated focal uptake in the anterior midventricular region of the heart. (**d**) to (**f**), At the last imaging time point (days 5 to 8), an anterior apical region of MSC uptake (arrow) is shown in three representative animals in the coronal view. This more anterior apical distribution was present independent of whether an early focal hot spot was observed (yellow arrowhead, f only). Reprinted from Kraitchman *et al.* (2003).[71]

prepare it to receive a transplant from the Rosa26-GFP mouse which constitutively expresses the green fluorescent protein gene in all of its cells.[73] In these experiments, MSC-like cells are referred to as mouse osteoprogenitor cells (MOPC). The irradiated mouse received a subcutaneous whole femur implant from the Rosa26 mouse on its back, and survived the irradiation due to the hematopoietic stem cells present in the femur. The recipient mouse now had GFP⁺ hematopoietic cells as well as a reservoir of GFP⁺ MOPCs — the bone under its skin. A parabiotic pairing was created with this mouse and a syngeneic wild-type partner and a collagen matrix containing bone morphogenetic protein 2 (BMP2) was inserted under the skin of the partner. BMP2 is a known inducer of bone formation by MOPCs (or MSCs) and represents an "injured bone" site in these experiments. Over one to three weeks, the BMP2-collagen pellet attracted the MOPCs, became positive for GFP, and showed histological evidence of new bone formation, demonstrating migration and differentiation of the MOPCs. Additional evidence suggested that HIF-1α and SDF-1 are factors involved in the migration and homing of MOPCs.

CLINICAL EXPERIENCE OF MSC INFUSION

MSCs have been brought to the clinic for several purposes; to repair damaged tissues, to promote repair of ischemia, to produce enzymes missing in patients with genetic metabolic disorders, to promote hematopoietic engraftment after allogeneic stem cell transplantation, and for immunosuppression in graft-versus-host disease or autoimmune disorders. Most of these studies have utilized BM-MSCs except for one below but more studies are planned for UC-MSCs, UCB-MSCs and placental-MSCs.

The first recipients of culture-expanded BM-MSCs were cancer patients in remission receiving autologous culture-expanded MSCs solely as a safety trial, and the i.v. infusions were safe with no untoward effects.[74] Subsequent trials in patients undergoing myeloablative therapy for breast cancer indicated that MSCs were not easily grown from the autologous bone marrow of such patients, likely due to the course of disease and medical treatment.[75]

In the genetic disease area, Horwitz *et al.* treated five children who had osteogenesis imperfecta (OI) — a rare genetic disorder in the coding sequence for type I collagen and resulting in brittle bones — with a bone marrow transplant from an HLA-identical sibling donor after full myeloablative conditioning. All five patients increased their total body mineral content and growth velocity.[76] In a later study, the same children received an infusion of *ex vivo* expanded MSCs derived from their donor.[77] Here, a low level of MSC engraftment was observed, particularly when the cells had been minimally expanded *in* vitro prior to infusion.

MSCs have also been used to treat inborn errors of metabolism, as they produce and secrete several enzymes, including arylsulphatase A and B, alpha-1-iduronidase, glucocerebrosidase and adrenoleukodystrophy protein.[55] Koc and colleagues reported that they cultured MSCs and infused them into patients with metachromatic leukodystrophy and Hurler's disease several years after the previous allogeneic stem cell transplantation. Four out of the five patients showed a promising improvement in the nerve conduction velocity.[55]

The first patient to receive mismatched BM-MSCs was an elderly woman with end stage severe aplastic anemia, who received MSCs derived from an HLA-haploidentical son on two occasions.[7] The engraftment of the donor MSCs was detected by PCR in the endosteum of a bone marrow biopsy, but not in marrow aspirates suggesting that the MSCs were engrafted in the bone tissue, and further indicating that human MSCs could be transplanted between HLA-disparate individuals without rejection.

However, experience in transplantation across HLA barriers in the absence of immunosuppression is limited. Fully mismatched allogeneic fetal liver-derived MSCs were transplanted into an immunocompetent fetus in the third trimester of gestation.[78] A bone marrow biopsy revealed 0.3%–7.4% Y-chromosome positive cells by fluorescent *in situ* hybridization, indicating engraftment of the donor MSCs. Prior to the transplantation, proliferation of donor lymphocytes against allogeneic peripheral blood lymphocytes was detected in MLC, confirming the immunocompetence of the fetus in the third trimester. In contrast, alloreactivity against donor MSCs was not detected *in vitro* either before or after transplant, further indicating that MSCs can be tolerated when transplanted across MHC barriers in humans.

MSCs Enhance Engraftment of Hematopoietic Stem Cells

Human MSCs promote engraftment of unrelated and umbilical cord-derived hematopoietic stem cells (HSCs) in NOD-SCID mice and in fetal sheep.[79–81] The enhancing effect is most prominent when the dose of HSCs is limiting. In the study by Maitra *et al.*, two out of ten mice engrafted when transplanted with a low dose of HSCs whereas eight out of ten mice showed persistent engraftment when co-transplanted with HSCs and MSCs. In another study, HLA-matched sibling-derived MSCs were co-infused with HSCs from the same donor in 46 patients to promote hematopoietic engraftment and to limit GVHD.[8] The MSCs were given in escalating doses from 1 to 5×10^6/kg. Biopsies demonstrated stromal cell chimerism in two out of 19 examined patients at six and 18 months post transplant. Moderate to severe acute GVHD was observed in 28% of the patients, and chronic GVHD was present in 61%. The MSC infusion caused no discernable acute or long-term MSC-associated adverse events.

In another study, HLA-identical or haploidentical MSCs were co-transplanted with HSCs in seven patients.[82] Three of the patients were re-transplanted for graft failure. Two of the patients suffered from severe combined immunodeficiency associated with high peri-transplant mortality. In all patients, co-transplantation of MSCs and HSCs resulted in fast engraftment of neutrophils and platelets and 100% donor chimerism. One of the patients, diagnosed with aplastic anemia had graft failure after her first transplantation and severe Henoch-Schönlein purpura. After a second co-transplantation, she recovered from both the Henoch-Schönlein purpura and aplasia.

In a further study, donor-derived MSCs were co-transplanted with HLA-disparate CD34[+] cells from a relative (haploidentical) in 14 children. Haploidentical hematopoietic stem-cell transplantation is associated with an increased risk of graft failure. While graft failure in 47 historic controls was 15%, all patients given MSCs showed sustained hematopoietic engraftment without any adverse reactions or increased number of infections.[82]

MSC Infusion in Inflammatory Disorders and Tissue Injury

In experimental animal models, infused MSCs improved the outcome of renal, neural and lung injury possibly by promoting a shift from the production of pro-inflammatory to anti-inflammatory cytokines at the site of injury.[31,45,83] In humans, as mentioned earlier, haploidentical MSCs were infused into a patient with treatment-resistance severe acute GVHD of the gut and liver, with the aim of harnessing the tissue-repair effect shown in animal models and the immunomodulatory effects on human lymphocytes seen *in vitro*.[21] Recovery in terms of improved liver values and intestinal recovery was prompt. Upon discontinuation of cyclosporine, the patient's acute GVHD recurred suggesting that tolerance was not induced. However, the GVHD was still responsive to a second MSC infusion.

Treatment of an additional eight patients with similar steroid-refractory grades III to IV acute GVHD have been reported.[23] Acute GVHD disappeared completely in six of the eight patients. Their survival rate was better than that of 16 patients who had steroid-resistant biopsy-proven GVHD during the same period that were not treated with MSC.[24] In one patient, DNA from both MSC donors (one haplo-identical and one mismatched unrelated) could be detected at low levels in the colon and lymph nodes of the gastrointestinal tract one month after infusion. Although the number of patients treated in this way is still low, MSC infusion seems a promising treatment.

MSCs have also been used to treat therapy-induced tissue toxicity after HSC transplantation.[84] Seven patients had hemorrhagic cystitis, two had a pneumomediastinum and one had perforated colon and peritonitis. The majority of MSC donors were third party, HLA-mismatched. In five patients, the severe hemorrhagic cystitis cleared after MSC infusion. The remaining two patients had reduced transfusion requirements after infusion, but died of multiorgan failure. In two patients, pneumomediastinum disappeared after HSC infusion. The perforated diverticulitis and peritonitis disappeared in the patient with gut GVHD. Further studies evaluating a role for MSCs in tissue toxicity are warranted in the future.

MSCs may also be used to treat autoimmune disorders. Systemic sclerosis is an autoimmune connective tissue disorder characterized by activation of the immune system, by vasculopathy due to dysfunction of endothelial cells and by an excessive deposition of collagen. A first report described the feasibility, safety and efficacy of transplanting bone marrow-derived human MSCs from allogeneic haploidentical-related donors after *in vitro* expansion into a patient suffering of systemic sclerosis with diffuse cutaneous involvement.[85] A 41-year-old female refractory to numerous immunosuppressive agents had painful ulcerations and rapidly progressive massive skin sclerosis. Three months after transplantation, a significant decrease in the patient's painful ulcerations was measured. After six months, all ulcerations but one had healed. Blood circulation improved in hands and fingers and transcutaneous partial oxygen pressure improved. The Rodnan skin score was reduced from 25 to 11, paralleled by a reduction of the visual analog scale for pain.

Four patients with Buerger's disease (thromboangiitis obliterans) were treated with UCB-MSCs by Kang and colleagues.[86] This disease is an inflammatory vaso-oclusive disorder affecting both arteries and veins of the feet and hands, and may lead to disuse and amputation as there is no curative treatment. The delivery of HLA-matched UCB-MSCs to the affected region rapidly relieved pain, and eliminated necrotic skin lesions at four weeks, and angiography revealed greater capillary density and improved overall vascular flow. As this study was small, we must await the results of a larger double blind study before drawing further conclusions.

Allogeneic BM-MSCs are currently under clinical evaluation for treatment of acute cardiac infarction and the Phase I study of 53 patients has been completed. The study protocol delivered the MSCs to the patients by a simple intravenous infusion in the hospital room and relied on the MSCs' ability to home to the sites of cardiac infarct. This double-blind, placebo controlled study involved three cohorts of escalating dose. The results indicate improvement in the treatment group versus placebo in heart ejection fraction (25%), improved lung function, fewer adverse events, fewer premature contractions and significantly improvement in overall condition, all reported at six months.[87]

CONCLUSIONS

MSCs have been known to have unusual interactions with allogeneic immune cells since the late 1990s. Many studies have shown that BM-MSCs, UC-MSCs, UCB-MSCs and placental-MSCs all have a low immunogenic surface phenotype and actively interact with immune cells to suppress a rejection response, and the cells can be transplanted across HLA barriers. Data indicate that the MSCs are able to migrate to sites of injury and inflammation, regulate inflammatory and immune responses depending on the mix of immune cells and factors present in local microenvironment at the time, and to induce immune tolerance. These characteristics distinguish MSCs from other cell types with immune modulating potential, making these cells an attractive candidate for development of clinical therapies for immune-mediated and inflammatory diseases as well as mesenchymal tissues damaged by injury, disease or genetic insufficiency.

The challenges of developing cellular therapies are only beginning to be addressed. While patients are currently undergoing treatment with allogeneic MSCs in response to many disorders including GVHD caused by treatment for hematopoietic cancers, genetic disorders such as osteogenesis imperfecta and glycogen storage diseases, acute cardiac infarction and chronic sclerotic disease, further understanding of the biology of MSCs, isolated from different tissues, will improve these cell-based therapies by continuing *in vitro* experimentation and *in vivo* animal studies. Only through continued *in vitro* experimentation and *in vivo* animal studies can the necessary questions be addressed, to improve the clinical use of MSCs.

References

1. Johnson A, Dorshkind K. Stromal cells in myeloid and lymphoid long-term bone marrow cultures can support multiple hemopoietic lineages and modulate their production of hemopoietic growth factors. *Blood* 1986 Dec;68(6): 1348–54.
2. Kincade PW. Molecular interactions between stromal cells and B lymphocyte precursors. *Semin Immunol* 1991 Nov;3(6):379–90.
3. McGlave P, Verfaillie C, Miller J. Interaction of primitive human myeloid and lymphoid progenitors with the marrow microenvironment. *Blood Cells* 1994;20(1):121–6; discussion 6–8.
4. Zipori D, Tamir M. Stromal cells of hemopoietic origin. *Int J Cell Cloning* 1989 Sep;7(5):281–91.
5. Pittenger MF, Mackay AM, Beck SC, Jaiswal RK, Douglas R, Mosca JD, *et al.* Multilineage potential of adult human mesenchymal stem cells. *Science* 1999;284(5411):143–7.

6. Klyushnenkova E, Mosca JD, Zernetkina V, Majumdar MK, Beggs KJ, Simonetti DW, *et al.* T cell responses to allogeneic human mesenchymal stem cells: immunogenicity, tolerance, and suppression. *J Biomed Sci* 2005; 12(1):47–57.

7. Fouillard L, Bensidhoum M, Bories D, Bonte H, Lopez M, Moseley AM, *et al.* Engraftment of allogeneic mesenchymal stem cells in the bone marrow of a patient with severe idiopathic aplastic anemia improves stroma. *Leukemia* 2003 Feb;17(2):474–6.

8. Lazarus HM, Koc ON, Devine SM, Curtin P, Maziarz RT, Holland HK, *et al.* Cotransplantation of HLA-identical sibling culture-expanded mesenchymal stem cells and hematopoietic stem cells in hematologic malignancy patients. *Biol Blood Marrow Transplant* 2005 May;11(5):389–98.

9. Bieback K, Kern S, Kluter H, Eichler H. Critical parameters for the isolation of mesenchymal stem cells from umbilical cord blood. *Stem Cells* 2004;22(4):625–34.

10. Romanov YA, Svintsitskaya VA, Smirnov VN. Searching for alternative sources of postnatal human mesenchymal stem cells: candidate MSC-like cells from umbilical cord. *Stem Cells* 2003;21(1):105–10.

11. Ennis J, Götherstrom C, Le Blanc K, Davies JE. *In vitro* immunologic properties of human umbilical cord perivascular cells. *Cytotherapy* 2008;10(2): 174–81.

12. Sarugaser R, Lickorish D, Baksh D, Hosseini MM, Davies JE. Human umbilical cord perivascular (HUCPV) cells: a source of mesenchymal progenitors. *Stem Cells* 2005 Feb;23(2):220–9.

13. Chang YJ, Shih DT, Tseng CP, Hsieh TB, Lee DC, Hwang SM. Disparate mesenchyme-lineage tendencies in mesenchymal stem cells from human bone marrow and umbilical cord blood. *Stem Cells* 2006 Mar;24(3): 679–85.

14. Kern S, Eichler H, Stoeve J, Kluter H, Bieback K. Comparative analysis of mesenchymal stem cells from bone marrow, umbilical cord blood, or adipose tissue. *Stem Cells* 2006 May;24(5):1294–301.

15. Wagner W, Wein F, Seckinger A, Frankhauser M, Wirkner U, Krause U, *et al.* Comparative characteristics of mesenchymal stem cells from human bone marrow, adipose tissue, and umbilical cord blood. *Exp Hematol* 2005 Nov;33(11):1402–16.

16. Li C, Zhang W, Jiang X, Mao N. Human-placenta-derived mesenchymal stem cells inhibit proliferation and function of allogeneic immune cells. *Cell Tissue Res* 2007 Dec;330(3):437–46.

17. Li CD, Zhang WY, Li HL, Jiang XX, Zhang Y, Tang PH, *et al.* Mesenchymal stem cells derived from human placenta suppress allogeneic umbilical cord blood lymphocyte proliferation. *Cell Res* 2005 Jul;15(7):539–47.

18. Bartholomew A, Sturgeon C, Siatskas M, Ferrer K, McIntosh K, Patil S, *et al.* Mesenchymal stem cells suppress lymphocyte proliferation *in vitro* and prolong skin graft survival *in vivo*. *Exp Hematol* 2002;30(1):42–8.

19. Di Nicola M, Carlo-Stella C, Magni M, Milanesi M, Longoni PD, Matteucci P, *et al.* Human bone marrow stromal cells suppress T-lymphocyte proliferation induced by cellular or nonspecific mitogenic stimuli. *Blood* 2002 May 15;99(10):3838–43.

20. Le Blanc K, Tammik L, Sundberg B, Haynesworth SE, Ringdén O. Mesenchymal stem cells inhibit and stimulate mixed lymphocyte cultures and mitogenic responses independently of the major histocompatibility complex. *Scand J Immunol* 2003 Jan;57(1):11–20.

21. Le Blanc K, Rasmusson I, Sundberg B, Götherström C, Hassan M, Uzunel M, *et al.* Treatment of severe acute graft-versus-host disease with third party haploidentical mesenchymal stem cells. *Lancet* 2004;363:1439–41.

22. Ball LM, Bernardo ME, Roelofs H, Lankester A, Cometa A, Egeler RM, *et al.* Cotransplantation of *ex vivo* expanded mesenchymal stem cells accelerates lymphocyte recovery and may reduce the risk of graft failure in haploidentical hematopoietic stem-cell transplantation. *Blood* 2007 Oct 1;110(7):2764–7.

23. Ringden O, Uzunel M, Rasmusson I, Remberger M, Sundberg B, Lonnies H, *et al.* Mesenchymal stem cells for treatment of therapy-resistant graft-versus-host disease. *Transplantation* 2006 May 27;81(10):1390–7.

24. van Laar JM, Tyndall A. Adult stem cells in the treatment of autoimmune diseases. *Rheumatology (Oxford)*. 2006 Oct;45(10):1187–93.

25. Tse WT, Pendleton JD, Beyer WM, Egalka MC, Guinan EC. Suppression of allogeneic T-cell proliferation by human marrow stromal cells: implications in transplantation. *Transplantation* 2003 Feb 15;75(3):389–97.

26. Augello A, Tasso R, Negrini SM, Amateis A, Indiveri F, Cancedda R, *et al.* Bone marrow mesenchymal progenitor cells inhibit lymphocyte proliferation by activation of the programmed death 1 pathway. *Eur J Immunol* 2005 May;35(5):1482–90.

27. Krampera M, Glennie S, Dyson J, Scott D, Laylor R, Simpson E, *et al.* Bone marrow mesenchymal stem cells inhibit the response of naive and memory antigen-specific T cells to their cognate peptide. *Blood* 2003 May 1;101(9):3722–9.

28. Aggarwal S, Pittenger MF. Human mesenchymal stem cells modulate allogeneic immune cell responses. *Blood* 2005 Feb 15;105(4):1815–22.

29. Ryan JM, Barry F, Murphy JM, Mahon BP. Interferon-gamma does not break, but promotes the immunosuppressive capacity of adult human mesenchymal stem cells. Clin Exp *Immunol* 2007 Aug;149(2):353–63.

30. Sato K, Ozaki K, Oh I, Meguro A, Hatanaka K, Nagai T, *et al.* Nitric oxide plays a critical role in suppression of T-cell proliferation by mesenchymal stem cells. *Blood* 2007 Jan 1;109(1):228–34.

31. Ortiz LA, Dutreil M, Fattman C, Pandey AC, Torres G, Go K, *et al.* Interleukin 1 receptor antagonist mediates the antiinflammatory and antifibrotic effect of mesenchymal stem cells during lung injury. *Proc Natl Acad Sci USA* 2007 Jun 26;104(26):11002–7.

32. Meisel R, Zibert A, Laryea M, Gobel U, Daubener W, Dilloo D. Human bone marrow stromal cells inhibit allogeneic T-cell responses by indoleamine 2,3-dioxygenase-mediated tryptophan degradation. *Blood* 2004 Jun 15;103(12): 4619–21.

33. Kadri T, Lataillade JJ, Doucet C, Marie A, Ernou I, Bourin P, *et al.* Proteomic study of Galectin-1 expression in human mesenchymal stem cells. *Stem Cells Dev* 2005 Apr;14(2):204–12.

34. Nasef A, Mathieu N, Chapel A, Frick J, Francois S, Mazurier C, *et al.* Immunosuppressive effects of mesenchymal stem cells: involvement of HLA-G. *Transplantation* 2007 Jul 27;84(2):231–7.

35. Rasmusson I, Ringden O, Sundberg B, Le Blanc K. Mesenchymal stem cells inhibit the formation of cytotoxic T lymphocytes, but not activated cytotoxic T lymphocytes or natural killer cells. *Transplantation* 2003 Oct 27;76(8): 1208–13.

36. English K, Barry FP, Field-Corbett CP, Mahon BP. IFN-gamma and TNF-alpha differentially regulate immunomodulation by murine mesenchymal stem cells. *Immunol Lett* 2007 Jun 15;110(2):91–100.

37. Glennie S, Soeiro I, Dyson PJ, Lam EW, Dazzi F. Bone marrow mesenchymal stem cells induce division arrest anergy of activated T cells. *Blood* 2005 Apr 1;105(7):2821–7.

38. Benvenuto F, Ferrari S, Gerdoni E, Gualandi F, Frassoni F, Pistoia V, *et al.* Human mesenchymal stem cells promote survival of T cells in a quiescent state. *Stem Cells* 2007 Jul;25(7):1753–60.

39. Plumas J, Chaperot L, Richard MJ, Molens JP, Bensa JC, Favrot MC. Mesenchymal stem cells induce apoptosis of activated T cells. *Leukemia* 2005 Sep;19(9):1597–604.

40. Potian JA, Aviv H, Ponzio NM, Harrison JS, Rameshwar P. Veto-like activity of mesenchymal stem cells: functional discrimination between cellular responses to alloantigens and recall antigens. *J Immunol* 2003 Oct 1;171(7):3426–34.

41. Beyth S, Borovsky Z, Mevorach D, Liebergall M, Gazit Z, Aslan H, *et al.* Human mesenchymal stem cells alter antigen-presenting cell maturation and induce T-cell unresponsiveness. *Blood* 2005 Mar 1;105(5):2214–9.

42. Groh ME, Maitra B, Szekely E, Koc ON. Human mesenchymal stem cells require monocyte-mediated activation to suppress alloreactive T cells. *Exp Hematol* 2005 Aug;33(8):928–34.

43. Maccario R, Podesta M, Moretta A, Cometa A, Comoli P, Montagna D, *et al.* Interaction of human mesenchymal stem cells with cells involved in alloantigen-specific immune response favors the differentiation of CD4$^+$ T-cell subsets expressing a regulatory/suppressive phenotype. *Haematologica* 2005 Apr;90(4):516–25.

44. Gerdoni E, Gallo B, Casazza S, Musio S, Bonanni I, Pedemonte E, *et al.* Mesenchymal stem cells effectively modulate pathogenic immune response in experimental autoimmune encephalomyelitis. *Ann Neurol* 2007 Mar;61(3):219–27.

45. Zappia E, Casazza S, Pedemonte E, Benvenuto F, Bonanni I, Gerdoni E, *et al.* Mesenchymal stem cells ameliorate experimental autoimmune encephalomyelitis inducing T-cell anergy. *Blood* 2005 Sep 1;106(5): 1755–61.

46. Corcione A, Benvenuto F, Ferretti E, Giunti D, Cappiello V, Cazzanti F, *et al.* Human mesenchymal stem cells modulate B-cell functions. *Blood* 2006 Jan 1;107(1):367–72.

47. Rasmusson I, Le Blanc K, Sundberg B, Ringden O. Mesenchymal stem cells stimulate antibody secretion in human B cells. *Scand J Immunol* 2007 Apr;65(4):336–43.

48. Zhang W, Ge W, Li C, You S, Liao L, Han Q, *et al.* Effects of mesenchymal stem cells on differentiation, maturation, and function of human monocyte-derived dendritic cells. *Stem Cells Dev* 2004 Jun;13(3):263–71.

49. Jiang XX, Zhang Y, Liu B, Zhang SX, Wu Y, Yu XD, *et al.* Human mesenchymal stem cells inhibit differentiation and function of monocyte-derived dendritic cells. *Blood* 2005 May 15;105(10):4120–6.

50. Nauta AJ, Kruisselbrink AB, Lurvink E, Willemze R, Fibbe WE. Mesenchymal stem cells inhibit generation and function of both CD34[+]-derived and monocyte-derived dendritic cells. *J Immunol* 2006 Aug 15;177(4):2080–7.

51. Ramasamy R, Fazekasova H, Lam EW-F, Soeiro Is, Lombardi G, Dazzi F. Mesenchymal stem cells inhibit dendritic cell differentiation and function by preventing entry into the cell cycle. *Transplantation* 2007;83(1):71–6.

52. Djouad F, Charbonnier LM, Bouffi C, Louis-Plence P, Bony C, Apparailly F, *et al.* Mesenchymal stem cells inhibit the differentiation of dendritic cells through an interleukin-6-dependent mechanism. *Stem Cells* 2007 Aug;25(8):2025–32.

53. Sotiropoulou PA, Perez SA, Gritzapis AD, Baxevanis CN, Papamichail M. Interactions between human mesenchymal stem cells and natural killer cells. *Stem Cells* 2006 Jan;24(1):74–85.

54. Perillo NL, Marcus ME, Baum LG. Galectins: versatile modulators of cell adhesion, cell proliferation, and cell death. *J Mol Med* 1998 May;76(6): 402–12.

55. Koç ON, Day J, Nieder M, Gerson SL, Lazarus HM, Krivit W. Allogeneic mesenchymal stem cell infusion for treatment of metachromatic leukodystrophy (MLD) and Hurler syndrome (MPS-IH). *Bone Marrow Transplant* 2002 Aug;30(4):215–22.

56. Eliopoulos N, Stagg J, Lejeune L, Pommey S, Galipeau J. Allogeneic marrow stromal cells are immune rejected by MHC class I- and class II-mismatched recipient mice. *Blood* 2005 Dec 15;106(13):4057–65.

57. Spees JL, Gregory CA, Singh H, Tucker HA, Peister A, Lynch PJ, *et al.* Internalized antigens must be removed to prepare hypoimmunogenic mesenchymal stem cells for cell and gene therapy. *Mol Ther* 2004 May;9(5):747–56.

58. Horwitz EM, Gordon PL, Koo WK, Marx JC, Neel MD, McNall RY, *et al.* Isolated allogeneic bone marrow-derived mesenchymal cells engraft and

stimulate growth in children with osteogenesis imperfecta: implications for cell therapy of bone. *Proc Natl Acad Sci USA* 2002 Jun 25;99(13):8932–7.

59. Sundin M, Ringden O, Sundberg B, Nava S, Gotherstromm C, Le Blanc K. No alloantibodies against mesenchymal stromal cells, but presence of anti-fetal calf serum antibodies, after transplantation in allogeneic hematopoietic stem cell recipients. *Haematologica* 2007 Sep;92(9):1208–15.

60. Cipriani P, Guiducci S, Miniati I, Cinelli M, Urbani S, Marrelli A, *et al.* Impairment of endothelial cell differentiation from bone marrow-derived mesenchymal stem cells: new insight into the pathogenesis of systemic sclerosis. *Arthritis Rheum* 2007 Jun;56(6):1994–2004.

61. Sun LY, Zhang HY, Feng XB, Hou YY, Lu LW, Fan LM. Abnormality of bone marrow-derived mesenchymal stem cells in patients with systemic lupus erythematosus. *Lupus* 2007;16(2):121–8.

62. Arinzeh TL, Peter SJ, Archambault MP, van den Bos C, Gordon S, Kraus K, *et al.* Allogeneic mesenchymal stem cells regenerate bone in a critical-sized canine segmental defect. *J Bone Joint Surg Am* 2003 Oct;85-A(10):1927–35.

63. De Kok IJ, Peter SJ, Archambault M, van den Bos C, Kadiyala S, Aukhil I, *et al.* Investigation of allogeneic mesenchymal stem cell-based alveolar bone formation: preliminary findings. *Clin Oral Implants Res* 2003 Aug;14(4): 481–9.

64. Toma C, Pittenger MF, Cahill KS, Byrne BJ, Kessler PD. Human mesenchymal stem cells differentiate to a cardiomyocyte phenotype in the adult murine heart. *Circulation* 2002 Jan 1;105(1):93–8.

65. Nishiyama N, Miyoshi S, Hida N, Uyama T, Okamoto K, Ikegami Y, *et al.* The significant cardiomyogenic potential of human umbilical cord blood-derived mesenchymal stem cells *in vitro*. *Stem Cells* 2007 Aug;25(8):2017–24.

66. Nunes VA, Cavacana N, Canovas M, Strauss BE, Zatz M. Stem cells from umbilical cord blood differentiate into myotubes and express dystrophin *in vitro* only after exposure to *in vivo* muscle environment. *Biol Cell* 2007 Apr;99(4):185–96.

67. Prat-Vidal C, Roura S, Farre J, Galvez C, Llach A, Molina CE, *et al.* Umbilical cord blood-derived stem cells spontaneously express cardiomyogenic traits. *Transplant Proc* 2007 Sep;39(7):2434–7.

68. Bittira B, Shum-Tim D, Al-Khaldi A, Chiu RC. Mobilization and homing of bone marrow stromal cells in myocardial infarction. *Eur J Cardiothorac Surg* 2003 Sep;24(3):393–8.

69. Pittenger MF, Martin BJ. Mesenchymal stem cells and their potential as cardiac therapeutics. *Circ Res* 2004 Jul 9;95(1):9–20.

70. Kraitchman DL, Heldman AW, Atalar E, Amado LC, Martin BJ, Pittenger MF, *et al. In vivo* magnetic resonance imaging of mesenchymal stem cells in myocardial infarction. *Circulation* 2003 May 13;107(18):2290–3.

71. Kraitchman DL, Tatsumi M, Gilson WD, Ishimori T, Kedziorek D, Walczak P, *et al.* Dynamic imaging of allogeneic mesenchymal stem cells trafficking to myocardial infarction. *Circulation* 2005 Sep 6;112(10):1451–61.

72. Amado LC, Saliaris AP, Schuleri KH, St John M, Xie JS, Cattaneo S, *et al.* Cardiac repair with intramyocardial injection of allogeneic mesenchymal stem cells after myocardial infarction. *Proc Natl Acad Sci USA* 2005 Aug 9;102(32):11474–9.

73. Otsuru S, Tamai K, Yamazaki T, Yoshikawa H, Kaneda Y. Circulating bone marrow-derived osteoblast progenitor cells are recruited to the bone-forming site by the CXCR4/stromal cell-derived factor-1 pathway. *Stem Cells* 2008 Jan;26(1):223–34.

74. Lazarus HM, Haynesworth SE, Gerson SL, Rosenthal NS, Caplan AI. *Ex vivo* expansion and subsequent infusion of human bone marrow-derived stromal progenitor cells (mesenchymal progenitor cells): implications for therapeutic use. *Bone Marrow Transplant* 1995 Oct;16(4):557–64.

75. Koç ON, Gerson SL, Cooper BW, Dyhouse SM, Haynesworth SE, Caplan AI, *et al.* Rapid hematopoietic recovery after coinfusion of autologous-blood stem cells and culture-expanded marrow mesenchymal stem cells in advanced breast cancer patients receiving high-dose chemotherapy. *J Clin Oncol* 2000;18(2):307–16.

76. Horwitz EM, Prockop DJ, Fitzpatrick LA, Koo WW, Gordon PL, Neel M, *et al.* Transplantability and therapeutic effects of bone marrow-derived mesenchymal cells in children with osteogenesis imperfecta. *Nat Med* 1999;5(3):309–13.

77. Horwitz EM, Prockop DJ, Gordon PL, Koo WW, Fitzpatrick LA, Neel MD, *et al.* Clinical responses to bone marrow transplantation in children with severe osteogenesis imperfecta. *Blood* 2001 Mar 1;97(5):1227–31.

78. Le Blanc K, Gotherstrom C, Ringden O, Hassan M, McMahon R, Horwitz E, *et al.* Fetal mesenchymal stem-cell engraftment in bone after *in utero* transplantation in a patient with severe osteogenesis imperfecta. *Transplantation* 2005 Jun 15;79(11):1607–14.

79. Almeida-Porada G, Flake AW, Glimp HA, Zanjani ED. Cotransplantation of stroma results in enhancement of engraftment and early expression of donor hematopoietic stem cells *in utero*. *Exp Hematol* 1999 Oct;27(10):1569–75.

80. in 't Anker PS, Noort WA, Kruisselbrink AB, Scherjon SA, Beekhuizen W, Willemze R, *et al.* Nonexpanded primary lung and bone marrow-derived mesenchymal cells promote the engraftment of umbilical cord blood-derived CD34(+) cells in NOD/SCID mice. *Exp Hematol* 2003 Oct;31(10): 881–9.

81. Maitra B, Szekely E, Gjini K, Laughlin MJ, Dennis J, Haynesworth SE, *et al.* Human mesenchymal stem cells support unrelated donor hematopoietic stem cells and suppress T-cell activation. *Bone Marrow Transplant* 2004 Mar;33(6):597–604.

82. Le Blanc K, Samuelsson H, Gustafsson B, Remberger M, Sundberg B, Arvidson J, *et al.* Transplantation of mesenchymal stem cells to enhance engraftment of hematopoietic stem cells. *Leukemia* 2007 Aug;21(8):1733–8.

83. Tögel F, Hu Z, Weiss K, Isaac J, Lange C, Westenfelder C. Administered mesenchymal stem cells protect against ischemic acute renal failure through differentiation-independent mechanisms. *Am J Physiol Renal Physiol* 2005 Jul;289(1):F31–42.

84. Ringden O, Uzunel M, Sundberg B, Lonnies L, Nava S, Gustafsson J, *et al.* Tissue repair using allogeneic mesenchymal stem cells for hemorrhagic cystitis, pneumomediastinum and perforated colon. *Leukemia* 2007 Nov;21(11):2271–6.

85. Christopeit M, Schendel M, Foll J, Muller LP, Keysser G, Behre G. Marked improvement of severe progressive systemic sclerosis after transplantation of mesenchymal stem cells from an allogeneic haploidentical-related donor mediated by ligation of CD137L. *Leukemia* 2008 May;22(5):1062–4.

86. Kim SW, Han H, Chae GT, Lee SH, Bo S, Yoon JH, *et al.* Successful stem cell therapy using umbilical cord blood-derived multipotent stem cells for Buerger's disease and ischemic limb disease animal model. *Stem Cells* 2006 Jun;24(6):1620–6.

87. Hare JM, Traverse JH, Henry TD, Dib N, *et al.* A randomized, double-blind placebo controlled, dose escalation study of intravenous adult human mesenchymal stem cells (prochymal) after acute myocardial infarction. *J Am Coll Cardiol* 2009 Nov;54(24):2277–86.

9

Expansion of Mesenchymal Stem Cells (MSCs) for Clinical Use

Marie Prat-Lepesant, Marie-Jeanne Richard
and Jean-Jacques Lataillade

ABSTRACT

Mesenchymal stem cells (MSCs) are considered as emergent "universal" cells and various tissue repair programs using MSCs are today in development. The use of MSCs in such therapeutic strategies requires a large number of expanded cells and relies on their ability to maintain their own multilineage differentiation potential. MSCs expansion up to now almost completely relied on open preparations with multiple cell culture flasks. Moreover, these production processes are based on the use of fetal calf serum (FCS) and exogenous growth factors such as PDGF, FGF-2 or IGF-1 in an attempt to increase MSC proliferation rate. Today, a major challenge in producing cells for clinical use is the need to implement GMP compliant production processes to satisfy regulatory requirements. Some strategies are developed in this way using single-use materials like blood bags and transfer tubing, as well as closed culture devices validated in transfusion medicine. Furthermore, many studies demonstrated the advantage to replace FCS and exogenous recombinant growth factors by human platelet-derived growth factors.

It has been shown that platelet-derived growth factors were more efficient for *ex-vivo* expansion of MSCs than FCS: they promote MSC proliferation rates and decrease the time required to reach confluency. This completely closed process in serum-free conditions abrogates most safety concerns regarding microbial and prion contamination, which represents the major risk in autologous cell transplantation.

In the present chapter, we first describe the prerequisites for MSC isolation, culture and *in vitro* expansion: various sources of collection and different technical procedures of culture are discussed. Then the regulatory restrictions for a clinical-scale production of MSCs are detailed based on the French experience (traceability of ancillary products, closed medical devices, safe area, quality controls). Finally, an innovative technical procedure of clinical scale MSC production using clinical grade platelet lysate is described and production data are compared to the standard procedure using FCS and FGF-2.

INTRODUCTION

The fibroblastic subset of cells in the bone marrow (BM) stromal system is enriched for multipotent mesenchymal stem cells (MSCs) capable of self-maintenance and extensive proliferation as well as differentiation capability into cell types of the connective tissues such as bone, cartilage, tendon and marrow stroma.[1] More recently, conflicting evidence suggested that MSCs could also differentiate into cardiac and skeletal muscles[2] and neural cells.[3] Very early, it has been demonstrated that MSCs can be easily recovered and enriched through their property of adhering to tissue culture surfaces.[4] MSCs can be readily expanded many fold *in vitro* and can have many orthopedic and hematologically relevant applications such as large bone defects,[5] osteogenesis imperfecta,[6,7] hematopoietic stroma deficiency[8,9] and graft *versus* host disease (GvHD) prevention.[10] More recently, the trophic properties of MSCs have been successfully used in the treatment of radiation burns[11] and chemo-irradio-therapy damages.[12]

Although various sources can be considered, in most clinical trials, BM aspirates have served as the principle source of MSCs. However, the amount of primary MSCs obtained from BM (<0.01% of nucleated cells) is far too small to be of clinical relevance, as the use of MSCs in human therapeutic strategies requires a large number of expanded cells to obtain clinical doses of at least 2×10^6 MSCs per kg body weight of a patient. Therefore, *in vitro* expansion is necessary and growth conditions have to be optimized to yield large quantities of cells in the shortest period of time. The use of cellular products for therapy has been generally challenged by the necessity of bovine-derived sera and/or serum-derived products in the culture media.[13] Major hurdles for clinical use of MSCs are related to the use of such bovine-derived products. Thus, replacing fetal calf serum (FCS) by autologous human serum or human growth factors such as platelet-derived growth factor (PDGF) has become a focus of several groups in order to adhere to good manufacturing practice (GMP) guidelines and improve safety and efficiency of cell therapy.[14–17] By using such conditions, MSCs can be readily expanded several hundredfold in primary culture. Today, regulatory standards for GMP cell production are not clearly defined. In order to conduct large controlled clinical trials, quality assurance management specific to MSC production has to be established ensuring the demonstration of a quality control system with a defined workflow, in-process controls, well-defined technical procedures, specific release criteria and controlled clean room facilities consistent with the standards provided by European GMP requirements (European Medical Agency). In respect to these GMP requirements, cells must be expanded in completely "closed" CE-marked systems using sterile single use consumables based on blood banking technologies.

Problems that arise in technologies using multiple passaging of cells are genomic instability and deregulation of proliferation.[18] Therefore, quality controls of cell products must not only consider the phenotype, the functional properties and the microbiological contamination of the cells, but also their karyotypic and telomerase activity status.

TECHNICAL PREREQUISITES FOR *IN VITRO* MSC EXPANSION

For clinical-scale production of MSCs several parameters could influence the process and the yield: the source of MSCs (cord blood, bone marrow, adipose tissue, or peripheral blood), isolation and purification procedures, and the expansion technical parameters.

Source of MSCs

Bone marrow is the source of MSCs mainly used for clinical applications.[6,9,11,12] MSCs represent less than 0.01% of the nucleated bone marrow cell fraction and CFU-F frequency was estimated between 0.01% and 0.001% of the mononuclear cells.[4] The gender and the age of the donor have been described to influence the availability of MSCs: the number, the differentiation potential, and the maximal life span decline with increasing age.[19–21]

Although bone marrow-derived MSCs could be obtained from spongeous bone, for clinical applications bone marrow aspirates should be privileged regarding the feasibility and the asepsis of the procedure; 20 to 100 ml of bone marrow is usually sufficient to achieve MSCs engraftment.[6,9,11,12]

Other alternative sources for MSCs are also being considered. Two of them are fetal tissues[22] including umbilical cord blood (UCB),[23,24] umbilical cord (Wharton's jelly), placenta and adipose tissue (AT) which both should be of clinical interest because of less invasive harvesting procedures, compared to bone marrow. However, lower success rates for MSC generation were reported, with only approximately 60% of the cord blood units giving rise to MSCs (100% for BM and AT).[25] This could be due to the lower MSC frequency in UCB ($1/10^5$ to $1/10^8$, respectively) versus BM.[26]

Despite this lower yield, it has been demonstrated that fetal MSCs exhibit a higher proliferation capacity than bone marrow and adipose tissue-derived MSCs,[25] making this source very attractive for clinical scale MSC production. Moreover, the increasing numbers of UCB banks around the world could contribute to the rising trend of using cord blood as a source for MSCs.

Finally, other MSCs sources need to be mentioned such as the adult peripheral blood[27] and the amniotic fluid[28] which however do not as of yet represent reliable sources of MSCs for clinical production.

Isolation of MSCs

Three methods are widely used for MSC isolation based on their antigenic characteristics, adherence and density properties.

The mostly used method for isolating MSCs is their plastic adherence property, preceded by a density centrifugation separation allowing the isolation of the mononuclear cell fraction in which normally MSCs are found.

Because other cell types such as macrophages and endothelial cells also adhere to the plastic surface, normally several culture passages are performed to obtain a high final MSC purity. Whereas no universal antigenic profile is established for MSCs, some determinants could be used for magnetic bead separation such as CD105 (endoglin),[29] Stro 1,[30] CD49a (α1-integrin)[31] and nerve growth factor receptor (LNGFR; CD271).[32] For clinical applications, the immunomagnetic-based selection is not advisable because it is an expensive and time consuming process. Furthermore, clinical grade antibodies are not yet available.

For clinical grade UCB-derived MSCs production, we recommend to select MSCs from total cells by plastic adherence. The non-adherent hematopoietic residual cells are removed from the culture supernatant after one to three days.

Expansion of MSCs

For MSC clinical use, an *in vitro* expansion step is essential for obtaining a large enough number of cells in a short period of time. High expansion rate of clinical grade MSCs depends on several technical parameters such as culture medium, cell plating density, and number of passages, while differentiation of the cells must be avoided.

Culture medium

Optimal conditions for MSCs expansion require basic medium such as Dulbecco's modified eagle medium (DMEM) or alpha-modified eagle medium (α-MEM) supplemented with 10% or 20% fetal calf serum (FCS). Major hurdles for clinical use of MSCs are due to the use of FCS and non-clinical grade growth factors. The best FCS batch, in terms of expansion yield and kinetic growth, must be carefully evaluated. Although MSCs can be expanded in medium containing only FCS, for clinical applications, growth factors can be added to the medium for increasing the expansion rate. Among these growth factors, basic fibroblast growth factor (FGF-2) and platelet-derived growth factor (PDGF) are prime candidates. However their use is compromised because clinical-grade growth factors are not yet broadly available. Concerning the safety of the medium, the most critical risk is related to FSC viral, bacterial and prion contamination. Furthermore, another potential problem of FSC is the possible immune reaction against animal-derived proteins.[6,33] This may lead to non-engraftment or rejection of transplanted cells, especially if MSCs are administrated repeatedly.[17]

Altogether, these issues lead several regulatory authorities to impose some restrictions for clinical use of MSCs produced in presence of FCS. Consequently, replacing FCS has become a challenge of several teams in order to adhere to GMP guidelines and improve the safety of the cellular products. Therefore, important efforts are made to develop FCS-free alternatives. Recently Stute *et al.* have reported that 10% human autologous serum is at least as good as 10% FCS with regard to MSC expansion.[17]

Another alternative is the use of platelet-rich plasma-derived products. Platelets are known to produce multiple growth factors[34] and therefore could be very attractive for MSC expansion. In 2005 we have proposed an innovative culture system in which serum and growth factors required for MSC expansion were provided by human platelet lysate (PL).[14] We demonstrated that growth factors delivered in the culture medium from 5% PL were able to promote MSC expansion and decreased time to reach confluency, as compared to a combination of FCS and FGF-2. Advantages of PL for MSC expansion to clinically relevant quantities have been further confirmed by several other groups.[15,16,35–37] Interestingly, PL has been recently used for propagating cord blood-derived MSCs demonstrating the capacity of PL to expand MSCs from various sources.[16]

For the first time, this FCS-free culture medium has been recently applied to GMP-compliant expansion of autologous MSCs before their local implantation for the treatment of radiation burns.[11]

Cell plating density

The density of cell plating is a key parameter for warranting high level expansion rates while maintaining multipotent properties of MSCs. In primary culture, 1.5×10^5 to 1×10^6 mononuclear cells/cm^2 are usually plated. It has been demonstrated by Prockop's group that high yields of early progenitors were promoted when MNCs were plated at the very low density of 1.5×10^5.[38] After three days of culture, the medium has to be replaced for removing non-adherent hematopoietic cells. The medium is changed twice weekly thereafter. Once 80% confluency is reached, adherent MSCs are reseeded at a density varying from 1×10^4 to 1×10^3 cells/cm^2. It is well established that the highest proliferation rates are obtained when cells are plated at low density. For clinical production, a plating density of 1×10^3 to 3×10^3 cells/cm^2 seems suitable, balancing the high cells numbers one needs to obtain and limiting space for culture vessels. For UCB-derived MSC isolation and expansion, the cell plating densities do not differ from the parameters used for BM.[25]

Passage number

MSCs are characterized, like other adherent cells, by growth inhibition when they reach confluency. In primary culture, MSCs are contaminated by hematopoietic cells such as monocytes, macrophages, lymphocytes and endothelial cells. Non-adherent cells can be removed with sequential changes of media. Subcultures are thus required for avoiding adherent residual hematopoietic-derived cells and obtaining large amounts of pure MSCs. High number of passages can, however, modify the hallmark features of MSCs. It has been demonstrated that, after numerous passages, late MSCs exhibit lower growth kinetics, and progressively lose their multipotentiality.[39,40] For clinical applications, MSCs from early passages (one to two passages) should be preferred even if it is necessary to accomplish several bone marrow harvests.

The impact of oxygen tension

Oxygen tension is a critical parameter for homeostatic regulation of stem cell niches. Therefore, variations in environmental oxygen concentrations could affect the expansion and differentiation properties of MSCs. MSCs cultured under hypoxia (2% O_2) in tridimensional constructs maintained their primitive characteristics exhibiting higher CFU-F capabilities and higher levels of osteoblastic and adipocitic differentiation markers than in normoxic conditions (20%).[24]

The suitability of cultured MSCs for use in tissue engineering depends not only on their lineage commitment, but also on their 3D organization. Similar to articular chondrocytes, a 3D culture environment (for example micromass or alginate gel) is necessary to maintain the cartilage or disc cell phenotype *in vitro*. Hyaluronic acid, collagen, and chitosan have been used as scaffold materials to provide immobilization of MSCs in a 3D framework that maintains their cartilage or Nucleus Pulposus-like phenotype.[41] These 3D *in vitro* culture techniques allow for stable cell anchorage and permit independent production of intercellular matrix, cytokines, and bioactive molecules.[42]

REGULATORY RECOMMENDATIONS FOR CLINICAL-SCALE PRODUCTION OF MSCs

Clinical grade MSCs are used after isolation, expansion and sometimes differentiation or other manipulations. Although the amount of product is limited compared to standard pharmaceutical production, expanded cells are considered drugs from a regulatory perspective. As a consequence, GMP-compliant MSC production has to be transferred from clinical research laboratories into specially equipped facilities (cell processing laboratories which are able to produce according to GMP standards).

GMP regulations apply to all phases of cell collection, processing and storage as well as documentation, training personnel and the laboratory facility. A number of national or international guidelines, regulations, and federal laws have to be considered (e.g., Directive 2004/23/EC of the European Parliament and of the European Council). One prerequisite for regulatory approval is the appointment of qualified individuals responsible for production, quality control, distribution, and monitoring of the product's clinical safety.

Besides specific personnel requirements, GMP conditions include building and facility requirements for cell processing. MSCs expanded for clinical use have to be processed in clean room facilities according to specific guidelines of GMP for the manufacturing of cells for use in human. Within the clean room facility, the air is filtered to meet criteria for a certain air cleanliness level. The clean room facility should fulfill the air cleanliness classification class C or B (10,000 to 1000 particles > 0.5 $\mu m/ft^3$, according to US federal standard No 209B) and of class A (100 according to 209B; Federal Standard No. 209B. 1992. Clean Room and Work Station Requirements, Controlled Environment. April 24, 1973). The aim is to avoid contamination and to ensure sterility and safety of the product. Finally, GMP regulations also define that within the clean room production unit only one production process is allowed at a time to avoid cross-contamination among different samples.

Another major area of GMP are standard operating procedures (SOP) that have to exist for all manufacturing stages as well as documentation for in-process controls, cleaning, training, and product release. Methods and procedures have to be described precisely in the SOP and personnel have to strictly follow these SOPs. This represents a major difference compared to basic research related expansion of cells.

To ensure the continued safety of the cell product and to prevent the release of unsuitable MSC preparations, a number of in-process control assays have to be performed. These include routine testing for sterility at different critical points, determination of cell number and viability, and flow cytometry analysis to characterize the cells in the preparation. Briefly, anti-CD45 and anti-CD14 antibodies are used to identify contaminating hematopoietic cells, and anti-CD105 (SH2), anti-CD73 (SH3, SH4) and anti-CD90 (Thy-1) are used as specific markers for MSCs. MSCs express other antigens, including cell surface markers such as CD49a, CD166, Stro-1 and regardless of the medium some growth factor receptors, which could be analyzed to complete the MSC phenotype profile.[43,44] A functional test is recommended. The standard *in vitro* assay for MSCs is the colony forming unit fibroblast assay (CFU-F) in which cells are plated at low density, expanded as an adherent population and quantified by scoring individual colonies presumed to derive from a single precursor (Table 1).[38,45,46] In addition, acceptable range limits for MSC-cell products (cell number, viability, phenotype) have to be specified. Some quality controls could be added to ensure safety of cells e.g., karyotype and telomerase activity of the stem cells (Table 2).

Table 1. Quality controls to perform regardless of the different steps of the process: the attempted specifications contribute to ensure that the *ex vivo* generated cells are safe and defined as MSCs.

Controls	Microbiological control	Cell number	Viability	Phenotypic characterization	CFU-F assay
Bone marrow	Sterility (absence of aerobic and anaerobic germs)	According to the experimental design	>90%		No established specifications 0.001%–0.01% of nucleated cells
MSC subculture 0	Sterility (absence of aerobic and anaerobic germs)	According to the experimental design	>80%	Less than 1% of CD45 and CD14 expressing cells More than 90% of CD73, CD105, CD90 expressing cells	No established specifications Colony number depends on plating density
MSC subculture 1	Sterility (absence of aerobic and anaerobic germs)	According to clinical trial: $2–5 \times 10^6$ MSCs/kg	>80%	More than 90% of CD73, CD105, CD90 expressing cells	No established specifications
Thawed cells before transplantation	Sterility (absence of aerobic and anaerobic germs)	According to clinical trial: $2–5 \times 10^6$ MSCs/kg	>60%	More than 90% of CD73, CD105, CD90 expressing cells	No established specifications

Table 2. Specific controls related to stem cell characteristics and performed on the final MSC product.

Specific controls	Specifications
Karyotype	No genetic alterations
Human Telomerase Reverse Transcriptase (h-TERT)	No expression

Based on the results of these assays, the release of the final cell product is decided. Additionally, the potential of cells to differentiate under defined conditions *in vitro* into chondrocytes, osteoblasts and adipocytes could be investigated[43,47] (Table 3). Regarding the immunomodulatory properties of MSCs, some controls could be added: expression of HLA-DR, inhibition of T-cell proliferation by MSCs in mixed lymphocytes reactions (MLR)[46,48] (Table 4).

Additional activities to comply with GMP regulations include: validation of all methods, survey of room surfaces and air to monitor the fulfillment of air cleanliness criteria, release procedure for lab material prior to use, regular self-inspections, documentation of each production and patient-related data. All raw data (e.g., results of in-process control, validation studies, storage) have to be archived. In addition, key equipment

Table 3. Specific controls related to differentiation capacity of MSCs *in vitro*: these controls contribute to validate the functionality of the *in vitro* generated MSCs.

Specific controls	Specifications
Osteogenic differentiation	Incubation of an MSC confluent monolayer with ascorbic acid, beta-glycerophosphate and dexamethasone for 2–3 weeks gives rise to osteoblasts[14]
Chondrogenic differentiation	Micro mass cultures in presence of TGF-β give rise to chondrocytes[14]
Adipogenic differentiation	Incubation of an MSC confluent monolayer with isobutyl methyl xanthine, indomethacin and beta-glycerophosphate for 2–3 weeks gives rise to adipocytes[14]

Table 4. Specific controls related to immune properties of allogenic MSC.

Specific controls	Specifications
HLA-DR	No expression on cells cultured without FGF-2
	The expression depends on FGF-2 concentration
	mean fluorescence intensity remains low
Reactivity in MLR	MSCs inhibit T lymphocytes proliferation
	No alloreactivity between the patient's PBMCs
	and the donors' MSCs

such as incubators, liquid or vapor nitrogen tanks should be consistently monitored for adequate functionality. The materials used (e.g., bags, culture medium, FCS, cytokines or growth factors) should be of GMP grade, whenever possible. Otherwise, certificates of analysis for the specific product lots used are required. Ideally, all animal and human-derived products should be excluded and synthetic recombinant alternatives used instead.

MSC CULTURE PROCESS: AN INNOVATIVE SERUM-FREE MEDIUM SUITED FOR CLINICAL APPLICATIONS

In an attempt to set up GMP-grade MSC production several key parameters have to be defined: the cell seeding concentration, the safety and traceability of ancillary products (medium, serum, growth factors), as closed as possible cell culture devices and the number of passages to perform. Although the majority of production protocols today use media containing FCS and cytokines in open culture systems, an increasing effort is made to develop serum-free cell culture conditions. These processes are mainly based on platelet-derived products and their related growth factors and closed culture devices, especially in France where regulatory requirements are very strict. We have developed an innovative MSC expansion procedure based on the replacement of FCS by 5% platelet lysate (PL).[14] Recently, we have scaled-up our protocol for large production of GMP-grade MSCs. For this purpose, we have devised a clinical-grade process of preparing PL by using a closed CE-marked system.

Very recently this serum-free culture process was used in a GMP-compliant production of MSCs for grafting in a severe accidental radiation burn syndrome.[11] In the following, we present the main steps of this culture process including the preparation and validation of the clinical-grade PL.

PREPARATION OF CLINICAL-GRADE PLATELET LYSATE FOR GMP-COMPLIANT MSC PRODUCTION

Clinical-grade platelet lysate (c-PL) is obtained from platelet apheresis collected with informed consent from selected regular donors. In accordance with French and European legislation, donors are tested for HIV, HCV, HBV, Syphilis, CMV, and HTLV infection. Platelet number of each platelet-rich plasma (PRP) product is automatically measured and only samples containing more than 1×10^9 platelets/ml are selected. Immediately after collection the PRP product is divided into several mini-bags containing 25 ml PRP (50 ml EVA bags, MacoPharma Biotech). Each mini-bag is equipped with connecting tubes for sterile transfer of PRP into the MEM medium containing bag (Fig. 1).

PRP mini-bags are then stored at −40°C to obtain release of platelet-derived growth factors. A biological qualification of the resulting c-PL is performed on one aliquot of each PRP product. This qualification includes a sterility testing and an MSC proliferation assay in which the ability of 5%

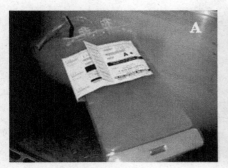

Fig. 1. PRP apheresis product (**A**) is divided in several mini-bags (**B**) containing 25 ml PRP (MacoPharma Biotech) and equipped with connecting tubes for sterile transfer of PRP in MEM medium bag.

c-PL to expand MSCs is compared to a medium containing 10% FCS. The batch of c-PL is validated for clinical use when serological testing of the donor is negative, tests of sterility are negative and when c-PL gives MSC proliferation rates at least as good as those obtained in FCS condition. This validation is performed on three independent proliferation assays for three MSC batches derived from three different sources. When a c-PL batch is functionally validated for MSC expansion, the resulting cells are phenotypically characterized and tested for their osteogenic, chondrogenic and adipogenic differentiation capability.

LARGE SCALE GMP-COMPLIANT MSC PRODUCTION

A GMP-manufactured large surface culture vessel (two-level 1,272 cm^2 CellStack$^®$; Corning, MacoPharma Biotech) is connected to a bag containing alpha-MEM medium in which 10 µg/ml ciprofloxacin (Ciflox$^®$ 400 mg/200 ml), 2 UI/ml heparin and thereafter 5% to 8% c-PL are added; 300 ml of this medium is then inoculated into the 1,272 cm^2 CellStack$^®$ (Fig. 2).

Finally, bone marrow nucleated cells are seeded in the cell culture device at a density of 200×10^3 cells/cm^2 and cultured at 37°C in 95% air and 5% CO_2. After three to four days, the non-adherent cells were removed and the culture is re-fed with fresh medium by means of bags equipped

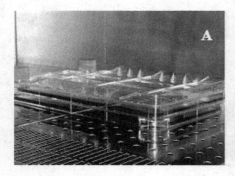

Fig. 2. Large scale GMP-compliant MSC production in a two-level 1,272 cm^2 CellStack$^®$ (**A**). A bag containing 300 ml alpha-MEM medium, 10 µg/ml ciprofloxacin, 2 UI/ml heparin, and 5%–8% c-PL is connected to the 1,272 cm^2 CellStack$^®$ (**B**).

Table 5. Large scale GMP compliant MSC production results.

Passage	Input of BMMNCs and MSCs/CellStack	Total MSC production/ CellStack	Culture duration (days)	Expansion rate
P0 (n = 5)	$63.8 \times 10^6 \pm$ 0.2×10^6	$34.49 \times 10^6 \pm$ 8.4×10^6	16.6 ± 1.2	0.54 ± 0.13
P1 (n = 4)	$2.5 \times 10^6 \pm 0$	$35.79 \times 10^6 \pm$ 17.6×10^6	12.75 ± 2.9	14.31 ± 7.04

with connecting tubes. On day 15–16, cells reach 80% confluence. Cells are harvested after trypsin-EDTA application for five minutes at 37°C and plated at 4×10^3 cells/cm^2 (first passage = P0). Harvesting and replating are performed by using connected bags. Seven to ten days later P0 cells reach confluence. Results of a large scale MSC production obtained in a 636 cm² CellStack® with a 8% c-PL medium are given in Table 5.

CONCLUSIONS

In conclusion, MSCs were first obtained from the bone marrow (BM) but they can also be isolated from umbilical cord blood and adipose tissue through *in vitro* expansion in medium containing either fetal calf serum (FCS), supplemented or not with fibroblast growth factor 2 (FGF-2), or platelet lysate (PL) as an FCS substitute. Their clinical-scale production according to GMP requirements is now possible. However, little is known about the consequences of donor selection, starting tissue material, or culture conditions on the functional properties and therapeutic potentials of clinical-grade MSCs and this lack of standardization is a major pitfall to address both the benefits and potential risks of this new therapeutic alternative. Before starting such a production, it is important to define precisely all of the parameters including controls, and to use traceable reagents and devices. A related concern is the capacity of MSCs for oncogenic transformation that requires specific controls such as karyotypes and tumorigenicity-related gene expression such as c-Myc, p16 and p21. Taking into account the previously described security requirements, it is both feasible and reasonable to begin production of MSCs for large Phase II clinical studies.

References

1. Caplan AI. Mesenchymal stem cells. *J Orthop Res* 1991 Sep;9(5):641–50.
2. Makino S, Fukuda K, Miyoshi S, Konishi F, Kodama H, Pan J, *et al.* Cardiomyocytes can be generated from marrow stromal cells *in vitro. J Clin Invest* 1999 Mar;103(5):697–705.
3. Hermann A, Gastl R, Liebau S, Popa MO, Fiedler J, Boehm BO, *et al.* Efficient generation of neural stem cell-like cells from adult human bone marrow stromal cells. *J Cell Sci* 2004 Sep 1;117(Pt 19):4411–22.
4. Friedenstein AJ, Gorskaja JF, Kulagina NN. Fibroblast precursors in normal and irradiated mouse hematopoietic organs. *Exp Hematol* 1976 Sep;4(5):267–74.
5. Quarto R, Mastrogiacomo M, Cancedda R, Kutepov SM, Mukhachev V, Lavroukov A, *et al.* Repair of large bone defects with the use of autologous bone marrow stromal cells. *N Engl J Med* 2001 Feb 1;344(5):385–6.
6. Horwitz EM, Gordon PL, Koo WK, Marx JC, Neel MD, McNall RY, *et al.* Isolated allogeneic bone marrow-derived mesenchymal cells engraft and stimulate growth in children with osteogenesis imperfecta: implications for cell therapy of bone. *Proc Natl Acad Sci USA* 2002 Jun 25;99(13):8932–7.
7. Horwitz EM, Prockop DJ, Fitzpatrick LA, Koo WW, Gordon PL, Neel M, *et al.* Transplantability and therapeutic effects of bone marrow-derived mesenchymal cells in children with osteogenesis imperfecta. *Nat Med* 1999 Mar;5(3):309–13.
8. Fouillard L, Bensidhoum M, Bories D, Bonte H, Lopez M, Moseley AM, *et al.* Engraftment of allogeneic mesenchymal stem cells in the bone marrow of a patient with severe idiopathic aplastic anemia improves stroma. *Leukemia* 2003 Feb;17(2):474–6.
9. Koc ON, Gerson SL, Phillips GL, Cooper BW, Kutteh L, Van Zant G, *et al.* Autologous CD34$^+$ cell transplantation for patients with advanced lymphoma: effects of overnight storage on peripheral blood progenitor cell enrichment and engraftment. *Bone Marrow Transplant* 1998 Feb;21(4):337–43.
10. Le Blanc K, Rasmusson I, Sundberg B, Gotherstrom C, Hassan M, Uzunel M, *et al.* Treatment of severe acute graft-versus-host disease with third party haploidentical mesenchymal stem cells. *Lancet* 2004 May 1;363(9419):1439–41.
11. Lataillade JJ, Doucet C, Bey E, Carsin H, Huet C, Clairand I, *et al.* New approach to radiation burn treatment by dosimetry-guided surgery combined with autologous mesenchymal stem cell therapy. *Regen Med* 2007 Sep;2(5):785–94.

12. Le Blanc K, Samuelsson H, Gustafsson B, Remberger M, Sundberg B, Arvidson J, *et al.* Transplantation of mesenchymal stem cells to enhance engraftment of hematopoietic stem cells. *Leukemia* 2007 Aug;21(8):1733–8.
13. Kuznetsov SA, Mankani MH, Robey PG. Effect of serum on human bone marrow stromal cells: *ex vivo* expansion and *in vivo* bone formation. *Transplantation* 2000 Dec 27;70(12):1780–7.
14. Doucet C, Ernou I, Zhang Y, Llense JR, Begot L, Holy X, *et al.* Platelet lysates promote mesenchymal stem cell expansion: a safety substitute for animal serum in cell-based therapy applications. *J Cell Physiol* 2005 Nov;205(2):228–36.
15. Lange C, Cakiroglu F, Spiess AN, Cappallo-Obermann H, Dierlamm J, Zander AR. Accelerated and safe expansion of human mesenchymal stromal cells in animal serum-free medium for transplantation and regenerative medicine. *J Cell Physiol* 2007 Oct;213(1):18–26.
16. Reinisch A, Bartmann C, Rohde E, Schallmoser K, Bjelic-Radisic V, Lanzer G, *et al.* Humanized system to propagate cord blood-derived multipotent mesenchymal stromal cells for clinical application. *Regen Med* 2007 Jul;2(4):371–82.
17. Stute N, Holtz K, Bubenheim M, Lange C, Blake F, Zander AR. Autologous serum for isolation and expansion of human mesenchymal stem cells for clinical use. *Exp Hematol* 2004 Dec;32(12):1212–25.
18. Rubio D, Garcia-Castro J, Martin MC, de la Fuente R, Cigudosa JC, Lloyd AC, *et al.* Spontaneous human adult stem cell transformation. *Cancer Res* 2005 Apr 15;65(8):3035–9.
19. Mueller SM, Glowacki J. Age-related decline in the osteogenic potential of human bone marrow cells cultured in three-dimensional collagen sponges. *J Cell Biochem* 2001;82(4):583–90.
20. Nishida S, Endo N, Yamagiwa H, Tanizawa T, Takahashi HE. Number of osteoprogenitor cells in human bone marrow markedly decreases after skeletal maturation. *J Bone Miner Metab* 1999;17(3):171–7.
21. Stenderup K, Justesen J, Clausen C, Kassem M. Aging is associated with decreased maximal life span and accelerated senescence of bone marrow stromal cells. *Bone* 2003 Dec;33(6):919–26.
22. Jo C, Kim O-S, Park E-Y, Kim B, Lee J-H, Kang S-B, *et al.* Fetal mesenchymal stem cells derived from human umbilical cord sustain primitive characteristics during extensive expansion. *Cell Tissue Res* 2008;334(3): 423–33.

23. Erices A, Conget P, Minguell JJ. Mesenchymal progenitor cells in human umbilical cord blood. *Br J Haematol* 2000 Apr;109(1):235–42.

24. Grayson WL, Zhao F, Izadpanah R, Bunnell B, Ma T. Effects of hypoxia on human mesenchymal stem cell expansion and plasticity in 3D constructs. *J Cell Physiol* 2006 May;207(2):331–9.

25. Kern S, Eichler H, Stoeve J, Kluter H, Bieback K. Comparative analysis of mesenchymal stem cells from bone marrow, umbilical cord blood, or adipose tissue. *Stem Cells* 2006 May;24(5):1294–301.

26. Bieback K, Kern S, Kluter H, Eichler H. Critical parameters for the isolation of mesenchymal stem cells from umbilical cord blood. *Stem Cells* 2004;22(4):625–34.

27. Roufosse CA, Direkze NC, Otto WR, Wright NA. Circulating mesenchymal stem cells. *Int J Biochem Cell Biol* 2004 Apr;36(4):585–97.

28. In 't Anker PS, Scherjon SA, Kleijburg-van der Keur C, Noort WA, Claas FH, Willemze R, *et al.* Amniotic fluid as a novel source of mesenchymal stem cells for therapeutic transplantation. *Blood* 2003 Aug 15;102(4):1548–9.

29. Majumdar MK, Banks V, Peluso DP, Morris EA. Isolation, characterization, and chondrogenic potential of human bone marrow-derived multipotential stromal cells. *J Cell Physiol* 2000 Oct;185(1):98–106.

30. Gronthos S, Simmons PJ. The growth factor requirements of STRO-1-positive human bone marrow stromal precursors under serum-deprived conditions *in vitro*. *Blood* 1995 Feb 15;85(4):929–40.

31. Deschaseaux F, Gindraux F, Saadi R, Obert L, Chalmers D, Herve P. Direct selection of human bone marrow mesenchymal stem cells using an anti-CD49a antibody reveals their CD45med, low phenotype. *Br J Haematol* 2003 Aug;122(3):506–17.

32. Quirici N, Soligo D, Bossolasco P, Servida F, Lumini C, Deliliers GL. Isolation of bone marrow mesenchymal stem cells by anti-nerve growth factor receptor antibodies. *Exp Hematol* 2002 Jul;30(7):783–91.

33. Spees JL, Gregory CA, Singh H, Tucker HA, Peister A, Lynch PJ, *et al.* Internalized antigens must be removed to prepare hypoimmunogenic mesenchymal stem cells for cell and gene therapy. *Mol Ther* 2004 May;9(5):747–56.

34. Sanchez AR, Sheridan PJ, Kupp LI. Is platelet-rich plasma the perfect enhancement factor? A current review. *Int J Oral Maxillofac Implants* 2003 Jan–Feb;18(1):93–103.

35. Bernardo ME, Avanzini MA, Perotti C, Cometa AM, Moretta A, Lenta E, *et al*. Optimization of *in vitro* expansion of human multipotent mesenchymal stromal cells for cell-therapy approaches: further insights in the search for a fetal calf serum substitute. *J Cell Physiol* 2007 Apr;211(1):121–30.

36. Capelli C, Domenghini M, Borleri G, Bellavita P, Poma R, Carobbio A, *et al*. Human platelet lysate allows expansion and clinical grade production of mesenchymal stromal cells from small samples of bone marrow aspirates or marrow filter washouts. *Bone Marrow Transplant* 2007 Oct;40(8):785–91.

37. Mirabet V, Solves P, Minana MD, Encabo A, Carbonell-Uberos F, Blanquer A, *et al*. Human platelet lysate enhances the proliferative activity of cultured human fibroblast-like cells from different tissues. *Cell Tissue Bank* 2008 Mar;9(1):1–10.

38. Sekiya I, Larson BL, Smith JR, Pochampally R, Cui JG, Prockop DJ. Expansion of human adult stem cells from bone marrow stroma: conditions that maximize the yields of early progenitors and evaluate their quality. *Stem Cells* 2002;20(6):530–41.

39. Muraglia A, Cancedda R, Quarto R. Clonal mesenchymal progenitors from human bone marrow differentiate *in vitro* according to a hierarchical model. *J Cell Sci* 2000 Apr;113(Pt 7):1161–6.

40. Vacanti V, Kong E, Suzuki G, Sato K, Canty JM, Lee T. Phenotypic changes of adult porcine mesenchymal stem cells induced by prolonged passaging in culture. *J Cell Physiol* 2005 Nov;205(2):194–201.

41. Risbud M, Ringe J, Bhonde R, Sittinger M. *In vitro* expression of cartilage-specific markers by chondrocytes on a biocompatible hydrogel: implications for engineering cartilage tissue. *Cell Transplant* 2001;10(8):755–63.

42. Risbud M, Albert T, Guttapalli A, Vresilovic E, Hillibrand A, Vaccaro A, *et al*. Differentiation of mesenchymal stem cells towards a nucleus pulposus-like phenotype *in vitro*: implications for cell-based transplantation therapy. *Spine* 2004;29(23):2627–32.

43. Chamberlain G, Fox J, Ashton B, Middleton J. Concise review: mesenchymal stem cells: their phenotype, differentiation capacity, immunological features, and potential for homing. *Stem Cells* 2007 Nov;25(11):2739–49.

44. Deans RJ, Moseley AB. Mesenchymal stem cells: biology and potential clinical uses. *Exp Hematol* 2000 Aug;28(8):875–84.

45. Horwitz EM, Le Blanc K, Dominici M, Mueller I, Slaper-Cortenbach I, Marini FC, *et al*. Clarification of the nomenclature for MSC: the

International Society for Cellular Therapy position statement. *Cytotherapy* 2005;7(5):393–5.

46. Sotiropoulou PA, Perez SA, Salagianni M, Baxevanis CN, Papamichail M. Characterization of the optimal culture conditions for clinical scale production of human mesenchymal stem cells. *Stem Cells* 2006 Feb;24(2):462–71.

47. Gregory CA, Prockop DJ, Spees JL. Non-hematopoietic bone marrow stem cells: molecular control of expansion and differentiation. *Exp Cell Res* 2005 Jun 10;306(2):330–5.

48. Plumas J, Chaperot L, Richard MJ, Molens JP, Bensa JC, Favrot MC. Mesenchymal stem cells induce apoptosis of activated T-cells. *Leukemia* 2005 Sep;19(9):1597–604.

PART III

Other Stem and Progenitor Cells in Cord Blood and Surrounding Tissue

10

The Unrestricted Somatic Stem Cell (USSC)

Gesine Kögler, Teja F. Radke and Rüdiger V. Sorg

ABSTRACT

Although unrestricted somatic stem cells (USSCs) from cord blood show a very homogeneous immunophenotype even after extensive amplification *in vitro*, our functional data suggest heterogeneity and the presence of a hierarchy of cells within such USSC populations and other non-adherent CD45⁻ populations. Certain USSC preparations reveal multipotency, associated with the lack of markers of senescence, long telomeres and high proliferative capacity. Others, in contrast, show a more restricted differentiation potential e.g., the cells can only form cell types of the mesenchymal lineage and are also limited in expansion capacity. We were able to define functional and transcriptome markers to dissect the hierarchy of USSCs and MSCs in cord blood and to highlight differences with bone marrow mesenchymal stroma cells. Based on these characterizations, USSC lines were generated and cultivated under GMP-conditions achieving 1.5×10^9 cells in passage 3 to 4 corresponding to ten to 20 population doublings. Possible clinical and preclinical applications towards cardiac regeneration and support of hematopoiesis are the focus of this chapter.

INTRODUCTION

In 1999 our group for the first time described osteoblast precursor cells derived from human umbilical cord blood (UCB).[1] In 2000, adherently growing, fibroblast-like cells derived from cord blood (UCB) were identified, which revealed an immunophenotype (CD45[-], CD13[+], CD29[+], CD73[+], CD105[+]) similar to bone marrow (BM)-derived mesenchymal stroma cells (MSCs).[2] Soon thereafter, Campagnoli et al. confirmed the MSC nature of such cells from fetal blood by showing that they have the potential to differentiate into osteocytes, adipocytes and chondrocytes.[3] The mesodermal differentiation potential of cord blood-derived MSCs has also been confirmed by other groups.[4-6] In addition to this classical three-mesodermal-lineage potential of typical MSCs (osteogenic, adipogenic and chondrogenic), UCB-derived MSC-like cells also differentiate to cells with features of skeletal muscle.[7] However, only about 50% of the cells became positive for myosin heavy chain during differentiation. This raises the question whether a common progenitor gives rise to the different lineage committed cells or distinct progenitors, which are either unrelated or part of a stem cell hierarchy consisting of cells with increasingly restricted differentiation potential, similar to what has been suggested for BM-MSCs by Charbord et al.[8] This issue becomes even more complex because other authors have described UCB-derived non-hematopoietic stem cell populations, which show an even wider differentiation spectrum that spans across different germ layers. Goodwin et al.[9] described a cell population, which can give rise to cells with features of adipocytes, osteocytes or neural cells, thus, cell types belonging to the mesodermal and ectodermal germ layer. Generation of cells with neuronal as well as glial marker gene expression has also been reported by others.[10-12] However, since in most studies only differentiation potentials for individual or few lineages were analyzed, no conclusions can be drawn on the overall differentiation potential of the particular cell population studied in individual labs. Interestingly, neural lineage commitment was high in one study (81% of cells showed features of neural differentiation[12] and in another study, differentiation along the adipocytic, osteogenic and neural pathways has been implicated on a clonal level,[10] suggesting that in UCB there is a common progenitor, which can give rise to cells of mesodermal as well as ectodermal lineages. Thus non-hematopoietic stem cells in UCB appear not to be simply MSCs as shown by our group.[13,14] Moreover, there was also evidence that UCB-derived cells differentiate to hepatocytes, which are normally a derivative of the endodermal germ layer. When mononuclear cells from UCB were transplanted into NOD/SCID mice, human hepatocytes, which were not a result of cell fusion were detected in the liver, producing human albumin detectable in the sera of the mice.[15,16] Although in these studies the identity of the hepatocyte progenitors in the MNC fraction of cord blood was not determined and it is unknown whether these progenitors might also have generated other tissue types, Hong *et al.*[17] reported hepatic *in vitro* differentiation (again only about 50% of cells acquired the features of hepatocytes) of a population of UCB-derived cells similar to those cells for which myogenic and neural differentiation[18] had been described by others. This suggests that in UCB there is a cell with the potential to generate tissues of endodermal, ectodermal and mesodermal origin. Indeed, we have identified such a multipotent cell, which we have termed unrestricted somatic stem cell (USSC) (Fig. 1).[19,20] USSCs have the potential to differentiate into mesodermal osteoblasts, chondroblasts, adipocytes, hematopoietic cells, and

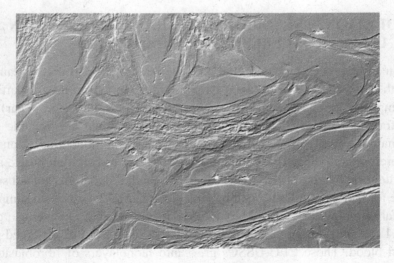

Fig. 1. Characteristics of USSCs: Spindle-shaped USSCs plated at low density.

cardiomyocytes, as well as into ectodermal derivatives such as all three neural lineages and into endodermal hepatic cells.[21,22] Chondrogenic, adipogenic, osteogenic, neural and hepatic differentiation potential of UCB-derived non-hematopoietic stem cells on a clonal level by limiting dilution analyses has also been shown by Lee *et al.*[23] USSCs can be differentiated easily towards MSCs with restricted differentiation potential, by specific cell culture conditions (see below). This spontaneous mesenchymal specification *in vitro* may be the explanation why several groups failed to detect a wider differentiation potential and have regarded non-hematopoietic UCB-derived stem cells as early mesodermal cells or simply as MSCs. Markers, which delineate such differences, are currently unknown and all immunophenotypic studies describe a very similar CD marker profile. Thus, the discrimination of unrestricted and mesodermally committed, or possibly cells committed to other lineages/germ layers, currently relies on the determination of the differentiation potential of the cells, similar to what is still true for the hematopoietic stem and progenitor cell compartment.[24,25] Nevertheless, these results suggest that there is a hierarchy in the non-hematopoietic stem cells from UCB, and it will be important to elucidate this hierarchy in more detail, which includes further delineation of the individual differentiation pathways.

ISOLATION, EXPANSION AND CHARACTERIZATION OF USSCs FROM FRESH CORD BLOOD

Collection, processing and initial characterization of UCB units obtained by the José Carreras cord blood bank in Düsseldorf from its 110 participating collection sites/maternity hospitals (8000 units per year) is performed routinely. Only about 30% of these units are suitable for hematopoietic banking, mainly due to cell number limitations.[26] Therefore, sufficient UCB units are available for research purposes, and the donor mothers consent to the use for research if donations are not suitable for banking. In USSC cultures initiated from 772 cord blood samples so far only 43% of the cord bloods gave rise to USSCs. On average, one to 11 colony forming units USSCs (CFU-USSCs) could be observed per cord blood. Those CFU-USSCs grew into monolayers of fibroblastoid, spindle-shaped cells within two to three weeks (Fig. 1).

Passage (% USSC lines)

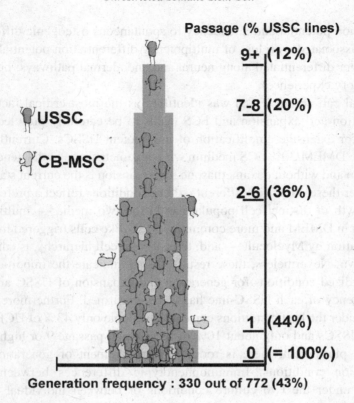

9+ (12%)

7-8 (20%)

USSC

CB-MSC

2-6 (36%)

1 (44%)

0 (= 100%)

Generation frequency : 330 out of 772 (43%)

Fig. 2. The generation frequency of USSCs is at present 43%. However, of all lines established (=100%), only 44% and 36% reach passages 1 and 2 to 6, respectively and only 20% passage 7 to 8; 12% can be expanded beyond passage 9. This may reflect a stem cell hierarchy (USSC vs. MSC) or biological aging of the cells.

Once established (if CFU were observed), 44%, 36%, 20% and 12% of USSCs reached one, two to six, seven to eight and beyond nine passages, respectively (Fig. 2). The colonies reaching passage 9, frequently could be expanded for more than 20 passages, yielding up to 10^{15} total cells.[20] Culture conditions influenced the frequency of generation and the overall expansion potential. In the presence of Myelocult (Stem Cell Technologies Inc., Vancouver, BC, Canada) and 10^{-7} M dexamethasone, USSC cultures were initiated from 57% of UCB samples, compared to 43% in DMEM/10^{-7}M dexamethasone. Although the frequency was higher in Myelocult medium, Myelocult grown cells showed a reduced

expansion potential and a tendency to spontaneous osteogenic differentiation, associated with loss of multipotency differentiation potential (cells
no longer differentiated along neural and endodermal pathways; our own
laboratory experience).

Fetal calf serum (FCS) was identified as the most critical factor for
generation and expansion and FCS needs to be carefully preselected to
allow for extensive amplification of multipotent USSCs. Currently, low
glucose DMEM/30% FCS medium with dexamethasone in the generation
medium and without dexamethasone for expansion is the current standard.
Whether these effects of different culture conditions reflect a preferential
outgrowth of distinct cell populations in the two media — multipotent
USSCs in DMEM and more committed MSC-like cells triggered into differentiation by Myelocult — and, thus, a stem cell hierarchy, is currently
unknown. Nevertheless, these results clearly indicate the importance of
standardized conditions for generation and expansion of USSC and that
the potency of each USSC-line has to be confirmed. Furthermore, since
even under the best conditions currently available only 43% of UCB units
yield USSCs and only about 10% of these reach passage 9 or higher and
remain pluripotent, there is room for improvement of generation and
expansion conditions. Immunophenotypic differences between cells
grown under the two culture conditions or between individual USSC
preparations have not been observed. USSCs lack expression of CD4,
CD8, CD11a, CD11b, CD11c, CD14, CD15, CD16, CD18, CD25, CD27,
CD31, CD33, CD34, CD40, CD45, CD49d, CD50, CD56, CD62E, CD62L,
CD62P, CD80, CD86, CD106, CD117, Cadherin V, glycophorin A and
HLA-class II, but are positive for CD10, CD13, CD29, CD44, CD49e,
CD58, CD71, CD73, CD90, CD105, CD146, CD166, vimentin, cytoker-
atin 8 and 18, as well as HLA-class I, and show only limited expression of
CD49b and CD123.[19,20]

GENERATION AND CULTURE OF GMP-GRADE USSCs

Due to their proliferation and differentiation capacity, USSCs are interesting candidates for the future development of cellular therapies as well as
for support systems for hematopoietic reconstitution. Since generation
and expansion under GMP conditions is mandatory for use in clinical

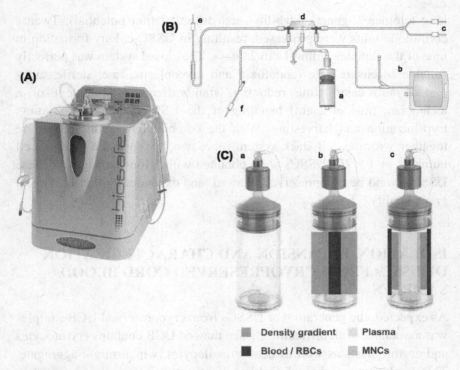

Fig. 3. Principle of the Sepax cell separating system. (**A**) The Sepax S-100 main processing unit allows the automated MNC isolation by Ficoll density gradient-based separation (DGBS). (**B**) Cord blood processing takes place in a closed and sterile environment using the Sepax single-use Ficoll kit CS-900. (**C**) Ficoll based separation principle.

applications, the automated cell processing system Sepax (BIOSAFE, Fig. 3) with the CS900 separation kit (Fig. 3) was used for mononuclear cell separation from cord blood in a similar way as it was described for bone marrow MNCs.[27] For the subsequent generation of USSC, 30% GMP-grade fetal calf serum, low-glucose DMEM-medium including 10^{-7} M dexamethasone was used. Expansion of USSCs was performed in a closed system applying cell stacks (Costar Corning).[28]

Results indicate that the frequency of generation is very similar under GMP conditions as compared to non-GMP conditions (45% and 43%, respectively). Those cells also had a similar quality (high population doubling, long telomeres, genetic stability, and differentiation potential). Twenty cord blood units were processed, resulting in USSC-colony formation in nine of the samples within 14 to 28 days. The closed system was perfectly suitable to ensure safe (endotoxin and mycoplasma-free, sterile, automated) and easy (time reducing, standardized, independent of a technician, time efficient) handling of the USSCs, including seeding, trypsinization and harvesting. With the combination of this procedure together with the cell stack system (one, two, five and ten layers), cell numbers of 1×10^9 USSCs were obtained within four passages.[28] These USSCs could be cryopreserved, thawed, and expanded further in clinical grade quality.

ISOLATION, EXPANSION AND CHARACTERIZATION OF USSCs FROM CRYOPRESERVED CORD BLOOD SAMPLES

As expected, the generation of USSCs from cryopreserved UCB samples was associated with difficulties, since thawed UCB contains erythrocytes and erythroblasts as well as dead granulocytes, which might aggregate. This is always a technical problem if hematopoietic cells are cryopreserved and thawed. Initial experiments performed with unseparated frozen UCB showed that after Ficoll isolation and erythrocyte lysis cells generally did not adhere to the culture flasks and as such the yield of MNCs was very low. Subsequent experiments were performed with volume-reduced UCB units, where the majority of erythrocytes were already depleted before cryopreservation. Only one experimental condition, namely the MNC isolation with erythrocyte lysis and further culture in the presence of Myelocult including dexamethasone allowed successful generation of USSCs in seven out of 36 UCB samples (19.5%). In DMEM/10^{-7} M dexamethasone (n = 10) USSCs could not be generated due to a low frequency in thawed blood. Due to these difficulties to generate USSCs from cryopreserved material, either technical improvements are necessary or, alternatively, therapeutic applications of USSCs will probably rely on the establishment of USSC banks, with the USSCs generated from fresh UCB.

DIFFERENTIATION OF USSCs TOWARDS
THE MESENCHYMAL LINEAGE

To test for multipotent differentiation capacity of USSCs, several *in vitro* differentiation assays were performed to document differentiation of cells towards all three germinal layers.[20] Differentiation of USSCs into *osteoblasts* was induced by culturing the cells in the presence of dexamethasone, ascorbic acid and β-glycerol phosphate. Cells formed nodules, which stained positive for Alizarin red, which is an indication for osteoblast-typical calcification and serves as a functional criterion for bone differentiation. Bone-specific alkaline phosphatase (ALP) activity was detected and continuous increase in Ca^{2+} release was documented.[20] Osteogenic differentiation was further confirmed by increased expression of ALP, osteonectin, osteopontin, Runx1, bone sialo-protein and osteocalcin as well as Notch 1 and 4 detected by RT-PCR and confirmed on protein level.[20]

A pellet culture technique in the presence of dexamethasone, proline, sodium pyruvat, ITS$^+$ premix and TGF-β1 was employed to trigger USSC differentiation towards the *chondrogenic* lineage.[20] The chondrogenic nature of differentiated cells was assessed by Alcian blue staining and by expression analysis of the cartilage extracellular protein collagen type II. Chondrogenesis was further confirmed by RT-PCR showing expression of the cartilage specific mRNAs encoding Cart-1, collagen type II and chondroadherin. All USSC lines tested were capable of differentiating along the osteogenic and chondrogenic differentiation pathways.[20] The *in vivo* regenerative capacity of USSCs for both bone and cartilage was confirmed in models for repair of critical size bone defects in athymic Harlan nude rats and the implantation of gelatin sponges into nude mice.[20]

For induction of *adipogenic* differentiation, USSCs were cultured with dexamethasone, insulin, IBMX and indomethacin. Adipogenic differentiation was demonstrated by Oil Red-O staining of intracellular lipid vacuoles. For adipogenic differentiation we were able to identify two subsets of USSC lines: cells which could be differentiated towards adipocytes (which are more likely MSCs based on the criteria in Table 1, Fig. 4) and USSC cell lines that could not be differentiated towards adipocytes but had otherwise a much better (Table 1) differentiation potential (Fig. 4).

Table 1.   Differences between USSC and CB MSC.

USSC	CB MSC
No adipogenic differentiation or only weak in late passages	Adipogentic differentiation
Strong expression of DLK-1	No expression of DLK-1
Neural differentiation	Restricted neural differentiation
Strong expression of SCRG 1, LIF, STAT2, NEUROG1, EPHA5, ZIC1	Weak expression of SCRG 1, LIF, STAT2, NEUROG1, EPHA5, ZIC1
Longer telomeres, higher proliferation	
Strong expression of CDKN1A, G0S2, CDK6, IL8, no expression of HOXA2, HOXB7, HOXC9, HOXC10, HOXD8, HOXD9	Weak expression of CDKN1A, G0S2, CDK6, IL8; strong expression of HOXA2, HOXB7, HOXC9, HOXC10, HOXD8, HOXD9

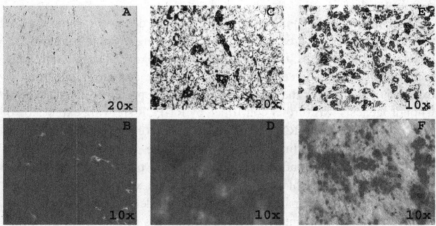

Fig. 4.   Adipogenic differentiation was performed in the presence of dexamethasone, indomethacin, IBMX and insulin. Oil Red-O staining of lipid vesicles performed two weeks after stimulation demonstrates an ongoing adipogenesis. We were able to identify two subsets of lines: USSC lines which could not be differentiated towards adipocytes (**A**), and MSC-like cell lines from cord blood that showed adipogenic differentiation (**C**). Bone marrow MSCs served as controls (**E**). (×20 magnification). (**B and C**) Despite their different adipogenic differentiation, all UCB lines tested were capable of osteogenic differentiation. For osteoblast differentiation, cells were cultured at 8000 cells/cm^2 in 24-well plates in DMEM, 30% FCS, 10^{-7} M dexamethasone, 50 μM ascorbic acid-2 phosphate and 10 mM β-glycerol phosphate (DAG). Differentiation to osteoblasts is shown by Alizarin red staining to determine calcium deposition (×10 magnification). (**D**) Bone marrow MSCs served as controls.

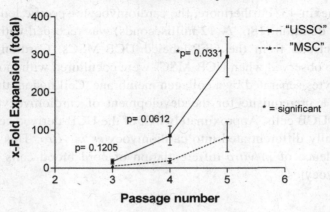

Fig. 5. USSC lines showed a significantly higher yield per passage as compared to UCB-MSCs. At passage 5, a mean of 82.87 ± 39.70-fold expansion was achieved for the more committed UCB-MSCs, while a 294.4 ± 53.00-fold expansion was observed for UCB-USSCs.

The lack of adipogenic differentiation capacity of UCB-derived MSC-like cells has also been reported by others.[4,29] In contrast, BM-MSCs always showed adipogenic differentiation. Therefore, we postulate two functional subsets of adherent non-hematopoietic cells in UCB and the hypothesis that USSCs and MSCs can be easily distinguished in cord blood by the adipogenic differentiation potential. Already the generation and expansion data (lower in UCB-MSCs as compared to USSCs, Fig. 5) suggested that there might be a hierarchy in the non-hematopoietic stem cells from cord blood.

DIFFERENTIATION OF USSCs INTO CARDIOMYOCYTES AND MYOCARDIAL APPLICATIONS

In Vitro Differentiation

As shown recently by Nishiyama *et al.*[30] cord blood adherent cells after coculture with fetal murine cardiomyocytes contracted rhythmically and synchronously, suggesting the presence of electrical communication between the fetal cardiomyocytes and the UCB cells. Cardiomyocytes derived from MSC-like UCB cells stained positive for cardiac troponin-I and connexin-43. Furthermore, the cardiomyocyte-specific long action potential duration (186 $^{+}/-$ 12 milliseconds) was recorded with a glass microelectrode from the GFP-labeled UCB-MSCs. Cardiomyocytes were also observed when UCB-MSCs were cocultured with mouse cardiomyocytes separated by a collagen membrane. Cell fusion, therefore, was not a prerequisite for the development of cardiomyocytic cells from the UCB cells. Approximately half of the UCB-derived cells were successfully differentiated into cardiomyocytes *in vitro*. This was the first evidence of *in vitro* differentiation of cord blood cells towards cardiomyocytes.

In Vivo Animal Models

Differentiation of USSCs into human cardiomyocytes was first described by our group in 2004 by transplanting USCCs into the preimmune fetal sheep model.[20] Eight months after transplantation USSCs were found to differentiate into cardiomyocytes in this model. Cells derived from USSCs stained positive with the human-specific anti-HSP27 antibody and were found in both atria and ventricles as well as in the septum of the heart. Double labeling showed that the USSCs stained also positive for the ryanodine receptor, myosin heavy chain and dystrophin. In this study, engraftment of USSCs into the Purkinje fiber system was also detected. Cells showed characteristic morphologies and were stained with the Purkinje fiber marker PGP 9.5. Compared to human MSCs from bone marrow there was a difference in the distribution of USSC-derived cells in the heart. In previous studies using the preimmune fetal sheep model, it was shown that human MSCs engraft predominantly in the Purkinje fiber system and that there are very few ventricular or atrial cardiomyocytes of human origin. In contrast, USSC-derived cells formed both, Purkinje fiber cells and cardiomyocytes. This indicates that USSCs are an earlier cell type than bone marrow MSCs, and possibly may represent an MSCs precursor. Since the preimmune model is a developmental model, following studies in porcine models addressed the question, whether USSCs can engraft in other more clinically relevant conditions such as post infarction. Kim *et al.* showed in a porcine model of chronic myocardial infarction that transplanted USSCs survived one month after intra-myocardial transplantation and were able to adopt cardiac phenotypes accompanied by improvement of myocardial function.[31]

We were able to show in a porcine model of acute myocardial infarction (MI) that after eight weeks USSCs were undetectable in the infarcted area as demonstrated by FISH and immunohistology.[32] Although human cells were not detectable two months after transplantation, the USSCs preserved the recipient myocardium and prevented scar formation after acute ischemia. This was associated with a significant improvement of left ventricular ejection fraction and the prevention of left ventricular dilation. In both studies side effects such as those described by Moelker *et al.*[33] which showed calcification after infusion of 1×10^8 USSCs into the infarct–related coronary artery, were not observed. The differences between Kim's study and ours could be explained by the different numbers of transplanted cells (1×10^8 versus 1.3×10^7) as well as by the difference in the observation time points (one versus two months). With regard to the results, there are many potential explanations for the observed functional recovery and prevention of scar formation: induction of cytokines can lead to the *de novo* formation of myocardium or the preservation of the recipient myocardium by paracrine effects. USSCs can release a variety of cytokines *in vitro* and, therefore, may accelerate regeneration after transplantation.[19]

NEURAL DIFFERENTIATION OF USSCs

Neural differentiation of USSCs[20,34] was induced by a defined combination of growth and differentiation factors. Neurofilament-positive (NF) cells were detected (70% of cells) and double immunostaining revealed colocalization of neurofilament and sodium channels in a small proportion of cells. Expression of synaptophysin was detected after four weeks of differentiation. USSC-derived neurons stained positive for TH, the key enzyme of the dopaminergic pathway. In approximately 30% of the cells, choline acetyltransferase was detected. Recent data demonstrated functional neuronal properties by applying patch-clamp analysis and high performance liquid chromatography (HPLC).[34] Although a rare event, voltage-gated sodium channels could be clearly identified and HPLC analysis confirmed synthesis and release of dopamine and serotonine by differentiated USSCs.[34] The *in vitro* data of USSC and MSC differentiation towards neural lineage has been reported by other groups[35] and is reviewed in more detail in Chapter 7. Sun *et al.*[36] described voltage-sensitive and ligand-gated channels in differentiating fetal neural cells isolated from the non-hematopoietic fraction of human cord blood.

The ability to migrate, integrate and differentiate into neuronal-like cells *in vivo* was shown by stereotactical implantation into the hippocampus region of an intact adult rat brain. Human Tau-positive cells persisted for up to three months.[20]

Taken together, there is clear evidence of USSC differentiation to cells of all three neural lineages and of functional activity of the differentiated cells. However, it is still unclear why transplantation of USSCs into ventricles of E16.5 rat embryos resulted in cluster formation in the striatum, cortex and hippocampus without expression of neural markers.[37]

ENDODERMAL DIFFERENTIATION OF USSCs

Transplantation of USSC in a non-injury model, the preimmune fetal sheep,[20] revealed more than 20% albumin-producing human parenchymal hepatic cells without any evidence of cell fusion. Meanwhile, we could observe endodermal differentiation of USSCs also *in vitro* by applying protocols described for both, embryonic and adult stem cells.[21,22] USSCs were negative for the majority of endodermal markers tested by RT-PCR. Only expression of HGF and cytokeratin 8, 18 and 19 was observed on RT-PCR level. After induction of differentiation, USSCs did not express NeuroD, HNF1, HNF3b, PDX-1, PAX4, insulin and α-fetoprotein, but did express, depending on the culture conditions,[21,22] the common endodermal precursor markers GSC, Sox-17, HNF4α, GATA4 as well as albumin, Cyp2B6, Cyp3A4, Gys2 (liver development) and Nkx6.1 and ISL-1 (pancreatic development). Furthermore, glycogen storage (PAS staining) and albumin secretion were detected. These results clearly indicate endodermal differentiation of USSCs *in vitro*.[21,22] Based on RT-PCR data as well as functional data, an endodermal differentiation pathway could be delineated for USSCs.[22]

HEMATOPOIESIS SUPPORTING ACTIVITY OF USSC

Recent experience showed that it is possible to reconstitute the adult hematopoietic system after myeloablative or non-myeloablative conditioning by applying one large UCB unit or by combining two cord blood units.[38,39] To date 626 UCB transplants (51% children, 49% adults) have been delivered from the Düsseldorf UCB bank to transplant centers worldwide for clinical transplantations. Many adult patients have received a double cord blood transplantation to increase hematopoietic progenitor and granulocyte numbers and to reduce the duration of post-transplant neutropenia. Recently, data presented by the EUROCORD registry in Paris showed excellent engraftment kinetics of the double UCB transplants and high survival rates. Other clinical approaches to improve reconstitution used the cotransplantation of a single UCB unit together with highly purified CD34+ mobilized peripheral blood stem cells or bone marrow from a haploidentical related donor. The specialized BM stroma environment consisting of extracellular matrix and stroma cells has been shown to be crucial for hematopoietic regeneration after any stem cell transplantation. Therefore, cotransplantation of UCB-USSCs may also improve engraftment. Our group has studied cytokine production and *in vitro* hematopoiesis-supporting stromal activity of USSCs *in vitro* and *in vivo* in comparison to BM-derived MSCs.[19] USSCs constitutively produced cytokines such as SCF, LIF, TGF-1β, M-CSF, GM-CSF, VEGF, IL-1β, IL-6, IL-8, IL-11, IL-12, IL-15, SDF-1α and HGF. When USSCs were stimulated with IL-1β, G-CSF was released. Production of SCF and LIF was significantly higher in USSCs compared to BM-MSCs.[19] In order to determine the hematopoiesis supporting stromal activity of USSCs as compared to BM-MSCs, UCB CD34+ cells were expanded in cocultures. At one, two, three and four weeks, coculture of CD34+ cells on USSC feeder layers resulted in a 14-fold, 110-fold, 151-fold and 183.6-fold amplification of total cells and a 30-fold, 101-fold, 64-fold and 29-fold amplification of CFCs (colony forming cells), respectively. LTC-IC expansion at one and two weeks was with two-fold and 2.5-fold significantly higher for USSCs than BM-MSCs (one-fold and one-fold), but declined after day 21. Transwell cocultures of USSCs did not significantly alter total cell or CFC expansion. In summary, USSCs produce significant amounts of supporting cytokines for hematopoiesis and are superior to BM-MSCs in supporting the expansion of CD34$^+$ cells from UCB. USSCs are therefore suitable candidates for stroma-driven *ex-vivo* expansion of hematopoietic stem cells for short-term reconstitution.

Recent *in vivo* data of Chan *et al.*[40] have shown that USSCs induced a significant enhancement of CD34$^+$ cell homing to both bone marrow and spleen. In the publication of Huang *et al.*[41] UCB–MSCs were used for the *ex vivo* expansion of UCB progenitors and resulted in a fast recovery after transplantation into NOD SCID mice.

COMPARISON OF USSCs AND MSCs FROM CORD BLOOD

USSCs share many overlapping features with MSCs from fetal tissue[42] such as immunophenotype and differentiation potential *in vitro* and *in vivo*. Based on the amplification kinetics (Fig. 5) certain authors who have described MSCs from UCB may possibly have obtained USSCs if appropriate fetal calf serum and medium conditions were applied.

USSCs can be easily triggered towards MSCs with restricted differentiation potential, depending on culture conditions. If a medium with horse serum is used and a fetal calf serum containing higher dosages of steroids, the differentiation towards MSCs is triggered.[19] Markers, which delineate such differences, are currently unknown and all immunophenotypic studies describe only homogenous populations. Thus, discrimination of unrestricted and mesodermally determined or possibly for other lineages/germinal layers committed cells currently relies on the determination of the functional differentiation potential of the cells, similar to what is still true for the hematopoietic stem and progenitor cell compartment. Nevertheless, current results[13,14] suggest that there is a hierarchy in the non-hematopoietic stem cells from UCB, and it will be important to

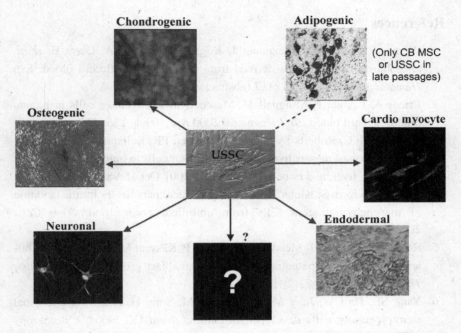

Fig. 6. Summary of the differentiation potential of USSCs *in vitro* and *in vivo*. Only very late passages of USSCs differentiate into adipocytes, whereas UCB-MSCs are easily triggered towards adipocytes in early passages.

elucidate this hierarchy, which includes further delineation of the individual differentiation pathways as shown in Fig. 6.

ACKNOWLEDGMENTS

The authors would like to thank all the co-authors of our jointly published papers that were cited here and the Deutsche Forschungsgemeinschaft (DFG) for funding the research group FOR 717 including the project Ko2119/6-1. Studies reported from the author's laboratory on hematopoiesis supporting activity, which were reviewed here, have been supported by the German José Carreras Leukemia Foundation grant DJCLS-R03/06; DJCLS-R07/05v and the EUROCORD III grant QLRT-2001-01918. Thanks to S. Maria Kluth, Aurelie Lefort and Anja Buchheiser, PhD., for their excellent technical support.

References

1. Wernet P, Callejas J, Enczmann J, Kogler G, Knipper A, Derra E, *et al.* Osteoblast precursor cells derived from human umbiilcal córd blood. *Exp Hematol* 1999;27(Suppl 1):117 (abstract).
2. Erices A, Conget P, Minguell JJ. Mesenchymal progenitor cells in human umbilical cord blood. *Br J Haematol* 2000 Apr;109(1):235–42.
3. Campagnoli C, Roberts IA, Kumar S, Bennett PR, Bellantuono I, Fisk NM. Identification of mesenchymal stem/progenitor cells in human first-trimester fetal blood, liver, and bone marrow. *Blood* 2001 Oct 15;98(8):2396–402.
4. Bieback K, Kern S, Kluter H, Eichler H. Critical parameters for the isolation of mesenchymal stem cells from umbilical cord blood. *Stem Cells* 2004;22(4):625–34.
5. Rosada C, Justesen J, Melsvik D, Ebbesen P, Kassem M. The human umbilical cord blood: a potential source for osteoblast progenitor cells. *Calcif Tissue Int* 2003 Feb;72(2):135–42.
6. Yang SE, Ha CW, Jung M, Jin HJ, Lee M, Song H, *et al.* Mesenchymal stem/progenitor cells developed in cultures from UC blood. *Cytotherapy* 2004;6(5):476–86.
7. Gang EJ, Jeong JA, Hong SH, Hwang SH, Kim SW, Yang IH, *et al.* Skeletal myogenic differentiation of mesenchymal stem cells isolated from human umbilical cord blood. *Stem Cells* 2004;22(4):617–24.
8. Charbord P, Oostendorp R, Pang W, Herault O, Noel F, Tsuji T, *et al.* Comparative study of stromal cell lines derived from embryonic, fetal, and postnatal mouse blood-forming tissues. *Exp Hematol* 2002 Oct;30(10):1202–10.
9. Goodwin HS, Bicknese AR, Chien SN, Bogucki BD, Quinn CO, Wall DA. Multilineage differentiation activity by cells isolated from umbilical cord blood: expression of bone, fat, and neural markers. *Biol Blood Marrow Transplant* 2001;7(11):581–8.
10. Yu M, Xiao Z, Shen L, Li L. Mid-trimester fetal blood-derived adherent cells share characteristics similar to mesenchymal stem cells but full-term umbilical cord blood does not. *Br J Haematol* 2004 Mar;124(5):666–75.
11. Bicknese AR, Goodwin HS, Quinn CO, Henderson VC, Chien SN, Wall DA. Human umbilical cord blood cells can be induced to express markers for neurons and glia. *Cell Transplant* 2002;11(3):261–4.

12. Buzanska L, Machaj EK, Zablocka B, Pojda Z, Domanska-Janik K. Human cord blood-derived cells attain neuronal and glial features *in vitro*. *J Cell Sci* 2002 May 15;115(Pt 10):2131–8.

13. Jansen BJ, Gilissen C, Roelofs H, Schaap-Oziemlak A, Veltman J, Raymakers RA, Jansen JH, Kögler G, Figdor CG, Torensma R, Adema GJ. Functional differences between mesenchymal stem cells populations are reflected by their transcriptome. *Stem Cells Dev* 2010 Apr;19(4): 481–90.

14. Kluth SM, Buchheiser A, Houben AP, Geyh S, Krenz T, Radke TF, Wiek C, Hanenberg H, Reinecke P, Wernet P, Kögler G. DLK-1 as a marker to distinguish unrestricted somatic stem cells and mesenchymal stromal cells in cord blood. *Stem cell Dev* 2010 Aug 28 [Epub ahead of print]. PubMed PMID: 20331358.

15. Kakinuma S, Tanaka Y, Chinzei R, Watanabe M, Shimizu-Saito K, Hara Y, *et al.* Human umbilical cord blood as a source of transplantable hepatic progenitor cells. *Stem Cells* 2003;21(2):217–27.

16. Newsome PN, Johannessen I, Boyle S, Dalakas E, McAulay KA, Samuel K, *et al.* Human cord blood-derived cells can differentiate into hepatocytes in the mouse liver with no evidence of cellular fusion. *Gastroenterology* 2003 Jun;124(7):1891–900.

17. Hong SH, Gang EJ, Jeong JA, Ahn C, Hwang SH, Yang IH, *et al. In vitro* differentiation of human umbilical cord blood-derived mesenchymal stem cells into hepatocyte-like cells. *Biochem Biophys Res Commun* 2005 May 20;330(4): 1153–61.

18. Jeong JA, Gang EJ, Hong SH, Hwang SH, Kim SW, Yang IH, *et al.* Rapid neural differentiation of human cord blood-derived mesenchymal stem cells. *Neuroreport* 2004 Aug 6;15(11):1731–4.

19. Kogler G, Radke TF, Lefort A, Sensken S, Fischer J, Sorg RV, *et al.* Cytokine production and hematopoiesis supporting activity of cord blood-derived unrestricted somatic stem cells. *Exp Hematol* 2005 May;33(5):573–83.

20. Kogler G, Sensken S, Airey JA, Trapp T, Muschen M, Feldhahn N, *et al.* A new human somatic stem cell from placental cord blood with intrinsic pluripotent differentiation potential. *J Exp Med* 2004 Jul 19;200(2):123–35.

21. Kogler G, Sensken S, Wernet P. Comparative generation and characterization of pluripotent unrestricted somatic stem cells with mesenchymal stem cells from human cord blood. *Exp Hematol* 2006 Nov;34(11):1589–95.

22. Sensken S, Waclawczyk S, Knaupp AS, Trapp T, Enczmann J, Wernet P, *et al. In vitro* differentiation of human cord blood-derived unrestricted somatic stem cells towards an endodermal pathway. *Cytotherapy* 2007;9(4):362–78.

23. Lee OK, Kuo TK, Chen WM, Lee KD, Hsieh SL, Chen TH. Isolation of multipotent mesenchymal stem cells from umbilical cord blood. *Blood* 2004 Mar 1;103(5):1669–75.

24. Geiger H, True JM, de Haan G, Van Zant G. Age- and stage-specific regulation patterns in the hematopoietic stem cell hierarchy. *Blood* 2001 Nov 15;98(10):2966–72.

25. Lemischka IR. Hematopoietic stem cells, explicity. *Blood* 2006;107(6):2216.

26. Kogler G, Tutschek B, Koerschgen L, Platz A, Bender HG, Wernet P. Die José Carreras Stammzellbank Düsseldorf im NETCORD/EUROCORD-Verbund: Entwicklung, klinische Ergebnisse, Perspektiven und Aufklärung werdender Eltern. *Gynäkologe* 2005;38:836–46.

27. Aktas M, Radke TF, Strauer BE, Wernet P, Kogler G. Separation of adult bone marrow mononuclear cells using the automated closed separation system SEPAX. *Cytotherapy* 2008;10(2):203–11.

28. Radke TF, Buchheiser A, Lefort A, Maleki M, Wernet P, Kogler G. GMP-conform generation and cultivation of USSC from cord blood using the SEPAX-separation method and a closed culture system applying cell stacks. *Blood* 2007;110(11):367 (abstract).

29. Kern S, Eichler H, Stoeve J, Kluter H, Bieback K. Comparative analysis of mesenchymal stem cells from bone marrow, umbilical cord blood, or adipose tissue. *Stem Cells* 2006 May;24(5):1294–301.

30. Nishiyama N, Miyoshi S, Hida N, Uyama T, Okamoto K, Ikegami Y, *et al.* The significant cardiomyogenic potential of human umbilical cord blood-derived mesenchymal stem cells *in vitro. Stem Cells* 2007 Aug;25(8):2017–24.

31. Kim BO, Tian H, Prasongsukarn K, Wu J, Angoulvant D, Wnendt S, *et al.* Cell transplantation improves ventricular function after a myocardial infarction: A preclinical study of human unrestricted somatic stem cells in a porcine model. *Circulation* 2005 Aug 30;112(9 Suppl):196–104.

32. Ghodsizad A, Niehaus M, Koegler G, Martin U, Wernet P, Bara C, *et al.* Transplanted human cord blood derived unrestricted somatic stem cells improve left-ventricular function and prevent left-ventricular dilation and scar formation after acute myocardial infarction. *Heart* 2009 Jan;95(1):27–35.

33. Moelker AD, Baks T, Wever KM, Spitskovsky D, Wielopolski PA, van Beusekom HM, *et al.* Intracoronary delivery of umbilical cord blood derived unrestricted somatic stem cells is not suitable to improve LV function after myocardial infarction in swine. *J Mol Cell Cardiol* 2007 Apr;42(4):735–45.

34. Greschat S, Schira J, Kury P, Rosenbaum C, de Souza Silva MA, Kogler G, *et al.* Unrestricted somatic stem cells from human umbilical cord blood can be differentiated into neurons with a dopaminergic phenotype. *Stem Cells Dev* 2008 Apr;17(2):221–32.

35. Wang TT, Tio M, Lee W, Beerheide W, Udolph G. Neural differentiation of mesenchymal-like stem cells from cord blood is mediated by PKA. *Biochem Biophys Res Commun* 2007 Jun 15;357(4):1021–7.

36. Sun W, Buzanska L, Domanska-Janik K, Salvi RJ, Stachowiak MK. Voltage-sensitive and ligand-gated channels in differentiating neural stem-like cells derived from the nonhematopoietic fraction of human umbilical cord blood. *Stem Cells* 2005 Aug;23(7):931–45.

37. Coenen M, Kogler G, Wernet P, Brustle O. Transplantation of human umbilical cord blood-derived adherent progenitors into the developing rodent brain. *J Neuropathol Exp Neurol* 2005 Aug;64(8):681–8.

38. Barker JN, Weisdorf DJ, DeFor TE, Blazar BR, McGlave PB, Miller JS, *et al.* Transplantation of 2 partially HLA-matched umbilical cord blood units to enhance engraftment in adults with hematologic malignancy. *Blood* 2005 Feb 1;105(3):1343–7.

39. Brunstein CG, Barker JN, Weisdorf DJ, DeFor TE, Miller JS, Blazar BR, *et al.* Umbilical cord blood transplantation after nonmyeloablative conditioning: impact on transplantation outcomes in 110 adults with hematologic disease. *Blood* 2007 Oct 15;110(8):3064–70.

40. Chan SL, Choi M, Wnendt S, Kraus M, Teng E, Leong HF, *et al.* Enhanced *in vivo* homing of uncultured and selectively amplified cord blood CD34+ cells by cotransplantation with cord blood-derived unrestricted somatic stem cells. *Stem Cells* 2007 Feb;25(2):529–36.

41. Huang GP, Pan ZJ, Jia BB, Zheng Q, Xie CG, Gu JH, *et al. Ex vivo* expansion and transplantation of hematopoietic stem/progenitor cells supported by mesenchymal stem cells from human umbilical cord blood. *Cell Transplant* 2007;16(6):579–85.

42. O'Donoghue K, Fisk NM. Fetal stem cells. *Best Pract Res Clin Obstet Gynaecol* 2004 Dec;18(6):853–75.

11

Endothelial Precursor Cells

Marcie R. Finney and Mary J. Laughlin

ABSTRACT

The initial isolation of endothelial precursor cells (EPCs) over ten years ago has led to vast amounts of research and literature regarding these cells and their capabilities in the vascular system to contribute to new vessel formation. EPCs are endothelial progenitors that express combinations of CD34, VEGFR-2 and CD133. These EPCs differentiate into mature endothelial cells *in vitro* with endothelial functionality. In addition, EPCs have the capacity to be expanded *in vitro* enabling the generation of adequate quantities of this rare circulating cell. Various animal models of peripheral and myocardial ischemia have been utilized to show the effectiveness of EPCs in increasing perfusion, reduction of ischemic tissue and some incorporation of cells into new vasculature. It has been shown that EPCs diminish in number and function with age. However, future translational therapies with EPCs would likely involve an older patient population. Therefore umbilical cord blood may provide a source of younger EPCs. Studies show that UCB-derived EPCs have equivalent functionality to those derived from bone marrow or peripheral blood. With such potent capabilities, additional investigation into the potential of EPCs as cellular therapy for augmentation of vascularization and the use of UCB as a readily available source of EPCs are warranted.

INTRODUCTION

Worldwide, heart failure remains the leading cause of death and current therapies merely delay progression of the disease. Coronary artery disease (CAD), is the most common type of heart failure, and is the leading fatality for both men and women in the United States. Patients with CAD have arteries narrowed by atherosclerotic plaques resulting in obstruction of blood flow to the heart, leading to both acute and chronic cardiac conditions. As narrowing or blockage of arterial vessels occurs, angina results from an acute loss of blood to an area of the heart. With reduced amounts of oxygen-rich blood in a chronic setting, the heart muscle will weaken, resulting in heart failure or arrhythmia. Patients suffering from heart failure have a deficient amount of blood being pumped by the affected heart whereas patients with arrhythmias have dysregulation of cardiac innervations leading to irregular heart rate or rhythm.

In addition to lifestyle changes, current therapies for CAD include pharmacological intervention and surgical procedures. Medical therapy for patients may consist of antiplatelet therapy, treatment with HMG-CoA reductase inhibitors (statins), angiotensin converting enzyme (ACE) inhibitors, nitrates, beta-blockers, and calcium channel blockers. Revascularization procedures include catheter-based or surgical approaches to reopen blocked vessels. Commonly performed percutaneous coronary interventions may consist of angioplasty, where a balloon is inflated to open a blocked artery, or stent deployment where a mesh cylinder is inserted into the artery to reinforce the artery and prevent reoccurrence of the blockage. Coronary artery bypass grafting (CABG) is the surgical procedure to revascularize blocked arteries. This invasive surgical procedure is performed when patients are not candidates for angioplasty due to the size or number of blocked arteries. In CABG healthy arteries/veins from the patient are grafted into the heart to bypass the blocked artery.

Although these surgical procedures are widespread, with approximately one million angioplasties and 500,000 CABG procedures performed each year in the US alone, to restore blood flow to ischemic myocardium, there are many patients in need of alternative cardiovascular therapies. Many patients are not suitable candidates for revascularization procedures or have undergone these procedures with unsatisfactory results. Patients with angina and advanced disease may be less responsive to anti-anginal medication and are not eligible for either percutaneous or surgical revascularization, leading to frequent symptoms, reduced exercise capacity, and poor quality of life. In a recent study that focused on the frequency of incomplete revascularization in CABG patients, approximately 30% of patients with multi-vessel disease undergoing elective surgery were incompletely revascularized. Not surprisingly, patients who had complete revascularization have improved rates of survival when compared to those incompletely revascularized.[1] Treatment of CAD by pharmacological methods or procedural interventions has limitations and those patients that are not responsive to, or are not candidates for those treatments, face limited options.

ANGIOGENESIS AND ENDOTHELIAL PROGENITOR CELLS

Age-related diminution of vascular endothelial cell number and function has been observed by several groups.[2,3] Related research indicated that in older patients, who are most likely to suffer from vascular problems, the number of responsive endothelial cells are reduced and the number of dysfunctional endothelial cells is increased.[4] Endothelial progenitor cells (EPCs) were studied in healthy patients and it was found that as risk of coronary heart disease increased, the number of circulating EPCs decreased. Reduced numbers of circulating EPCs correlated with reduced endothelial function. Wojakowski *et al.* investigated peripheral blood mononuclear cells in patients with a myocardial infarction (MI), stable angina and healthy controls.[5] Circulating stem cells were highest in patients following MI. The high levels decreased steadily over time and returned to normal by seven days. These data taken together with the findings by Hill *et al.* suggest that the body possesses circulating progenitor cells, and that endogenous stem cells respond immediately after myocardial injury.[4] However, total numbers of EPCs and their proliferative capacity are reduced in patients who need them most — those at highest cardiovascular risk. Furthermore, the progenitor cell response following myocardial injury is not sustained days after onset of symptoms, during which adverse myocardial remodeling is known to occur.

The determination of loss of function of endothelial cells with increased age and endogenous recruitment of progenitor cells has spurred development of an entire arena of research into the transfer or transplantation of endothelial cells to treat patients whose own cells have reduced capacity to repair cardiac damage and may not be eligible for other traditional cardiovascular treatment.

The production of new blood vessels from pre-existing vasculature is a potent mechanism for maintaining perfusion of ischemic tissue. Classically, this process termed angiogenesis, was understood to be the result of recruitment and proliferation of mature endothelial cells[6] that sprout capillary vessels from existing vessels. However, investigations have identified circulating EPCs that migrate from bone marrow in adults to areas of ischemia via cytokine signaling, to participate in neovascularization.[7,8] This process of vasculogenesis was previously considered to occur only in the embryo and not in adults.[9] In the embryo, blood islands or cell clusters form, proliferate and fuse with other islands to form the capillary structures. Hematopoietic stem cells are located in the center of these blood islands and EPCs are located in the perimeter. The discovery by Asahara and colleagues that circulating EPCs exist in adult peripheral blood, started intensive investigation into neovascularization in adults.

EPC PHENOTYPE

Initial reports of the existence of EPCs in the bone marrow described the participation of these cells in new blood vessel formation and showed homing to vasculature.[10] To determine if peripheral blood contains cells capable of differentiating into mature endothelial cells, the antigens CD34 and Flk-1 were utilized as markers. These markers are shared by mature endothelial cells and hematopoietic stem cells. Flk-1 is a receptor for vascular endothelial growth factor 2 (VEGF-2). Cells were initially selected for CD34, labeled with a fluorescent dye, and cocultured with the negative fraction of the mononuclear cells. After a week in culture on fibronectin-coated culture dishes, fluorescent cells became spindle shaped and had proliferated. The CD34$^+$ labeled cells that had been 1% of the cells, now accounted for 60% of the total cells attached to the cell culture plates. When isolating and culturing Flk-1$^+$ cells, they behaved similarly, by growing to spindle-shaped, attached cells. These studies went on to show that the cells differentiated into mature endothelial cells *in vitro* and that these EPCs homed to sites of injury in animal models. The studies by Asahara *et al.* clearly showed that EPCs are a distinct population from mature endothelial cells, but their derivation and characteristics are still a vibrant discussion in the literature today, and ten years later there is still debate as to the exact population of cells contributing to these observations.[11]

ISOLATION AND CULTURE OF EPCs

It has been shown that EPCs are rare in unmobilized peripheral blood (0.04 ± 0.2%).[12] Due to the scarcity of these cells in circulation, methods were developed to expand or select EPCs *in vitro*.[13,14]

Attempts to directly isolate EPCs have utilized various surface markers, including CD34, CD133 (AC133), and VEGFR2.[10,12,15–17] Notably, these markers are also expressed on hematopoietic stem cells and methods to isolate EPCs from whole cellular preparations would not exclude HSCs.[18–20]

The literature in the last decade has been filled with many techniques to isolate EPCs directly, select EPCs following short-term culture and to attempt to expand EPCs *in vitro*. It is clear that these methods result in cells with vascular potential; however it is not clear what the exact nomenclature or composition of these EPCs is.

Two types of *in vitro* selection of EPCs are mainly found in the literature and produce early outgrowth and late outgrowth EPCs. Asahara's original method to generate early outgrowth EPCs involved culture of adult peripheral blood or umbilical cord blood mononuclear cells on fibronectin-coated culture dishes.[10] In this method a pre-plating step removed mature endothelial cells and monocytes. Short-term culture of these cells (seven days) resulted in spindle-shaped individual EPCs that were positive for endothelial markers such as CD31, Flk-1 and Tie-2. Another major finding of this work was that the spindle-shaped cells developing from mononuclear cells were derived from CD34$^+$ cells. Inclusion of the CD34$^-$ cells resulted in a ten-fold increase in the number of spindle-shaped attached EPCs, showing a synergistic effect of the whole mononuclear cell preparation on the emerging EPCs. A variation of this method produced colony forming unit-endothelial cells, CFU-EC. Hill *et al.* isolated mononuclear cells from peripheral blood and plated them on fibronectin-coated dishes for 48 hours to remove contaminating mature endothelial cells. The non-adherent cells were then replated on new fibronectin-coated plates and cultured for seven days.[4] This cell selection method was commercialized and is utilized in several clinical studies of various disease states including cardiovascular risk and diabetes. Another early EPC derivation method utilizes short-term culture without the adherence-depletion step. Mononuclear cells are cultured for four days on fibronectin, non-adherent cells are discarded and adherent cells are cultured short term (less than ten days).[13,21,22] Several groups have used these cells successfully for regeneration in animal models of ischemia and refer to the cells with various nomenclature including EPCs, CFU-ECs and circulating angiogenic cells, CACs. The main unifying properties of these early endothelial outgrowth cells are spindle-shaped morphology, limited proliferative capacity, and short-term culture duration. These early EPCs showed expression of endothelial (VE-cadherin, VEGFR2), myeloid (CD14) and pan-leukocytic (CD45) markers.

The culture selection of late outgrowth EPCs involves culture of cells for two to three weeks and develops a population of EPCs that rapidly grows with extensive population doublings.[23,24] These cells have also been termed endothelial colony-forming cells (ECFCs) or outgrowth endothelial cells (OECs) in the literature and have been described as displaying a cobblestone appearance after this long term culturing procedure.[23,25] When transplanted into *in vivo* models, ECFC receiving animals had increased blood flow in injury models and these studies have shown that ECFCs have the ability to form perfused vessels *in vitro*. This lead to the continued development of these cells for potential use in ischemic repair.[24,26,27] In addition, this population has such expansion capacity that

ECFCs have been utilized in preclinical studies due to the abundance of cells and uniformity of the resultant cell population.

ANIMAL MODELS OF NEOVASCULARIZATION

In their seminal report, Asahara and colleges utilized the hind limb ischemia model to study the contributions of culture-generated EPCs to angiogenesis.[8,10] Several weeks following ligation, injected cells were located in capillary vessel walls and arranged in preserved muscle structures. Importantly, no transplanted cells were found in non-injured legs, highlighting the homing capacity of these cells to ischemic sites. These studies challenged the paradigm that vasculogenesis was not possible after embryogenesis and opened the field for extensive investigations of neoangiogenesis.

Since then, rodent models of both myocardial and peripheral ischemia have been utilized to examine the effect of administration of EPCs on neovascularization.[21,22,26,28–30] The murine femoral ligation model has been used extensively with EPCs from multiple sources. In many cases, athymic animals or NOD/SCID mice were used to minimize rejection of transplanted cells. Figure 1 shows an example of such an experiment. A baseline image was taken (pre-ligation), then the ligation surgery was performed and another image was taken (post-ligation). Following recovery from surgery, mice were injected with 0.5×10^6 CD133$^+$ EPCs derived from umbilical cord blood. Mice were then imaged each week for three weeks. The expression of the blood flow data as a ratio of the injured versus the uninjured limb reduces data variables that include normal growth, temperature and ambient light. The particular animals in Fig. 1 showed a ratio of 0.48 (CD133$^+$ EPCs) and 0.23 (control mouse) on Day 21 following the ligation, indicating a revascularization of the ligated limb.

In vivo effectiveness was compared between early and late EPCs using this hind limb ischemia model. Hur *et al.* cultured both late and early EPCs from the same donor.[26] Mononuclear cells were plated on gelatin-coated wells in common endothelial growth medium containing fetal bovine serum, hEGF, VEGF, FGF-B, IGF-1, ascorbic acid and heparin. Late EPC colonies were harvested from within the early EPC spindle-shaped cells. For this, late EPCs were isolated based upon the

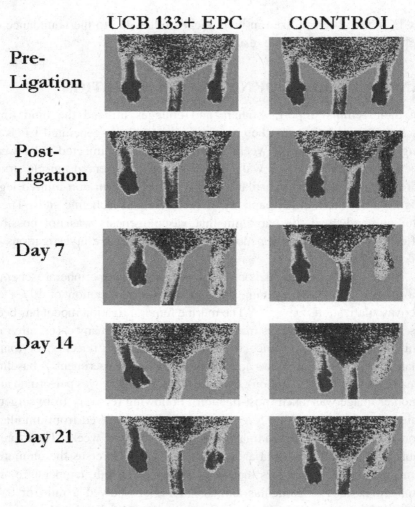

Fig. 1. CD133+ EPCs increase blood flow recovery ratios after injection into NOD/SCID mice with induced hind limb ischemia. Prior to injection of EPC, mice were sublethally irradiated and anesthetized. A baseline measurement of blood flow of the hind limbs was measured using a laser Doppler imager (pre-ligation). The right femoral artery was exposed, transected and ligated. Blood flow measurements on both feet were taken to determine that the ligation was successful in reducing blood flow. Blood flow recovery following ligation and injection of 0.5×10^6 CD133+ EPCs from umbilical cord blood (left panels) was then monitored over time (post-ligation, Days 7, 14 and 21) by laser Doppler. Control animals (Control, right panels) received culture medium injection without cells. Analysis of the images utilizes an arbitrary Perfusion Unit (PU) calculation with the color scale range from 0 PU to red 1,000 (PU). As can be seen, control animals also recovered, but only to a certain extent.

smooth shape and cobble stone morphology and were replated on new plates and cultured for 12 weeks.

Mononuclear cells prior to culture expressed Flt-1, eNOS and VWF. Early EPCs showed weak expression of VE-cadherin and KDR with increased expression of Flt-1, pointing to differentiation to endothelial lineage. In contrast, late EPCs showed elevated expression of endothelial-specific VE-cadherin, KDR Flt-1, eNOS and VWF. Late EPCs showed increased *in vitro* vascular functionality including nitric oxide production, capillary tube formation and integration into endothelial monolayer. Conversely, early EPCs produced increased angiogenic cytokines including vascular endothelial growth factor and interleukin-8 (IL-8). When injected into the hind limb ischemia model, both cell types promoted recovery and limb salvage. However, despite the differences in cell phenotype, there was no difference in blood flow recovery. This study did confirm prior findings that treatment with EPCs, regardless of type, resulted in limb salvage.

Yoon and colleagues continued on this theme and reported a synergistic effect of a combination of early and late EPCs in promoting neovascularization *in vivo*.[27] This research focused on impact of cytokine production and matrix metalloproteinases, both known to be instrumental in augmenting vasculogenesis. Early EPCs secreted large amounts of VEGF and IL-8, cytokines instrumental in endothelial proliferation, migration and tube formation. This group also determined that early EPCs secreted MMP-9 where late EPCs secreted MMP-2. Injured mice recovered when either early or late EPCs were transplanted, however the mice with greatest improvement over control were those injected with a mixed population of both types of cells. The mice given both cells also showed greater limb salvage and increased capillary density. These findings highlighted the importance of paracrine and autocrine mechanisms working to augment neovascularization.

Ligation of the left anterior descending coronary artery (LAD) provides an ideal animal *in vivo* model for myocardial ischemia that has been used extensively for determination of efficacy of transplanted EPCs.[28,31,32] Left ventricular (LV) function was determined using echocardiography and fractional shortening, regional wall motion, and scarring were parameter compared between treatment groups. When compared to animals receiving control medium, transplantation of EPCs preserved left ventricular function and incorporated into foci of myocardial neovascularization.[28,31]

EPCs FROM UCB

Several research groups have shown the potential of umbilical cord blood as a source for EPCs and the effectiveness of these cells in augmenting neovascularization in preclinical animal studies.[22,30,33–37] Several qualities of UCB render it more advantageous for culturing EPCs when compared to bone marrow, including a higher concentration of stem cells, and greater proliferation capacity. Indeed, when comparing the EPC colony formation frequency between UCB and peripheral blood, the number of colonies was greater in UCB.[30,34] In addition, it was determined that the EPCs grown from UCB have significantly higher growth potential.

Murohara *et al.* determined that a greater number of EPC could be derived from umbilical cord blood (UCB) than from the same amount of peripheral blood.[30] These UCB-derived EPCs possessed characteristics associated with EPCs such as incorporation of acetylated-LDL, release of nitric oxide and expression of surface molecules characteristic of EPCs including CD34, Tie-2, KDR and VE-cadherin. In addition, the efficacy of transplantation of EPC in the unilateral hind limb ischemia model was quantified via regional blood perfusion ratios. Histological studies showed that EPCs from UCB homed to the sites of injury and that the injected cells were viable. Laser Doppler perfusion showed augmentation of blood flow in all animals treated with EPCs derived from UCB. This study showed that EPCs derived from UCB were effective in treatment of ischemia by utilizing methods to quantify augmentation of neovascularization.

Our group compared UCB and bone marrow as sources for EPCs in the hind limb ischemia model.[22] Umbilical cord blood and bone marrow MNCs were cultured for seven days on fibronectin coated tissue culture plates. After this short-term culture, the cells exhibited surface markers considered specific to endothelial cells including CD31 and P1H12. The majority of the cells endocytosed acetylated-LDL and a minority exhibited lectin binding, two important cytochemical characteristics of endothelial cells. Infusions of EPCs culture-expanded from non-selected UCB versus adult marrow had comparable biologic effect in a NOD/SCID model of hind limb vascular injury. The data showed that in a direct comparison study, UCB and marrow-derived EPCs cell infusions both significantly increased blood flow in the ischemic leg by day 14 post-injury/cell infusion above that of cytokine infusions alone.

The effectiveness of umbilical cord blood-derived EPCs was also determined in the LAD ligation myocardial ischemia model.[34] Following ligation, UCB EPCs were transplanted into the ischemic myocardium. One week later, areas just outside the ischemic zone developed vascular structures that were positive for human CD31. When examined after 14 days, cells positive for human CD31 coexpressed Ki67, a known proliferation marker. These cells also reduced ischemia-induced myocardial damage enumerated by ejection fraction obtained by electrocardiography.

A novel *in vivo* experimental system that uses implantation of matrigel plugs seeded with EPCs subcutaneously into immunodeficient mice, showed that UCB-derived EPCs formed extensive networks of vascular structures.[38] Recent work shows that EPCs from UCB formed normally functioning, persistent blood vessels that survived for four weeks. In contrast, vessels derived from peripheral blood EPCs were unstable and began deterioration after three weeks.[39] These findings point to the potential use of EPC from UCB in engineered organs or tissue regeneration in addition to use in ischemic disease.

CONCLUSIONS

UCB is a valuable stem cell source due to its high content of early CD133[+] stem cells capable of EPC differentiation, as well as its robust proliferative capacity and "off the shelf" clinical application potential as it can be stored with ease. UCB collection poses no imposition on the normal birthing process and has not been associated with ethical concerns that surround the use of embryonic stem cells.[40] UCB stem cell expansion has been routinely performed for hematology clinical use and to date no malignant transformation have been observed in more than 8,000 patients treated with the exception of rare reported cases of donor-derived malignancy.[41]

Obviously, utilization of autologous cells remains the ideal scenario for application of cellular therapy. However, the rarity of EPCs in circulation and the concept of reduced capacity of EPCs with increased age has prompted investigations of UCB as a allogeneic source. Currently the potency of UCB-derived EPCs has been well described, however the immunological implications of transplantation of an allogeneic cell must be clarified before UCB can be used large scale in the clinic for neovasculogenesis. Laboratory work reported to date supports the use of UCB as a potential source of EPCs for therapeutic vasculogenesis.

Extensive studies have examined the characteristics of EPCs homing capacity and ability to stimulate neovascularization in animal models of ischemia. However, persistence and mechanism of influence in *in vivo* models still need to be addressed in more detail. Preclinical studies reported support the hypothesis that paracrine mechanisms account for EPC efficacy with minimal anatomic incorporation of EPCs infused into injured vasculature. These observations further support the use of allogeneic UCB as therapeutic agent to augment vasculogenesis in patients failing conventional revascularization therapies.

References

1. Jones EL, Weintraub W. The importance of completeness of revascularization during long-term follow-up after coronary artery operations. *J Thorac Cardiovasc Surg* 1996;112(2):227–37.
2. Chauhan A, More RS, Mullins PA, Taylor G, Petch C, Schofield PM. Aging-associated endothelial dysfunction in humans is reversed by L-arginine. *J Am Coll Cardiol* 1996;28(7):1796–804.
3. Tschudi MR BM, Bersinger NA, Moreau P, Cosentino F, Noll G, Malinski T, Luscher TF. Effect of age on kinetics of nitric oxide release in rat aorta and pulmonary artery. *J Clin Invest* 1996;98(4):899–905.
4. Hill JM, Zalos G, Halcox JP, Schenke WH, Waclawiw MA, Quyyumi AA, *et al.* Circulating endothelial progenitor cells, vascular function, and cardiovascular risk. *N Engl J Med* 2003 Feb 13;348(7):593–600.
5. Wojakowski W, Tendera M, Michalowska A, Majka M, Kucia M, Maslankiewicz K, *et al.* Mobilization of CD34/CXCR4+, CD34/CD117+, c-met+ stem cells, and mononuclear cells expressing early cardiac, muscle,

and endothelial markers into peripheral blood in patients with acute myocardial infarction. *Circulation* 2004 Nov 16;110(20):3213–20.

6. Folkman J. Angiogenesis in cancer, vascular, rheumatoid and other disease. *Nat Med* 1995 Jan;1(1):27–31.

7. Shi Q, Rafii S, Wu M. Evidence for circulating bone marrow-derived endothelial cells. *Blood* 1999;92:362–7.

8. Asahara T, Masuda H, Takahashi T. Bone marrow origin of endothelial progenitor cells responsible for postnatal vasculogenesis in physiological and pathological neovascularization. *Circ Res* 1999;85(3):221–28.

9. Isner JM, Asahara T. Angiogenesis and vasculogenesis as therapeutic strategies for postnatal neovascularization. *J Clin Invest* 1999 May;103(9):1231–6.

10. Asahara T, Murohara T, Sullivan A, Silver M, van der Zee R, Li T, *et al.* Isolation of putative progenitor endothelial cells for angiogenesis. *Science* 1997 Feb 14;275(5302):964–7.

11. Asahara T. Stem cell biology for vascular regeneration. *Ernst Schering Res Found Workshop* 2005(54):111–29.

12. Peichev M, Naiver AJ, Pereira D, Zhu Z, Lane WJ, Williams M, Oz MC, Hicklin DJ, Witte L, Moore MA, Rafii S. Expression of VEGFR-2 and CD133 by circulating human CD34(+) cells identifies a population of functional endothelial precursors. *Blood* 2000;95(3):952–8.

13. Dimmeler S, Zeiher AM. Endothelial cell apoptosis in angiogenesis and vessel regression. *Circ Res* 2000 Sep 15;87(6):434–9.

14. Schatteman GC, Hanlon HD, Jiao C, Dodds SG, Christy BA. Blood-derived angioblasts accelerate blood-flow restoration in diabetic mice. *J Clin Invest* 2000 Aug;106(4):571–8.

15. Salven P, Mustjoki S, Alitalo R, Alitalo K, Rafii S. VEGFR-3 and CD133 identify a population of CD34+ lymphatic/vascular endothelial precursor cells. *Blood* 2003 Jan 1;101(1):168–72.

16. Timmermans F, Van Hauwermeiren F, De Smedt M, Raedt R, Plasschaert F, De Buyzere ML, *et al.* Endothelial outgrowth cells are not derived from CD133+ cells or CD45+ hematopoietic precursors. *Arterioscler Thromb Vasc Biol* 2007 Jul;27(7):1572–9.

17. Gehling UM ES, Schumacher U, Wagener C, Pantel K, Otte M, Schuch G, Schafhausen P, Mende T, Kilic N, Kluge K, Schafer B, Hossfeld DK, Fiedler W. *In vitro* differentiation of endothelial cells from CD133-positive progenitor cells. *Blood* 2000;95(10):3106–12.

18. Yin AH, Miraglia S, Zanjani ED, Almeida-Porada G, Ogawa M, Leary AG, *et al.* CD133, a novel marker for human hematopoietic stem and progenitor cells. *Blood* 1997 Dec 15;90(12):5002–12.

19. Matthews W, Jordan CT, Gavin M, Jenkins NA, Copeland NG, Lemischka IR. A receptor tyrosine kinase cDNA isolated from a population of enriched primitive hematopoietic cells and exhibiting close genetic linkage to c-kit. *Proc Natl Acad Sci USA* 1991 Oct 15;88(20):9026–30.

20. Katoh O, Tauchi H, Kawaishi K, Kimura A, Satow Y. Expression of the vascular endothelial growth factor (VEGF) receptor gene, KDR, in hematopoietic cells and inhibitory effect of VEGF on apoptotic cell death caused by ionizing radiation. *Cancer Res* 1995 Dec 1;55(23):5687–92.

21. Kalka C, Masuda H, Takahashi T, Kalka-Moll WM, Silver M, Kearney M, *et al.* Transplantation of *ex vivo* expanded endothelial progenitor cells for therapeutic neovascularization. *Proc Natl Acad Sci USA* 2000;97:3422–7.

22. Finney MR, Greco J, Haynesworth S, Martin J, Hedrick D, Swan J, *et al.* Direct comparison of umbilical cord blood versus bone marrow-derived endothelial precursor cells in mediating neovascularization in response to vascular ischemia. *Biol Blood Marrow Transplant* 2006 May;12(5):585–93.

23. Lin Y, Weisdorf DJ, Solovey A, Hebbel RP. Origins of circulating endothelial cells and endothelial outgrowth from blood. *J Clin Invest* 2000;105:17–9.

24. Yoder MC, Mead LE, Prater D, Krier TR, Mroueh KN, Li F, *et al.* Redefining endothelial progenitor cells via clonal analysis and hematopoietic stem/ progenitor cell principals. *Blood* 2007 Mar 1;109(5):1801–9.

25. Ingram DA, Mead LE, Moore DB, Woodard W, Fenoglio A, Yoder MC. Vessel wall-derived endothelial cells rapidly proliferate because they contain a complete hierarchy of endothelial progenitor cells. *Blood* 2005 Apr 1;105(7):2783–6.

26. Hur J, Yoon CH, Kim HS, Choi JH, Kang HJ, Hwang KK, *et al.* Characterization of two types of endothelial progenitor cells and their different contributions to neovasculogenesis. *Arterioscler Thromb Vasc Biol* 2004 Feb;24(2):288–93.

27. Yoon CH, Hur J, Park KW, Kim JH, Lee CS, Oh IY, *et al.* Synergistic neovascularization by mixed transplantation of early endothelial progenitor cells and late outgrowth endothelial cells: the role of angiogenic cytokines and matrix metalloproteinases. *Circulation* 2005 Sep 13; 112(11):1618–27.

28. Kawamoto A, Gwon H, Iwaguro H, Yamaguchi J, Uchida S, Masuda H, *et al.* Therapeutic potential of *ex vivo* expanded endothelial progenitor cells for myocardial ischemia. *Circulation* 2001;103:634–7.

29. Masuda H, Kalka C, Asahara T. Endothelial progenitor cells for regeneration. *Hum Cell* 2000 Dec;13(4):153–60.

30. Murohara T, Ikeda H, Duan J, Shintani S, Sasaki K, Eguchi H, *et al.* Transplanted cord blood-derived endothelial precursor cells augment post-natal neovascularization. *J Clin Invest* 2000;105(11):1527–36.

31. Kawamoto A, Tkebuchava T, Yamaguchi J, Nishimura H, Yoon Y, Milliken C, *et al.* Intramyocardial transplantation of autologous endothelial progenitor cells for therapeutic neovascularization of myocardial ischemia. *Circulation* 2003 Jan 28;107(3):461–8.

32. Orlic D, Kajstura J, Chimenti S, Jakoniuk I, Anderson S, Li B, *et al.* Bone marrow cells regenerate infarcted myocardium. *Nature* 2001 Apr 5;410(6829):701–5.

33. Kim BO, Tian H, Prasongsukarn K, Wu J, Angoulvant D, Wnendt S, *et al.* Cell transplantation improves ventricular function after a myocardial infarction: a preclinical study of human unrestricted somatic stem cells in a porcine model. *Circulation* 2005 Aug 30;112(9 Suppl):I96–104.

34. Ott I, Keller U, Knoedler M, Gotze K, Doss K, Fischer P, *et al.* Endothelial-like cells expanded from CD34+ blood cells improve left ventricular function after experimental myocardial infarction. *FASEB J* 2005 Jun;19(8):992–4.

35. Wu KH, Zhou B, Mo XM, Cui B, Yu CT, Lu SH, *et al.* Therapeutic potential of human umbilical cord-derived stem cells in ischemic diseases. *Transplant Proc* 2007 Jun;39(5):1620–2.

36. Yamahara K, Sone M, Itoh H, Yamashita JK, Yurugi-Kobayashi T, Homma K, *et al.* Augmentation of neovascularizaiton in hindlimb ischemia by combined transplantation of human embryonic stem cells-derived endothelial and mural cells. *PLoS ONE* 2008;3(2):e1666.

37. Yang C, Zhang ZH, Li ZJ, Yang RC, Qian GQ, Han ZC. Enhancement of neovascularization with cord blood CD133+ cell-derived endothelial progenitor cell transplantation. *Thromb Haemost* 2004 Jun;91(6):1202–12.

38. Melero-Martin JM, Khan ZA, Picard A, Wu X, Paruchuri S, Bischoff J. *In vivo* vasculogenic potential of human blood-derived endothelial progenitor cells. *Blood* 2007 Jun 1;109(11):4761–8.

39. Au P, Daheron LM, Duda DG, Cohen KS, Tyrrell JA, Lanning RM, *et al.* Differential *in vivo* potential of endothelial progenitor cells from human umbilical cord blood and adult peripheral blood to form functional long-lasting vessels. *Blood* 2008 Feb 1;111(3):1302–5.

40. Ghen MJ, Roshan R, Roshan RO, Blyweiss DJ, Corso N, Khalili B, *et al.* Potential clinical applications using stem cells derived from human umbilical cord blood. *Reprod Biomed Online* 2006 Oct;13(4):562–72.

41. Sevilla J, Querol S, Molines A, Gonzalez-Vicent M, Balas A, Carrio A, *et al.* Transient donor cell-derived myelodysplastic syndrome with monosomy 7 after unrelated cord blood transplantation. *Eur J Haematol* 2006; 77(3):259–63.

12

Mesenchymal Stromal Cells Derived from Wharton's Jelly

Mark L. Weiss and Kiranbabu Seshareddy

ABSTRACT

The umbilical cord is a source of therapeutic cells. As described in other chapters in this book, umbilical cord blood is used clinically for reconstitution of the hematopoietic system following a number of diseases such as cancers, and certain congenital or metabolic diseases. We suggest that mesenchymal stromal cells derived from the umbilical cord may be useful therapeutic tools for a variety of diseases and for tissue engineering. While mesenchymal stromal cells are found infrequently in umbilical cord blood, they are found in abundance in the loose connective tissue surrounding the umbilical vessels, i.e., Wharton's jelly. In this chapter the characterization of Wharton's jelly-derived mesenchymal stromal cells, their status of banking for clinical evaluation, and preclinical work performed with these cells are reviewed and discussed.

INTRODUCTION

Embryonic stem cells, e.g., stem cells derived from the inner cell mass of early embryos, have received much attention due to their pluripotency and unlimited self-renewal potential.[1] To date, however, these cells have not been translated into the clinic. In contrast, adult-derived cells such as mesenchymal cells derived from the bone marrow, have clinical value, despite their limited self-renewal ability and limited differentiation potential.[2–7] Contrary to embryonic stem cells, many clinical trials are on-going for adult-derived cells, and for some adult-derived cells clinical use is already a routine.[8]

One problem ascribed to cell therapies derived from adult tissues is cell collection which may involve hospitalization and a painful, invasive procedure for the donor. Additionally, there may be a waiting period for locating the donor and arranging for the collection. Other problems are that the cells must be isolated and expanded using established good tissue manufacturing practices (GMP). However, adult-derived cells have limited expansion ability *in vitro* and require tissue matching to avoid rejection and to limit graft versus host disease (GVHD). Cryogenic stability is another concern with adult-derived therapeutic cells.[9] These problems are significant barriers preventing extensive cell banking for adult-derived therapeutic cells. The ability to bank cells for therapeutic use is a prerequisite for off-the-shelf cell therapy to produce commercial success and a reason for consideration of umbilical cord blood or Wharton's jelly-derived mesenchymal stromal cells for cellular therapies.

One appealing aspect of umbilical cord-derived cells and tissue is the fact that it is derived from a fetal deciduous tissue, e.g., a tissue rendered superfluous at birth. The collection of blood and cells from discarded fetal tissue is free of ethical controversy. Since these materials are discarded at birth, collection of blood and tissue from the cord offers an once-in-a-life-time opportunity to collect and store potentially valuable cells for future use. Finally, the collection process itself is simple and inexpensive, safe and pain-free to mother and child.

A number of reports published over the past ten years indicate that umbilical cord blood is a valuable source of hematopoietic stem cells.[10,11] Umbilical cord blood collection and cryogenic storage methods are known and are relatively established to date. Thus, generation of a bank of potentially useful cells from cord blood is not technically daunting. The sharing of haplotype and other data via cord blood bank networks permits rapid release of umbilical cord blood units that meet treatment requirements.[12] This allows rapid access to umbilical cord blood for cell therapy. Transplantation of cord blood has a lower incidence and lower severity of GVHD, too, compared to bone marrow transplant.[13,14] Additionally, umbilical cord blood has a lower incidence of viral contamination than adult blood. In summary, umbilical cord blood banking makes good sense because umbilical cord blood can be used as an off-the-shelf cell product for therapeutic use.

Our group has focused on a population of primitive cells that can be isolated and enriched from the loose connective tissue surrounding the umbilical vessels, called Wharton's jelly. As shown in Fig. 1, cells of

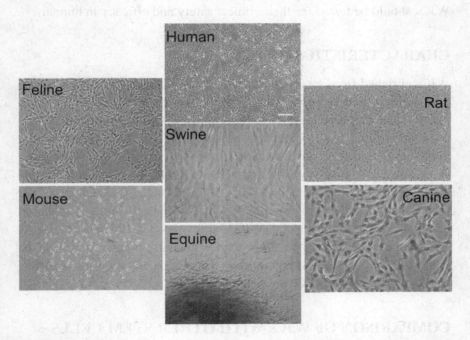

Fig. 1. Wharton's jelly-derived cells (WJCs) from seven species: human, rat, canine, equine, mouse, feline and swine. Note that the Wharton's jelly-derived cells have a similar morphology to mesenchymal stromal cells derived from other tissues.

similar morphology can be derived from Wharton's jelly from human, rodents, swine, equine, dog.[15–19] In addition, Wharton's jelly-derived cells (WJCs) have been isolated from bovine and rabbit (not shown).

These cells, which are referred to here as WJCs, have the properties of primitive mesenchymal stromal cells since they grow as plastic adherent cells with a cell surface marker expression profile similar to mesenchymal stromal cells (MSCs) derived from other tissues such as the bone marrow. Moreover, WJCs can differentiate along mesenchymal lineages.[20–22] WJCs can be extensively expanded *in vitro* and have shown therapeutic benefits in several preclinical models of human disease. In contrast to other sources of MSCs, WJCs are collected easily, and in large numbers from a seemingly inexhaustible source. Published information about WJCs characteristics, their isolation, and preclinical results are reviewed here. WJCs are effective for cellular therapy in preclinical models of human diseases, tissue engineering, and biotechnology. This suggests that WJCs should be tested for their clinical safety and efficacy in humans.

CHARACTERISTICS OF WJCs

WJCs enriched from rat, mouse, swine, equine, bovine, human umbilical cord have similar morphologies (see Fig. 1). The cells are adherent to plastic, they are phase bright and either fibroblastic or rounded in appearance (suggesting two distinct cell populations), similar to adult MSCs. WJCs have a similar surface marker profile as compared to MSCs. As shown by flow cytometry, WJCs are CD34, CD45, MHC class II, CD31, CD14, and c-kit negative and CD105, CD73, CD44, CD13, CD29, CD90 CD49c and CD49e positive (with over 90% of the cells in the population being positive).[20–22] In general, WJCs express less CD105, CD106, HLA-ABC and more CD117 and CD146 than bone marrow-derived MSCs (see Table 1). Note: in Table 1, eight papers that compared two or more different tissue sources of MSCs are listed.

COMPARISON OF WJCs WITH OTHER STEM CELLS

WJCs, like hematopoietic stem cells, express CD90 (Thy-1) and CD10. Few WJCs express c-Kit (<3%). WJCs may express low levels of Oct4,

Table 1. Comparison of the phenotype of MSC-like cells from various tissues.

Reference #	70	70	70	20	20	29	29	22	22	22	22	71	71	71	30	30	30	28	28	72	72	72
Cell type	AEC	A-MSC	C-MSC	HUCPVC	BM-MSC	WJC	BM-MSC	WJC	PTC	BM-MSC	ADF	Umbilical fibroblast-like cell	BM-MSC	Skin fibroblast	BM-MSC	HUVEC	WJC	WJC	BM-MSC	WJC	BM-MSC	Skin fibroblast
Marker																						
CD10	pos	pos	pos			pos	pos	pos	pos			++	++	++						pos	pos	pos
CD13	pos	pos	N/D			pos	pos	pos	pos	N/D		++	++	+++				N/D	N/D	pos	pos	pos
CD14	N/D	N/D	pos			N/D	N/D			N/D						pos	N/D	++	++			
CD29	pos	pos	pos			pos	pos			pos					N/D	pos	N/D					
CD31	pos	N/D				N/D	N/D	N/D	N/D						N/D	pos	N/D					
CD33	N/D	N/D	N/D			N/D	N/D	N/D	N/D						N/D	N/D	N/D	N/D	N/D			
CD34	N/D	N/D				N/D	N/D	N/D	N/D	N/D		+/-	+/-	+/-	N/D	N/D	N/D	N/D	N/D			
CD38	pos	pos				N/D	N/D	pos	pos	pos												
CD44	pos	pos	pos			pos	pos	N/D	N/D	N/D		++	++	++	N/D	N/D	N/D	+++	+++	pos	pos	pos
CD45	N/D	N/D	N/D			N/D	N/D	N/D	N/D	N/D		+/-	+/-	+/-				N/D	N/D	low	low	N/D
CD49b	pos	pos										+	+	+				++	N/D	low	low	N/D
CD49c	pos	pos																				
CD49d	pos	pos		pos																		
CD49e	pos	pos		pos	pos													N/D	N/D			
CD54	pos	pos	pos	pos	pos	pos	pos	pos				++	++	+						pos	pos	
CD73	pos	pos	pos		pos	pos	pos	pos	pos		pos	++	++	+				+++	++++	pos	pos	N/D
CD80																		N/D	N/D			
CD86																		N/D	N/D			
CD90	pos	pos	pos		pos	pos	pos	pos	pos			++	++	++++	moderate	moderate	pos	+++	++++	pos	pos	pos
CD105	pos	pos	pos		pos	pos	pos	pos		pos		++	++	++	moderate	pos	moderate pos	++++	++++	low	low	low

(*Continued*)

Table 1. (*Continued*)

Reference #	70	70	70	20	20	29	29	22	22	22	22	71	71	71	30	30	30	28	28	72	72	72
Cell type	AEC	A-MSC	C-MSC	HUCPVC	BM-MSC	WJC	BM-MSC	WJC	PTC	BM-MSC	ADF	Umbilical fibroblast-like cell	BM-MSC	Skin fibroblast	BM-MSC	HUVEC	WJC	WJC	BM-MSC	WJC	BM-MSC	Skin fibroblast
CD106	+/-					+	+++					N/D	++	N/D						N/D	pos	N/D
CD117 c-kit		N/D	N/D	3%	0.94%			N/D	N/D			+	+	+						pos	low	N/D
CD133/AC133		N/D	N/D															N/D	N/D			
CD146				52%	15%																	
CD166	pos	pos	pos	<1%	<1%	pos	pos			pos					3%	moderate	moderate	++	+++			
STRO-1										pos												
CD140b	pos	pos	low																			
CD271	low	pos	N/D																			
CD324	pos	N/D	pos																			
CD349 Frizzled-9	N/D	pos	pos																			
HLA-ABC	pos	pos	pos		+	+++	++++	pos	pos			pos	+	+++				+	++			
HLA-DR	N/D	N/D	N/D			N/D	N/D	N/D	N/D			N/D	N/D	N/D				N/D	N/D			
CFU-F				++		++	+	+++	+++	+	++											
Growth potential				++						+	++				+		+					

Note: Grey highlight indicates markers that may be different between cell types.[20,22,28-30,70-72] AEC: Amniotic epithelial cells, A-MSC: amniotic mesenchymal stromal cells, C-MSC: chorionic mesenchymal stromal cells, HUCPVC: human umbilical cord perivascular cells, BM-MSC: bone marrow-derived mesenchymal stromal cells, WJC: Wharton's jelly-derived cells, HUVEC: human umbilical vein-derived endothelial cells, PTC: placental-derived cells, ADF: adult dermal fibroblasts, CFU-F: colony forming unit, fibroblast. N/D: Not detected, pos: positive.

Reference # in citation list	Reference
70	Parolini O *et al.*, Stem Cells 2008
20	Baksh D *et al.*, Stem Cells, 2007
29	Lu L-L, *et al.*, Haematologica, 2006
22	Lund RD *et al.*, Stem Cells, 2007

Reference # in citation list	Reference
71	Lupatov AY *et al.*, Bull Exp Biol Med 2006
30	Magin S *et al.*, SCDev 2008
28	Friedman R *et al.*, BBMT 2007
72	Suzdal'tseva YG *et al.*, Bull Exp Biol Med 2007

Sox2, Nanog, ESG-1, FGFR4, and Rex-1,[15] with Oct4 expressed at lower levels than in embryonic stem cells but at higher levels than in adult-derived cells.[15] It was recently reported that induction of high levels of expression of four genes, Oct4, Sox2, Nanog and c-myc in fully differentiated mouse or human fibroblasts induces pluripotency.[23,24] Owing to their young age and their higher growth potential than adult-derived cells, WJCs might be induced to pluripotency more efficiently than adult-derived cells; however, other than expression of early genes, there is no evidence of WJCs pluripotency. Additionally, the promoter of the pluripotency gene Oct4 is methylated in WJCs,[25] suggesting that the Oct4 promoter is inactive in WJCs.

WJCs are most similar to MSCs and a recent review makes the case that WJCs are a primitive MSC.[26] Thus, WJCs, like bone marrow-derived MSCs, meet the International Society for Cellular Therapies (ISCT) criteria for MSCs.[27] Several points, however distinguish WJCs from adult-derived MSCs. For example, WJCs express higher levels of FLK-1, ABCG2, CXCR4, TERT and CD44 than bone marrow-derived MSCs.[26] In contrast, bone marrow-derived MSCs express collagen X at higher levels than WJCs. Furthermore, WJCs, like MSCs derived from, e.g., early fetal bone marrow, have a faster doubling time and a greater expansion capacity prior to senescence than adult bone marrow-derived MSCs.[20,22] WJCs can also be distinguished from bone marrow-derived MSCs based upon their cytokine profile,[28,29] where WJCs release more G-CSF, GM-CSF, LIF, IL-6, IL-8, etc. into the medium than bone marrow-derived MSCs. Despite the higher cytokine levels, there are no obvious differences in WJCs as a stromal support (feeder layer) compared to bone marrow-derived MSCs.[29,30]

Cells derived from Wharton's jelly may be similar to MSCs isolated from other fetal structures, e.g., placenta, amniotic fluid, umbilical cord blood, bone marrow, liver, etc.; this point however has not been confirmed experimentally yet.[31] During development, hematopoietic cells are detected first in the yolk-sac region. These HSCs do not persist and the first HSCs made in the embryo originate from the dorsal aorta region. Differences are found in the transcription factor regulation of hematopoiesis between yolk sac and fetal HSCs. Analogous to the changes observed in the hematopoietic system, we speculate that WJCs are the first mesenchymal stromal cells. It is possible that WJCs migrate into the developing embryo along with the embryonic germ cells and the developing blood islands. The WJCs might take up residence in the AGM region to contribute stromal support for blood formation in that niche during embryonic development.

WJCs ISOLATED FROM OTHER SPECIES

Since WJCs can theoretically be isolated from any amniote, investigators from Kansas State University derived WJCs from swine, bovine, equine, rat, mouse and human. It was noted that differentiation potential, growth characteristics, isolation and enrichment protocols and expression of certain genes is similar between species (swine, equine, rat and human have been most closely evaluated). The ability to easily and inexpensively collect WJCs from animals suggests that this may be a means to store genetic material from endangered species, or as cell source for cell therapy for veterinary medicine, or for biotechnology applications.

ISOLATION METHODS FOR WJCs

Several general methods for isolation of WJCs have been described.[17,32,33] In the first method, small pieces (explants) of Wharton's jelly are cut free from the cord and plated. In the second method, explants are treated enzymatically to free the cells from Wharton's jelly (collagenase 3000 U/ml and hyaluronidase 1 mg/ml for one hour; 0.1% trypsin/EDTA for 30 minutes), followed by crushing of the cord tissue to release the cells. The third general method involves dissecting the vessels free from the cord, ligating them and exposing the outer surface of the cord to enzymes to free the cells from the Wharton's jelly in the perivascular zone.[33] Recently, WJCs were enzymatically extracted with collagenase following mincing of the entire umbilical cord without removing the blood vessels.[22] This method reduces manipulation of the tissue and speeds up the extraction process. Other modifications include slitting the cord down its length, inverting the cord, and opposing the outer edges and suturing the outer edges to hold the cord inside-out. This is followed by enzyme treatment to release the WJCs from the collagen and hyaluronic acid extracellular matrix[20] or using mechanical aids to enhance tissue disruption (Seshareddy, Davis, Weiss, unpublished).

In the initial isolation, WJCs, red blood cells, and non-adherent cells are isolated. Plating on tissue culture treated plates and subsequent attachment of the cells results in enrichment of WJCs, as does the use of specific medium which favors WJC propagation over endothelial cells or HSCs. After two or three passages, a uniform population of WJCs is derived, based upon flow cytometric characterization of surface markers.[34]

For commercial banking, it has been possible to freeze minced cord piece. For freezing, the cord pieces must be finely minced (approximately 2 mm), immersed in cryoprotectant and frozen using a controlled rate freezing device for best viability and start-up post thaw (Pojda Z, personal communication). It is also possible to reanimate vitrified cord pieces (Davis D and Troyer D, personal communication). To date, a commercial process for isolating WJCs has not been published. It is known that collection of cord blood does not impair the ability to also isolate WJCs (Seshareddy and Weiss, unpublished data).[30]

IN VITRO EXPANSION CONDITIONS

WJCs can be easily adapted to a variety of standard cell culture media. For example, WJCs have been expanded in Dulbecco's modified Eagle's medium (DMEM-high glucose) with 10%–20% fetal bovine serum, 0.1% non-essential amino acids, 2-mercaptoethanol with or without antibiotics and antimycotics, conditions used to expand bone marrow-derived MSCs.[17,22,30,33–35] WJCs also grow well in low serum (2%) medium described by the Verfaillie lab used for expansion of multipotent adult progenitor cells (MAPCs).[17,22,26] Morphological changes in WJCs have been reported following tranfer from fetal bovine serum to human serum.[28] WJCs survive[30] and expand in serum-free medium at the same rate as in medium with 2% serum and growth factors (Seshareddy *et al.*, abstract ISSCR meeting 2008).

WJCs should be grown in a HEPA-filtered incubator at 37°C with 5% CO_2 in saturating humidity. In general, WJCs are passaged every three to five days when the population reaches 70%–80% confluency. Cells lose their contact inhibition if left too long in culture and when becoming over-confluent. Three-dimensional colonies form in some over-confluent cultures or the sheet of cells may slip, e.g., slide off the substrate, and form three-dimensional tubular structures. These conditions should be avoided since it is more difficult to produce single cell suspensions and passage cells from over-confluent cultures. WJCs should be fed every two to three days, depending upon passage schedule. Normally, half of the spent medium volume is removed and replaced with fresh medium. For passaging, cells can be lifted by first washing with phosphate buffered saline, followed by a short trypsinization of less than five minutes. Coating of tissue culture plates is not required for attachment of WJCs, although coating of tissue culture plates with gelatin or hyaluronic acid has been reported and may aid in passaging the cells.[34]

WJCs expand rapidly with short doubling times after the first three passages, and the doubling rate is stable for about 20 passages producing greater than 10^{16} cells in approximately 80 days. WJCs grow faster than human adult dermal fibroblasts and human bone marrow-derived MSCs, and are comparable in expansion potential to placenta-derived cells.[22] WJCs have been shown to be karyotypically stable through ten to 12 passages.[21,22] Clonal expansion of WJCs, has not been reported yet. WJCs can be cryogenically stored, thawed and expanded. To freeze WJCs, after lifting cells in the log growth phase and following centrifugation, the cell pellet is suspended in cryoprotectant solution (usually 10% dimethylsulfoxide (DMSO), 90% fetal bovine serum or 10% DMSO in growth medium), and frozen using a controlled rate freezer ($-1°C$/min) to $-80°C$ for 24 hours prior to transfer to the vapor phase of a liquid nitrogen tank. For unknown reasons, there is significant variability in the viability of WJCs samples after thawing (viability reported to be 50%–98%).[21,28,30,33]

In our hands, no changes were observed between the surface marker phenotype of WJCs that had been frozen and thawed and those that had never been frozen.[34] In contrast, Saraguser et al. reported a loss of HLA class I positive WJCs resulting from freeze/thaw and enrichment of HLA class I and class II negative cells by repeated freeze/thaw cycles.[33] WJCs express the consensus set of MSC markers including CD44, CD73, and CD105 and are negative for CD34 and CD45.[36] There may be differences between WJCs and bone marrow MSCs in surface markers: Lu et al. reported that WJCs express less CD106 and HLA class I;[29] Friedman et al.

reported expression of CD49b in WJCs and not in bone marrow-derived MSCs;[28] Baksh *et al.* reported higher expression of CD117 and CD146 in WJCs compared to bone marrow-derived cells.[20]

IN VITRO DIFFERENTIATION OF WJCs

Multiple reports show that human WJC can undergo differentiation to mesenchymal lineages, e.g., bone, cartilage and adipose cells,[37] in contrast to umbilical cord blood-derived MSCs which have been reported to have less capability of differentiating to adipose cells.[38] The literature on WJCs, however is inconsistent: one group reported a decreased potential for WJCs to form adipose cells[39] while other groups did not report any difficulties.[20,40] Similarly, there are several reports for a neural differentiation potential.[41] while one group was unable to differentiate to neural cells.[33] Possibly differences in protocols, growth factors or other *in vitro* conditions could account for these differences. As discussed by Magin *et al.*, for WJCs to move into the clinics, good manufacturing processes need to be used for expansion, which might include serum-free expansion medium.[30]

PRECLINICAL WORK USING WJCs

WJCs have proven useful for combating diseases in animal models based upon three properties that they share in common with other MSCs: (1) homing, (2) immune modulation, and (3) stromal support. WJCs have been tested and found to be useful in neurological diseases such as Parkinson's disease,[26,41] stroke[41–45] and retinal degeneration,[22] and in myocardial infarction,[46,47] cancer[48–51] and reconstitution of the hematopoietic system.[28–30,52] Thus, WJCs have therapeutic efficacy similar to bone marrow-derived MSCs. While bone marrow-derived MSCs have certain advantages such as autologous use with minimal manipulation, WJCs may be useful for allogenic transplantation once a significant public bank of transplant-grade WJCs is available. Furthermore, since exogenous genes can be introduced easily and foreign proteins can be overexpressed in WJCs, WJCs may be suitable for applications in gene therapy.[51] These features, coupled with their ability to home to sites of pathology, suggest that WJCs could be used to deliver a therapeutic payload to eradicate certain cancers.[50,51]

The ease of expansion and *in vitro* differentiation of WJCs makes them an easy choice for tissue engineering. Accordingly, WJCs have been evaluated for tissue-engineered heart valves, pulmonary conduits, cartilage and bone constructs.[53-60] To summarize this work, WJCs are a useful, plentiful source of cells for tissue engineering especially for bone and cartilage. WJCs have also been evaluated as feeder cell layers for blood cell expansion,[28-30,52,61] and for embryonic stem cell expansion,[62,63] showing that WJCs support the *ex vivo* expansion of umbilical cord blood cells and embryonic stem cells. While no difference was found between the expansion potential of WJCs and other MSCs for HSC expansion,[29,30] WJCs are thought to be the best source of autologous stromal cells for cord blood expansion.[30] Since WJCs and umbilical cord blood are easily collected at birth and have a variety of uses both in tissue engineering and as cells for cellular therapy, there has been a push for banking of both WJCs and cord blood.[31,64]

UNRESOLVED ISSUES OF WJC CHARACTERIZATION

There are still several unresolved problems related to the characterization of WJCs which are discussed below:

(1) Two morphologically distinct cell types are found in the WJC cultures. Several groups have reported clonal expansion, however, these groups have not described multipotency of the expanded clonal cells.[19,43,65] To rule out the possibility that the observed multipotency of the WJC population is due to the coexistence of several subpopulations with unipotential differentiation potential, Sarugaser and colleagues derived clonal populations of WJCs that demonstrated self-renewal potential and multipotency.[66]

(2) Are WJCs stromal cells or true stem cells? To determine whether WJCs are mesenchymal stromal cells or true stem cells, transplantation experiments have to be performed in which WJCs are tested for long-term engraftment and contribution to multiple mesenchymal tissues. While such transplantations experiments have been performed, evaluations have only been performed short-term and engraftment incompletely evaluated. A second test is serial transplantation. To

date, WJCs have not been used in serial transplantation experiments. This is in contrast to the mouse hematopoietic stem cells, where demonstration of reconstitution of the lympho-hematopoietic system has been shown following transplantation of a single hematopoietic stem cell. A similar demonstration would provide convincing data that WJCs are true stem cells.

(3) The underlying mechanism(s) of therapeutic efficiency of WJCs. While WJCs have been shown to have therapeutic effects and be safe in certain animal models (as discussed above) the mechanisms are a matter of speculation.[18,22,67] Two mechanisms have been proposed so far.[26,36,68] Firstly, WJCs may act via stromal support by releasing cytokines and growth factors that rescue sick and dying cells. Secondly, WJCs may act via immune modulation by suppressing the inflammatory responses to reduce or eliminate collateral damage. There is evidence supporting both mechanisms. For example, cytokines and growth factors are released by WJCs into the medium[19,28,29] and the immune properties of WJCs indicate immune modulation and inhibition of activated T-cell proliferation.[68,69] Whether one mechanism or the other is most important is unknown and whether each property can be manipulated to improve therapeutic efficacy has not been evaluated to date.

TRANSLATION OF WJCs FROM THE LABORATORY BENCH TO THE CLINIC

Introduction of a new drug into clinical trials has a well-defined pathway. This includes several specific areas such as manufacturing or production of the drug and safety and efficacy testing of the drug for the particular uses. These will not be discussed further here.

For cell-based therapies, two strategies are used for clinical translation. The first strategy is that of an autologous product with minimal manipulation. In this scenario, the WJC cell-product is derived from the patient (and thus, perfectly matched to the patient). This scenario fits with the private banking of WJCs, which is just now becoming available. One might imagine using WJCs as a feeder layer for expanding umbilical cord blood for improving and accelerating, perhaps, hematopoietic stem cell engraft-ment[28,30] or use WJCs as a feeder to prepare a particular blood cells for immune therapy.[52] In contrast to the autologous cell-product, the allogeneic WJC cell-product would contain WJCs with matching MHC from another individual. Examples of allogeneic cell products are umbilical cord blood, Apligraf (www.apligraf.com), MultiStem from Athersys (www.athersys.com) and Prochymal from Osiris (www.osiristx.com).

It is anticipated that the first-generation of WJC therapeutics will be an allogenic product where WJCs are used as a feeder for *ex vivo* expansion of umbilical cord blood and coinfused for allogeneic hematopoietic transplantation. Similar clinical trials using haploidentical MSCs to expand umbilical cord blood are ongoing. As suggested by Magin *et al.*, WJCs would be a superior source of MSCs for autologous cord blood expansion.[30] Before such a trial can be considered, a commercial process to collect and store WJCs has to be developed, as well as clinical-grade expansion protocols. To our knowledge, two companies offer commercial collection and storage of WJCs in the USA: Auxocell (www. Auxocell.com) and HealthBank (www.HealthBank.com).

ACKNOWLEDGMENTS

This work was funded by NIH NS-34160, State of Kansas Legislature, Targeted Excellence Grant from KSU Provost's office, K-INBRE, KSU Developing Scholar's program, KSU Howard Hughes Undergraduate Research Support, KSU CVM Dean's office and Terry C Johnson Center for Basic Cancer Research. Dr. S. Bennett and the OB/GYN staff at Mercy Hospital are thanked for umbilical cord collection. The Weiss, Troyer and Davis labs are thanked for supplying the images used in Fig. 1. Dr. B. J. Lutjemeier is thanked for critically reading of the manuscript. B. G. Weiss is thanked for her support. Work is dedicated to David K. Weiss (father of Mark L. Weiss).

References

1. Keller GM. *In vitro* differentiation of embryonic stem cells. *Curr Opin Cell Biol* 1995;7(6):862–9.

2. Burt RK, Traynor AE, Craig R, Marmont AM. The promise of hematopoietic stem cell transplantation for autoimmune diseases. *Bone Marrow Transplant* 2003;31(7):521–4.

3. Chen J, Li Y, Wang L, Lu M, Zhang X, Chopp M. Therapeutic benefit of intracerebral transplantation of bone marrow stromal cells after cerebral ischemia in rats. *J Neurol Sci* 2001;189(1–2):49–57.

4. Hogge DE, Sutherland HJ, Lansdorp PM, Phillips GL, Eaves CJ. The elusive peripheral blood hemopoietic stem cell. *Semin Hematol* 1993;30 (4 Suppl 4):82–9.

5. Jackson KA, Majka SM, Wang H, Pocius J, Hartley CJ, Majesky MW, *et al*. Regeneration of ischemic cardiac muscle and vascular endothelium by adult stem cells. *J Clin Invest* 2001;107(11):1395–402.

6. Koc ON, Lazarus HM. Mesenchymal stem cells: heading into the clinic. *Bone Marrow Transplant* 2001;27(3):235–9.

7. Pittenger MF, Mackay AM, Beck SC, Jaiswal RK, Douglas R, Mosca JD, *et al*. Multilineage potential of adult human mesenchymal stem cells. *Science* 1999;284(5411):143–7.

8. Giordano A, Galderisi U, Marino IR. From the laboratory bench to the patient's bedside: an update on clinical trials with mesenchymal stem cells. *J Cell Physiol* 2007;211(1):27–35.

9. Bruder SP, Jaiswal N, Haynesworth SE. Growth kinetics, self-renewal, and the osteogenic potential of purified human mesenchymal stem cells during extensive subcultivation and following cryopreservation. *J Cell Biochem* 1997;64(2):278-94.

10. Frassoni F, Podesta M, Maccario R, Giorgiani G, Rossi G, Zecca M, *et al*. Cord blood transplantation provides better reconstitution of hematopoietic reservoir compared with bone marrow transplantation. *Blood* 2003;102(3):1138–41.

11. Kurtzberg J. Cord blood transplantation in genetic disorders. *Biol Blood Marrow Transplant* 2004;10(10):735–6.

12. Gluckman E, Broxmeyer HA, Auerbach AD, Friedman HS, Douglas GW, Devergie A, *et al*. Hematopoietic reconstitution in a patient with Fanconi's anemia by means of umbilical-cord blood from an HLA-identical sibling. *N Engl J Med* 1989;321(17):1174–8.

13. Broxmeyer HE, Kurtzberg J, Gluckman E, Auerbach AD, Douglas G, Cooper S, *et al*. Umbilical cord blood hematopoietic stem and repopulating cells in human clinical transplantation. *Blood Cells* 1991;17(2):313–29.

14. Risdon G, Gaddy J, Broxmeyer HE. Allogeneic responses of human umbilical cord blood. *Blood Cells* 1994;20(2–3):566–70.

15. Carlin R, Davis D, Weiss M, Schultz B, Troyer D. Expression of early transcription factors Oct-4, Sox-2 and Nanog by porcine umbilical cord (PUC) matrix cells. *Reprod Biol Endocrinol* 2006;4:8.

16. Hoynowski SM, Fry MM, Gardner BM, Leming MT, Tucker JR, Black L, *et al.* Characterization and differentiation of equine umbilical cord-derived matrix cells. *Biochem Biophys Res Commun* 2007;362(2):347–53.

17. Mitchell KE, Weiss ML, Mitchell BM, Martin P, Davis D, Morales L, *et al.* Matrix cells from Wharton's jelly form neurons and glia. *Stem Cells* 2003; 21(1):50–60.

18. Weiss ML, Mitchell KE, Hix JE, Medicetty S, El-Zarkouny SZ, Grieger D, *et al.* Transplantation of porcine umbilical cord matrix cells into the rat brain. *Exp Neurol* 2003;182(2):288–99.

19. Weiss ML, Troyer DL. Stem cells in the umbilical cord. *Stem Cell Rev* 2006;2(2):155–62.

20. Baksh D, Yao R, Tuan RS. Comparison of proliferative and multilineage differentiation potential of human mesenchymal stem cells derived from umbilical cord and bone marrow. *Stem Cells* 2007;25(6):1384–92.

21. Karahuseyinoglu S, Cinar O, Kilic E, Kara F, Gumus AG, Ozel DD, *et al.* Biology of the stem cells in human umbilical cord stroma: *in situ* and *in vitro* surveys. *Stem Cells* 2007;25(2):319–31.

22. Lund RD, Wang S, Lu B, Girman S, Holmes T, Sauve Y, *et al.* Cells isolated from umbilical cord tissue rescue photoreceptors and visual functions in a rodent model of retinal disease. *Stem Cells* 2007;25(3):602–11.

23. Yamanaka S. Induction of pluripotent stem cells from mouse fibroblasts by four transcription factors. *Cell Prolif* 2008;41(Suppl 1):51–6.

24. Yu J, Vodyanik MA, Smuga-Otto K, Antosiewicz-Bourget J, Frane JL, Tian S, *et al.* Induced pluripotent stem cell lines derived from human somatic cells. *Science* 2007;318(5858):1917–20.

25. He H, McHaney M, Hong J, Weiss ML. Cloning and characterization of 3.1kb promoter region of the Oct 4 gene from the Fischer 344 rat. *Open Stem Cell J* 2009;1(1): 30–39.

26. Weiss ML, Medicetty S, Bledsoe AR, Rachakatla RS, Choi M, Merchav S, *et al.* Human umbilical cord matrix stem cells: preliminary characterization and effect of transplantation in a rodent model of Parkinson's disease. *Stem Cells* 2006;24(3):781–92.

27. Dominici M, Le BK, Mueller I, Slaper-Cortenbach I, Marini F, Krause D, *et al.* Minimal criteria for defining multipotent mesenchymal stromal cells. The International Society for Cellular Therapy position statement. *Cytotherapy* 2006;8(4):315–7.

28. Friedman R, Betancur M, Boissel L, Tuncer H, Cetrulo C, Klingemann H. Umbilical cord mesenchymal stem cells: adjuvants for human cell transplantation. *Biol Blood Marrow Transplant* 2007;13(12):1477–86.

29. Lu LL, Liu YJ, Yang SG, Zhao QJ, Wang X, Gong W, *et al.* Isolation and characterization of human umbilical cord mesenchymal stem cells with hematopoiesis-supportive function and other potentials. *Haematologica* 2006;91(8):1017–26.

30. Magin AS, Koerfer NR, Partenheimer H, Lange C, Zander A, Noll T. Primary cells as feeder cells for coculture expansion of human hematopoietic stem cells from umbilical cord blood: a comparative study. *Stem Cells Dev* 2009;18(1):173–86.

31. Secco M, Zucconi E, Vieira NM, Fogaca LL, Cerqueira A, Carvalho MD, *et al.* Multipotent stem cells from umbilical cord: cord is richer than blood! *Stem Cells* 2008;26(1):146–50.

32. McElreavey KD, Irvine AI, Ennis KT, McLean WH. Isolation, culture and characterisation of fibroblast-like cells derived from the Wharton's jelly portion of human umbilical cord. *Biochem Soc Trans* 1991;19(1):29S.

33. Sarugaser R, Lickorish D, Baksh D, Hosseini MM, Davies JE. Human umbilical card perivascular cells (HUCPV): a source of mesenchymal progenitors. *Stem Cells* 2005;23:220–9.

34. Seshareddy K, Troyer D, Weiss ML. Method to isolate mesenchymal-like cells from Wharton's Jelly of umbilical cord. *Methods Cell Biol* 2008;86:101–19.

35. Ennis J, Sarugaser R, Gomez A, Baksh D, Davies JE. Isolation, characterization, and differentiation of human umbilical cord perivascular cells (HUCPVCs). *Methods Cell Biol* 2008;86:121–36.

36. Troyer DL, Weiss ML. Wharton's jelly-derived cells are a primitive stromal cell population. *Stem Cells* 2008;26(3):591–9.

37. Wang HS, Hung SC, Peng ST, Huang CC, Wei HM, Guo YJ, *et al.* Mesenchymal stem cells in the Wharton's jelly of the human umbilical cord. *Stem Cells* 2004;22(7):1330–7.

38. Kern S, Eichler H, Stoeve J, Kluter H, Bieback K. Comparative analysis of mesenchymal stem cells from bone marrow, umbilical cord blood, or adipose tissue. *Stem Cells* 2006;24(5):1294–301.

39. Kim JW, Kim SY, Park SY, Kim YM, Kim JM, Lee MH, *et al.* Mesenchymal progenitor cells in the human umbilical cord. *Ann Hematol* 2004; 83(12):733–8.
40. Karahuseyinoglu S, Kocaefe C, Balci D, Erdemli E, Can A. Functional structure of adipocytes differentiated from human umbilical cord stroma-derived stem cells. *Stem Cells* 2008;26(3):682–91.
41. Fu YS, Cheng YC, Lin MY, Cheng H, Chu PM, Chou SC, *et al.* Conversion of human umbilical cord mesenchymal stem cells in Wharton's jelly to dopaminergic neurons *in vitro*: potential therapeutic application for Parkinsonism. *Stem Cells* 2006;24(1):115–24.
42. Ding DC, Shyu WC, Chiang MF, Lin SZ, Chang YC, Wang HJ, *et al.* Enhancement of neuroplasticity through upregulation of beta1-integrin in human umbilical cord-derived stromal cell implanted stroke model. *Neurobiol Dis* 2007;27(3):339–53.
43. Jomura S, Uy M, Mitchell K, Dallasen R, Bode CJ, Xu Y. Potential treatment of cerebral global ischemia with Oct-4+ umbilical cord matrix cells. *Stem Cells* 2007;25(1):98–106.
44. Koh SH, Kim KS, Choi MR, Jung KH, Park KS, Chai YG, *et al.* Implantation of human umbilical cord-derived mesenchymal stem cells as a neuroprotective therapy for ischemic stroke in rats. *Brain Res* 2008;1229:233–48.
45. Wu KH, Zhou B, Mo XM, Cui B, Yu CT, Lu SH, *et al.* Therapeutic potential of human umbilical cord-derived stem cells in ischemic diseases. *Transplant Proc* 2007;39(5):1620–2.
46. Wu KH, Yang SG, Zhou B, Du WT, Gu DS, Liu PX, *et al.* Human umbilical cord derived stem cells for the injured heart. *Med Hypotheses* 2007;68(1):94–7.
47. Wu KH, Zhou B, Yu CT, Cui B, Lu SH, Han ZC, *et al.* Therapeutic potential of human umbilical cord derived stem cells in a rat myocardial infarction model. *Ann Thorac Surg* 2007;83(4):1491–8.
48. Ayuzawa R, Doi C, Rachakatla RS, Pyle MM, Maurya DK, Troyer D, *et al.* Naive human umbilical cord matrix derived stem cells significantly attenuate growth of human breast cancer cells *in vitro* and *in vivo*. *Cancer Lett* 2009;280(1):31–7.
49. Ganta C, Chiyo D, Ayuzawa R, Rachakatla R, Pyle M, Andrews G, *et al.* Rat umbilical cord stem cells completely abolish rat mammary carcinomas with no evidence of metastasis or recurrence 100 days post-tumor cell inoculation. *Cancer Res* 2009;69(5):1815–20.

50. Rachakatla RS, Marini F, Weiss ML, Tamura M, Troyer D. Development of human umbilical cord matrix stem cell-based gene therapy for experimental lung tumors. *Cancer Gene Ther* 2007;14(10):828–35.

51. Rachakatla RS, Pyle MM, Ayuzawa R, Edwards SM, Marini FC, Weiss ML, *et al.* Combination treatment of human umbilical cord matrix stem cell-based interferon-beta gene therapy and 5-fluorouracil significantly reduces growth of metastatic human breast cancer in SCID mouse lungs. *Cancer Invest* 2008;26(7):662–70.

52. Boissel L, Tuncer HH, Betancur M, Wolfberg A, Klingemann H. Umbilical cord mesenchymal stem cells increase expansion of cord blood natural killer cells. *Biol Blood Marrow Transplant* 2008;14(9):1031–8.

53. Bailey MM, Wang L, Bode CJ, Mitchell KE, Detamore MS. A comparison of human umbilical cord matrix stem cells and temporomandibular joint condylar chondrocytes for tissue engineering temporomandibular joint condylar cartilage. *Tissue Eng* 2007;13(8):2003–10.

54. Breymann C, Schmidt D, Hoerstrup SP. Umbilical cord cells as a source of cardiovascular tissue engineering. *Stem Cell Rev* 2006;2(2):87–92.

55. Ferguson VL, Dodson RB. Bioengineering aspects of the umbilical cord. *Eur J Obstet Gynecol Reprod Biol* 2009;144(Suppl 1):S108–13.

56. Hou T, Xu J, Wu X, Xie Z, Luo F, Zhang Z, *et al.* Umbilical cord Wharton's jelly: a new potential cell source of mesenchymal stromal cells for bone tissue engineering. *Tissue Eng Part A* 2009;15(9): 2325–34.

57. Schmidt D, Asmis LM, Odermatt B, Kelm J, Breymann C, Gossi M, *et al.* Engineered living blood vessels: functional endothelia generated from human umbilical cord-derived progenitors. *Ann Thorac Surg* 2006;82(4): 1465–71.

58. Schmidt D, Mol A, Neuenschwander S, Breymann C, Gossi M, Zund G, *et al.* Living patches engineered from human umbilical cord derived fibroblasts and endothelial progenitor cells. *Eur J Cardiothorac Surg* 2005;27(5):795–800.

59. Wang L, Seshareddy K, Weiss ML, Detamore MS. Effect of initial seeding density on human umbilical cord mesenchymal stromal cells for fibrocartilage tissue engineering. *Tissue Eng Part A* 2009;15(5):1009–17.

60. Wang L, Tran I, Seshareddy K, Weiss ML, Detamore MS. A comparison of human bone marrow-derived mesenchymal stem cells and human umbilical cord-derived mesenchymal stromal cells for cartilage tissue engineering. *Tissue Eng Part A* 2009;15(8): 2259–66.

61. Bakhshi T, Zabriskie RC, Bodie S, Kidd S, Ramin S, Paganessi LA, *et al.* Mesenchymal stem cells from the Wharton's jelly of umbilical cord segments provide stromal support for the maintenance of cord blood hematopoietic stem cells during long-term *ex vivo* culture. *Transfusion* 2008;48(12):2638–44.

62. Hiroyama T, Sudo K, Aoki N, Miharada K, Danjo I, Fujioka T, *et al.* Human umbilical cord-derived cells can often serve as feeder cells to maintain primate embryonic stem cells in a state capable of producing hematopoietic cells. *Cell Biol Int* 2008;32(1):1–7.

63. Saito S, Sawai K, Minamihashi A, Ugai H, Murata T, Yokoyama KK. Derivation, maintenance, and induction of the differentiation *in vitro* of equine embryonic stem cells. *Methods Mol Biol* 2006;329:59–79.

64. Klingemann H. Discarded stem cells with a future? *Expert Opin Biol Ther* 2006;6(12):1251–4.

65. La RG, Anzalone R, Corrao S, Magno F, Loria T, Lo IM, *et al.* Isolation and characterization of Oct-4+/HLA-G+ mesenchymal stem cells from human umbilical cord matrix: differentiation potential and detection of new markers. *Histochem Cell Biol* 2009;131(2):267–82.

66. Sarugaser R, Hanoun L, Keating A, Stanford WL, Davies JE. Human mesenchymal stem cells self-renew and differentiate according to a deterministic hierarchy. *PLoS One* 2009 Aug 4;4(8):e6498.

67. Medicetty S, Bledsoe AR, Fahrenholtz CB, Troyer D, Weiss ML. Transplantation of pig stem cells into rat brain: proliferation during the first 8 weeks. *Exp Neurol* 2004;190(1):32–41.

68. Weiss ML, Anderson C, Medicetty S, Seshareddy KB, Weiss RJ, VanderWerff I, *et al.* Immune properties of human umbilical cord Wharton's jelly-derived cells. *Stem Cells* 2008;26(11):2865–74.

69. Ennis J, Gotherstrom C, Le BK, Davies JE. *In vitro* immunologic properties of human umbilical cord perivascular cells. *Cytotherapy* 2008;10(2):174–81.

70. Parolini O, Alviano F, Bagnara GP, Bilic G, Buhring HJ, Evangelista M, *et al.* Concise review: isolation and characterization of cells from human term placenta: outcome of the First International Workshop on Placenta Derived Stem Cells. *Stem Cells* 2008;26(2):300–11.

71. Lupatov AY, Karalkin PA, Suzdal'tseva YG, Burunova VV, Yarygin VN, Yarygin KN. Cytofluorometric analysis of phenotypes of human bone

marrow and umbilical fibroblast-like cells. *Bull Exp Biol Med* 2006; 142(4):521–6.

72. Suzdal'tseva YG, Burunova VV, Petrakova NV, Vakhrushev IV, Yarygin KN, Yarygin VN. Comparatine analysis of cytophenotypes of cells of mesenchymal lineage isolated from human tissues. *Bull Exp Biol Med* 2007;143(1):147–54.

PART IV

Future Outlook

13

Banking of Cord Blood

Shin Y. Ong and William Y. K. Hwang

ABSTRACT

The storage of publicly banked cord blood (UCB) units could eventually allow every patient who needs a hematopoietic stem cell transplant (HSCT) access to life-saving blood stem cells because of enhanced ability of UCB transplantation (CBT) to tolerate HLA mismatches compared to bone marrow transplantation (BMT), where stricter HLA matching requirements exist. Enhancements in CBT protocols, double CBT, *ex vivo* UCB expansion and homing technologies, together with a special effort to bank the higher cell dose units could overcome the problem of low cell dose for larger patients and adults. A worldwide network of UCB banks currently exists to help ensure that patients of different ethnicities in need of HSCT have maximal access to matching UCB cells, regardless of which part of the world they are in. Strict regulations in UCB collection, testing, processing, storage and distribution exist to protect the interests of both the transplant patients and UCB donors. More than 15,000 CBT have been performed around the world, using UCB units from public UCB banks. The storage of UCB units for personal use by private UCB banks, while rarely used and not encouraged by organizations like the American Academy of Pediatrics (AAP), is a personal decision made by families based on the remote likelihood of usage and the possibility that science could evolve in future to make such treatments

291

more commonplace. Meanwhile public UCB banks, which provide the vast majority of UCB units for transplantation, have been serving as a common resource for the population for an increasing number of HSCT. Community and governmental support to increase the inventory of these public banks is critical to ensure the growth of this vital resource.

INTRODUCTION

In 1988, Eliane Gluckman and colleagues transplanted the cord blood (UCB) of an HLA-identical sibling to a six-year-old boy with Fanconi anemia. Prenatal testing showed that the sibling was unaffected by the disorder, and her UCB was thus collected at birth and cryopreserved. Upon transplantation, engraftment occurred on day 22, and complete hematologic reconstitution was achieved without graft-versus-host disease (GVHD).[1] Following this, the first series of sibling matched and one-antigen mismatched UCB transplants in 44 children with inherited diseases and hematologic malignancies was reported by Wagner *et al.*[2] High rates of engraftment (85% at 50 days) and low risk of acute or chronic GVHD were observed in this series. The confirmation that a small volume of UCB provides a sufficient source of HSCs to reconstitute hematopoiesis in children led to interest in banking unrelated donor UCB.

The trend of increasing use of UCB for transplantations over the years is multi-factorial. Firstly, UCB units are collected and stored before they are needed, allowing assured and rapid availability for transplantation. Studies have shown that patients can receive UCB transplantation (CBT) at a median of 25 to 35 days faster than BMT,[3,4] which is beneficial for patients in need of an urgent transplantation. Secondly, data has demonstrated that well-matched CBT may lead to better outcomes than equally matched bone marrow (BM) in children, which have encouraged its use over BMT in pediatric patients at some transplant centers, particularly for patients in need of an urgent transplant.[5] Broader use of CBT in adults is being made possible with new non-myeloablative conditioning regimens that allow engraftment of lower-dosed UCB grafts, and the use of double unit CBT.[6,7] The indications for CBTs are continually growing (Table 1).

Of great importance, likely because of the lower incidence of GVHD, CBT will tolerate up to two mismatches of HLA antigens. This has greatly expanded the donor pool, especially for ethnic and racial minorities with

Table 1. Diseases for which cord blood transplants have been performed.

Malignant diseases

Acute lymphoblastic leukemia
Juvenile myelomonocytic leukemia
Acute myeloid leukemia
Lymphoma
Chronic lymphocytic leukemia
Myelodysplastic syndrome
Chronic myelogenous leukemia
Neuroblastoma
T-cell leukemia
Common variable immunodeficiency-myelodysplastic syndrome

Non-malignant Diseases

Adrenoleukodystrophy
Krabbe's disease
Amegarkayocytic thrombocytopenia
Langerhans cell histiocytosis
Bare-leukocyte syndrome
Lesh-Nyhan disease
Blackfan-Diamond anemia
Leukocyte adhesion defect
Dyskeratosis congenital
Neuronal ceroid lipofuscinosis
Familial erythrophagocytic lymphohistiocytosis
Osteopertrosis
Fanconi anemia
Severe aplastic anemia
Global cell leukodystrophy
Severe combined immunodeficiency
Gunter disease
Sickle cell disease
Hunter's syndrome
Thalassemia
Hurler's syndrome
Wiskott-Aldrich syndrome
Kostmann's syndrome
X-linked lymphoproliferative disorder

rare alleles, because search for a compatible UCB unit means finding a unit of good cell dose ($>3 \times 10^7$ total nucleated cells (TNC)/kg) with mismatches of up to two of six HLA antigens. In contrast, for unrelated BMT, donor-recipient allele matching at HLA-A, -B, -C, and DRB1 is recommended, with additional matching at HLA-DQB1 leading to the highest survival probability.[8] It has been estimated that, for any given ethnic group, a 300,000 cord unit inventory may be sufficient to enable identification of a suitably matched UCB graft for every pediatric patient, and for more than 90% of adult patients.[9] On the other hand, it is impossible to identify an eight out of eight high resolution allele matched BM for every patient even if the registry size is infinitely large, due to the extensive polymorphism of HLA alleles.[10,11] Furthermore, minorities are under-represented in all BM registries, and identifying a marrow donor for ethnic minorities remains a challenge.[12] This imbalance can potentially be corrected by collecting UCB stem cells from minorities in proportion to their representation among live births. In fact, better minority representation has been achieved in many UCB banks, due to the establishment of collection centers in new sites of ethnic diversity, more willingness to donate UCB versus BM and an increased awareness of the benefits of UCB donation.[13–15]

The ability to use UCB from a sibling donor is dependent on the mother delivering a sibling of acceptable HLA-matching before the transplant, by chance or through the controversial practice of pre-implantation genetic testing and HLA typing.[16] However, not all patients in need of CBT have an available sibling donor who can donate UCB. It thus became apparent that establishing a public bank of frozen unrelated UCB units is critical for the wider clinical application of UCB stem cells.

FIRST PUBLIC CORD BLOOD BANK
AND UNRELATED DONOR TRANSPLANTS

The first public umbilical UCB bank was set up by Rubinstein and colleagues at the New York Blood Center in 1992 with funding support from the National Heart, Lung, and Blood Institute of the National Institutes of Health.[17] Rubinstein and colleagues also reported the method of volume reduction to conserve storage space, where red blood cells are removed and only a small volume of mononuclear cells are retained for storage, a technique still used by most public UCB banks today.[18] Considerable efforts have since been made to standardize methods of collection, processing, and cryopreservation of UCB stem cells.

Kurtzberg and colleagues at Duke University (Durham, NC) performed the first unrelated donor CBT in a child with recurrent T-cell acute lymphoblastic leukemia using an unrelated donor UCB unit from the New York Cord Blood Bank in 1993. In 1996, they reported the results of the first 25 unrelated, partially HLA-mismatched transplants in children, demonstrating 48% disease-free survival and less acute and chronic GVHD compared to better HLA-matched BM stem cells.[19]

Further clinical data from single institutions and various registries observed that the level of HLA matching and cell dose of the graft have significant impact on the outcomes of transplantation (Table 2).[20–22]

Table 2. Selection of cord blood units.

Program	Cell dose	HLA match
Eurocord	$> 3 \times 10^7$ NC/kg	Hematological malignancy • No effect of 1–2 antigen HLA mismatch on LFS • Mismatch adversely affects engraftment, but reduces relapse Non-malignant disease • Survival reduced with mismatch
New York Blood Center Cord Blood Program	• Cell dose (range: $0.7 – 10 \times 10^7$ NC/kg) does not affect outcome for 6/6 matched • Twice the cell dose needed to compensate for adverse effect on TRM and survival in 4/6 compared with 5/6 transplants	Mismatch adversely affects engraftment, GVHD, TRM, LFS
University of Minnesota Blood and Marrow Transplant Program	• $> 2.5 \times 10^7$ NC/kg for 5/6–6/6 matched single cord transplants • At least 3×10^7 NC/kg for double cord transplant	• Partial HLA matching required between the 2 units for double cord transplants

The above summarizes previously published data. Current recommendations are constantly evolving.
TRM: Transplant related mortality; LFS: leukemia free survival.

These insights have prompted efforts to increase the inventory of UCB banks, so as to improve the probability of identifying a donor unit of good cell dose (at least 3×10^7 NC/kg) with no more than two mismatches at the HLA-A, B antigens and DRB1 alleles.

SIBLING DONOR CORD BLOOD BANKS

A joint study by Eurocord and the International Bone Marrow Transplant Registry (IBMTR) comparing outcomes in 113 children who received CBT versus BMT from HLA-identical siblings found a lower risk of acute and chronic GVHD with CBT as well as no significant difference in relapse-related deaths and overall survival, indicating that CBT is as useful as HLA-identical sibling BMT.[23] Thus, in families affected by genetic diseases that can be corrected by allogeneic HSCT, it may be recommended to collect and freeze UCB units of healthy newborn siblings to allow earlier transplant rather than wait for the donor sibling to be old enough to donate BM.

To date, sibling UCB donations have been used for transplants for acute lymphoblastic leukemia, thalassemia major, Fanconi anemia, Diamond-Blackfan anemia and neutrophil dysplasia.[24–26] Before transplantation, each sibling UCB unit undergoes extensive individualized testing to exclude the specific genetic disease for which the transplant was indicated. A sibling UCB donor collection program initiated by the National Institutes of Health also waits until the UCB donor is of a certain age prior to transplantation before using the UCB unit, so that longitudinal assessment of the donor's clinical and hematological characteristics can help to exclude clinically significant genetic diseases.[26] For genetic diseases where graft-versus-tumor (GVT) reaction can have no benefit, CBT from an HLA-identical sibling offers the benefit of a reduced risk of potentially fatal GVHD after transplantation from an unrelated donor. However, in the case of leukemia and other malignancies, the GVT effect that occurs with allogeneic transplants can be essential to the success of the transplant, rendering sibling-directed CBT a less desirable treatment option if a suitably matched, unrelated UCB unit of adequate cell dose is available.

PRIVATE CORD BLOOD BANKS

Private UCB banks are now offering expectant parents the option to store UCB for exclusive possible future use by the donating child (autologous transplant) for a fee. Parents have been encouraged to bank their children's UCB for autologous use as a form of biological insurance. The validity of directed UCB storage in low risk families, however, has been challenged by some organizations like the American Academy of Pediatrics (AAP) for some of the reasons enumerated below.

Firstly, the likelihood of an individual in low-risk families using autologous UCB for hematopoietic disorders before the age of 20 is low (estimates range from one in 20,000 to one in 2700).[27,28] Furthermore, autologous UCB cells are not often felt to be the best treatment for acute leukemia, the most likely indication for HSC transplantations, since there may be no graft versus tumor (GVT) effect, and pre-leukemic mutations may be present in the UCB. Speculations that UCB can treat non-hematopoietic diseases are still of theoretical nature, although it can be argued that rapid advancements in the field of stem cell therapy may bring breakthroughs. If autologous stem cell transplantation for non-hematopoietic disease proves to be safe and effective in the future, but UCB stem cells were not stored at birth, stem cells can still be harvested from BM or peripheral blood. In addition, multipotent stem cells that are isolated from other tissue sources or generated through the reprogramming of somatic cells into a pluripotent state may also be used in cell and tissue regeneration in the future.[29,30]

Despite this, some parents who can afford the costs connected to private storage have selected this option to avoid any later regret of not having opted for this resource of stem cells for their children. The likelihood of a person developing a disorder that is treatable using UCB stem cells could increase over the lifetime of that individual, as science advances and the promises of regenerative medicine and gene therapy are realized. Since it is impossible to predict the results of research that may affect the future potential use of these cells, parents making this important one time decision sometimes decide against altruistic donation to public UCB banks. To-date not many UCB units have been used for the donor itself and whether the majority of these autologous treatments have been necessary or effective is still unclear. The stored UCB is available exclusively for sick siblings, while in a public bank such a unit might have been released to somebody else.

Recently, a private-public hybrid UCB bank was set up in the United Kingdom to cater to public donors and high-risk families.[31] The Virgin Health bank stores the child's UCB in two portions — one as a private sample for sole use of the child or family, and the second as a public sample. However, the cell dose for each of the units would be lower than if one uses the entire unit and could reduce the public unit beyond usability due to low cell dose. Nevertheless, the issue of hybrid banks is being discussed at international meetings and one possible model is to have UCB banks that strongly encourage public UCB banking, but which allow the parents to make the option of private banking if they have read and signed the literature in favor of public banking, especially if the UCB unit has fallen below the criteria given for the public bank.

WORLDWIDE CORD BLOOD BANKING NETWORKS

There are currently more than 400,000 banked UCB units throughout the world, and it is estimated that more than 15,000 UCB transplants have been performed, mainly in pediatric patients.[20] Japan leads the world in the number of CBT performed in adults; there are more than two times as many adult UCB recipients as pediatric recipients (https://www.netcord.org/inventory.html). Currently, transplant physicians have to approach multiple UCB banks to identify the best source of stem cells (Fig. 1), and there is a need to improve coordination to accelerate the procurement of stem cells.

Networks of UCB banks and registries are listed below.

BMDW (http://www.bmdw.org/)

The Bone Marrow Donors Worldwide (BMDW) is the biggest international UCB registry that provides centralized information on the HLA phenotype and cell dose of unrelated UCB units from many UCB registries around the world. All registries participating in BMDW are expected to adhere to the guidelines of the WMDA (World Marrow Donor Association), which promotes the definition and standardization of ethical, technical, medical

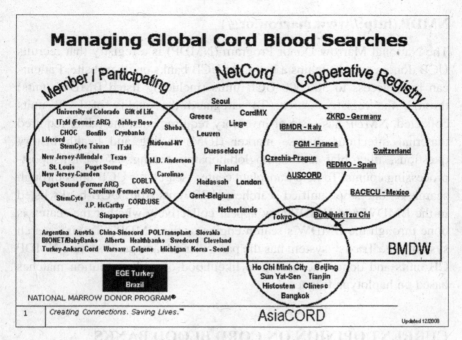

Fig. 1. Cord blood banks and registries. *Figure courtesy of the NMDP.*

and financial aspects of international UCB transplants. In 2007, the BMDW registry contained data on 280,000 samples in about 95 registries in different countries around the world.

NETCORD (https://www.netcord.org/index.html)

Netcord is an international network of UCB banks, and contains data on about 150,000 UCB units in 23 UCB banks around the world (as of September 2007). It provides a "virtual office" or online search system for real-time searches of compatible and available unrelated UCB units, and has established quality standards in collaboration with the Foundation for the Accreditation of Cellular Therapy (FACT) on the collection, cryopreservation, storage and release of units. All NETCORD banks must obtain accreditation by FACT and thus comply with quality standards of FACT/NETCORD.

NMDP (http://www.marrow.org/)

The National Marrow Donor Program (NMDP) is a registry that recruits UCB donors and maintains a listing of UCB banks on its website. Patients can have access to 300,000 UCB units (including listed BMDW units) through the registry search. To safeguard the quality of UCB units collected, NMDP has a set of inventory requirements, including required maternal infectious disease marker (IDM) testing and health history questions, microbial and hemoglobinopathy testing, as well as post-processing counts. To help transplant centers search for UCB units through seamless "one-stop" unified searches, UCB units from UCB banks listed in the BMDW and NMDP now appear collectively whether the search is done through the BMDW's search engine or the NMDP's Traxis® search system. The Traxis® system has the particular advantage of sorting NMDP CB units and donors according to likelihood of high-resolution matches based on haplotype frequencies.

CURRENT OPINION ON CORD BLOOD BANKS

There is now general consensus on the utility of UCB storage, and many agencies support the growth and expansion of UCB banks to increase the probability of finding better matches and improving transplantation outcomes. However, the establishment of UCB banks has raised a number of ethical issues that are challenging to resolve. Some of the most crucial of these are informed consent for the donation of UCB for banking and commercialization of UCB banks.

Informed consent is required by policy makers and ethical review boards for the banking of UCB because it should be a voluntary process and because sensitive medical information and labeled specimens are collected and tested for diseases. Institutional Review Boards or Ethical Committees should approve the process of UCB collection and informed consent. However, the consent policy is not consistent, and depends on whether UCB is collected for private or public banking, whether *in utero* and *ex utero* collection methods are used, and whether the collected samples remain traceable. Policies also differ in different countries. Most consent forms include elements of the risks and benefits of UCB collection, the possible future uses of UCB, and a discussion of confidentiality in the consent form. Although an extensive informed consent is desirable, consent procedures should take into account practical staffing issues.

Several governing bodies, notably the European Union (EU)[a] and some professional organizations [such as the AAP[32] and the Royal College of Obstetricians and Gynecologists (RCOG)][b] have adopted policy statements about the ethics of private storage of UCB. The general conclusion is that since the likelihood of the use of autologous stem cells is very low, UCB banking for one's own use is not encouraged. The RCOG opinion states, "There is still insufficient evidence to recommend directed commercial UCB collection and stem-cell storage in low-risk families." Similarly, the AAP policy recommends that "private storage of UCB as 'biological insurance' is unwise" unless there is a "full sibling in the family with a medical condition (genetic/malignant) that could potentially benefit" from CBT. The American Medical Association (AMA) also adopted guidelines in Nov 2007, arguing that "physicians should encourage donation to public cord blood banks," while the WMDA discourages funding of autologous UCB storage by national governments. A balanced perspective on the different options for banking should be provided to expectant parents to allow them to make an informed decision about whether to store the UCB in a public or private bank. It is important to realize, however, that these are very private, individual decisions that also need to take into account the possibility (however remote or unlikely) of evolution in the science of UCB banking with time.

CORD BLOOD COLLECTION PROCESS

Proper collection of umbilical UCB for banking purposes involves multiple steps to ensure safe, ethical and judicious UCB collection while maintaining the highest quality in collection, temporary storage and transport of the collected UCB unit (CBU). These processes start prior to birth of the baby and continue with the long-term follow-up of the baby after birth.

---

[a] http://ec.europa.eu/european_group_ethics/docs/avis19_de.pdf
[b] http://parentsguidecordblood.org/content/media/m_pdf/RCOG_opinion2006.pdf

Medical History

A thorough medical history from the mother is required to minimize the risk of transmission of genetic and infectious disease to the transplant recipient. This includes the mother's sexual and travel history as well as the history of any genetic diseases in first-degree relatives. There should be definite and clear criteria for the inclusion or exclusion of donors based on international guidelines, which currently do not exist. However, UCB banks also have to take into account the requirements for import or export of products into or out of specific countries. A case in point is the importation of UCB units collected in the UK and shipped for transplantation into the United States. In view of current criteria in the US, which stipulates that there is a risk of variant Creutzfeldt-Jakob disease (vCJD) for donors who have lived for a certain period of time in the UK, such UCB units could still be used in the US, but would be labeled as FDA-ineligible units.

Informed Consent

The informed consent process should include the UCB collection procedure, the donation of the UCB unit for use in transplantation, possible risks and benefits to the mother or infant donor including medical and ethical concerns, possible alternatives, the right of the mother to refuse without prejudice, the need to take personal and family medical history, the need to review the medical records of the mother and infant donor, as well as the need to take blood samples from the mother and UCB unit for infectious and genetic disease testing. The amount of blood required for testing, as well as the timing of these blood tests should be clearly stipulated. It is also necessary to reassure the mother of confidentiality while providing information that there will be storage of reference samples from the mother and the UCB unit for future testing, maintenance of linkage for the purpose of notifying infant donor/family of infectious or genetic diseases, and a possible need to contact the mother at any time after UCB donation. Confidentiality is particularly important because of the need for infectious disease testing and also when there is a possibility that the UCB units could also be used for laboratory-based research. The name of the donor should be on a "need-to-know" basis and strictly limited only to personnel in the UCB bank who must have this information to execute their tasks effectively. The mother should also know the policies of the UCB bank with regards to disposal of the UCB unit and whether she will be informed if the UCB unit is to be disposed.

Collection Procedures

Safety of the donor and mother is crucial and emergency medical care should be available for the mother and the infant donor. However, if protocols are followed and if the process of birth and delivery of the baby is not altered to facilitate UCB collection, there should be little or no risk to mother or baby (Fig. 2A).

When it comes to UCB collection, strict adherence to existing protocols is necessary, with adequate training of staff, aseptic techniques and measures in place to protect mother and infant donor. *In utero* UCB collection takes place from the placental umbilical vessels after the infant

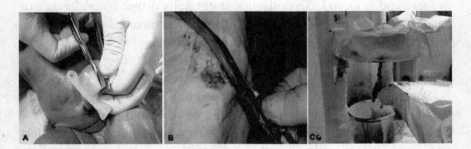

Fig. 2. Cord blood collection. (**A**) After birth of the baby, the umbilical cord is clamped and cut in the usual way. (**B**) The cord blood collection procedure should not interfere with the usual delivery process. If *in utero* collection is carried out, as in this example, then collection of the cord blood is carried out while awaiting delivery of the placenta. The surface of the umbilical cord should be thoroughly cleaned according to each cord blood bank's protocols. A needle (in this case from a blood collection bag) is inserted to collect the cord blood. Some other devices are available which collect the cord blood differently and some also incorporate a second needle to allow for further collection or flushing of the umbilical cord. (**C**) If *ex utero* collection is carried out, then the placenta is first delivered and the placenta and umbilical cord brought to another area or room where the collection will then be carried out. *Photos courtesy of the Singapore Cord Blood Bank.*

donor has been delivered and has been separated from the umbilical cord, but before the placenta has been delivered (Figs. 2A and 2B). Delivery practices should not be modified in attempt to increase UCB volume, and, when *in utero* UCB collection is performed, additional safeguards should be in place to ensure safety of mothers and infant donors. For example, we feel that *in utero* collections should be performed only in uncomplicated deliveries and only on singleton deliveries. The concern over collecting UCB in multiple pregnancies arises not only because of concerns of mixing of UCB between twins, but also because of concerns that UCB collection could start for the first twin before delivery of the second twin. *Ex utero* UCB collection takes place from the placental umbilical vessels after the placenta has been delivered. This usually takes place in a separate room and the placenta is usually suspended in a bowl or other device with a hole at the bottom to allow the umbilical cord to pass through (Fig. 2C). Because the UCB may have to pass through a few hands before it reaches the processing facility, procedures should also be in place to ensure confidentiality and yet traceability for purposes of correlation with medical history and follow-up. The choice between *ex utero* and *in utero* UCB collection is normally driven by preference of the obstetricians and collection centers, as well as financial considerations and staffing presence, as data has indicated that there is little significant difference between both methods when it comes to cell recovery and volume of collection.[33]

Several methods have been developed to increase the yield of collection of UCB. These include flushing or placental perfusion of the umbilical cord with heparinized saline[34] as well as strategies to collect additional UCB directly from the placenta.[35] Special equipment to help extract extra blood from an *ex utero* harvest have been devised as well as some novel devices in the pipeline (Fig. 3).[36] Some other procedures like placing the baby on the mother's abdomen prior to UCB collection and before clamping the umbilical cord are not advocated or practiced by the Singapore Cord Blood Bank, but investigators have shown that these methods can further enhance UCB yield without endangering the newborn.[37] However, the process of delivery of the newborn and clamping of the umbilical cord should not be altered to facilitate UCB collection.

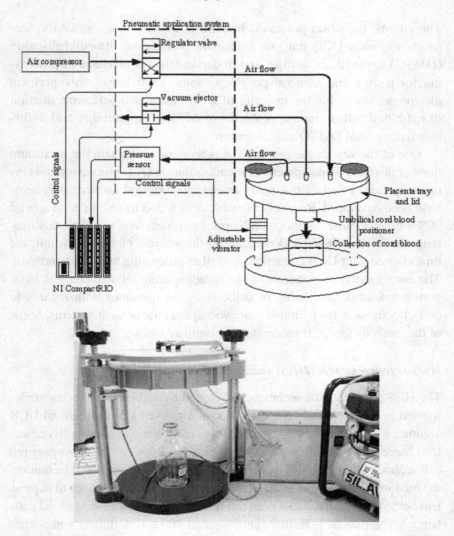

Fig. 3. Early prototype of enhanced *ex-utero* cord blood collection system. Figure courtesy of Tan Kok-Zuea, Electrical and Computer Engineering Department, National University of Singapore.

Processing

The processing of the UCB unit prior to cryopreservation essentially serves two purposes: volume reduction and the addition of a cryoprotectant.

The volume reduction process is helpful to reduce storage space requirements for each UCB unit, to reduce the volume of dimethylsulfoxide (DMSO) required for storage, and to derive leftover products of processing for testing and archival purposes. Some UCB banks only perform plasma depletion, but the majority of UCB banks also perform reduction of red blood cells to reduce problems of ABO incompatibility and to further reduce total DMSO concentration.

One of the key challenges in UCB processing is to retain the maximum number of viable hematopoietic progenitors during UCB processing and cryopreservation. To attain this goal, several methods and automated systems have been developed. Recovery of viable cells seems to be limited to around 90% of the original number of cells prior to processing, although this may range between 70%–80% depending on the volume of the UCB unit, the time elapsed after UCB collection, as well as processing technique and skill. The automated systems contribute to speedier, more reliable and less manpower-dependent processing of cells, however, questions with regards to cost-effectiveness have limited more widespread use of such systems. Some of the methods for UCB processing are outlined below.

Hydroxyethyl starch (HES) sedimentation

The HES sedimentation technique developed by Rubinstein *et al.* is implemented in many banks.[18] HES is added at a ratio of 1:5 of the initial UCB volume, which decreases the zeta potential (a measure of the repulsive electric charge between cells, especially red blood cells) of cells to allow easy red cell sedimentation and buffy coat formation. In one version of this technique, the mixture is centrifuged at 50 *g* and placed in a plasma extractor to express leukocyte-rich plasma into a connecting bag, leaving excess red blood cells behind. The plasma is further centrifuged at 400 *g* to sediment leukocytes, allowing excess plasma to be expressed and leaving the nucleated cells in a standard volume ready for cryopreservation (Fig. 4A). This is a labor-intensive process, and takes approximately an hour to process a single unit.

Top-and-bottom (T&B)

The top-and-bottom process uses differential centrifugation, followed by automatic expression (e.g., using Optipress II Blood component

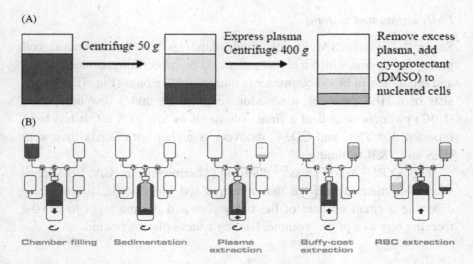

Fig. 4. Cord blood processing systems. **(A)** Hydroxyethyl starch sedimentation for volume reduction of cord blood units is the standard routine method for cord blood processing. **(B)** Other methods, including automated ones like the one depicted in the flowchart for cord blood processing in Sepax™. Figure courtesy of Biosafe SA.

extractor) and a triple blood bag system to contain separated red cells, plasma, and stem cell-rich buffy coat.[33] Although this is more efficient than the HES technique (~30 minutes), red blood cell depletion is not as effective.

Filtration

Filters that trap hematopoietic progenitors have been developed (e.g., StemQuick™ E filters from Asahi, Tokyo, Japan).[38] These devices filter UCB by gravity, trapping nucleated cells on the filter while allowing red blood cells and platelets to flow into the drain bag. The trapped nucleated cells are recovered from the filter by flushing with a reverse flow of protein rich recovery solution. This process takes only 12 minutes to complete, and has no negative effect on the engraftment potential of UCB cells.[38] However, recovery levels for TNC is approximately 25%–30% less than with the HES and T&B methods.

Fully automated systems

Sepax™ (Biosafe SA, Eysins, Switzerland) is a fully automated cell processing device, which includes automatic chamber filling, centrifugation, and separation of blood components into respective bags (Fig. 4B). It consists of a HES protocol, a no additive protocol, and a low buffy coat (LBC) protocol to collect a final volume of as low as 8 ml. It has been reported that TNC and CD34[+] recovery is higher with Sepax than with HES and T&B techniques.[39]

The AXP™ AutoXpress™ platform (Thermogenesis, CA, USA) consists of a metering device that removes red blood cells, followed by diverting a small amount of the buffy coat and plasma layer to fill the freezing bag to a preset volume, leaving excess plasma behind.

Cryopreservation

Cryopreservation of UCB units usually requires the controlled addition of a cryoprotectant such as DMSO and regulated freezing in automated freezing devices. Different concentrations of cryoprotectants have also been explored with variable results. The addition of osmotic agents such as dextran at the same time as DMSO appears to be necessary. A variety of membrane stabilizers like trehalose and catalase, as well as commercial preparations like Cryostor™, seem to have possible additional benefits in protecting cells against hypothermic stress.[37,38] The frozen UCB units are usually kept in liquid nitrogen as this provides the safest long term storage due to the relative independence from an uninterrupted power supply as well as the ability to keep temperatures of less than −190°C, offering a safe margin from the gas-phase transition temperature of −130°C. Maternal samples as well as other non-viability dependent samples may be kept in conventional −80°C freezers. There should be segments attached to the UCB bag (usually sealed segments of tubing), which can be used for confirmatory testing (CT) of the HLA type of the UCB unit.

The cord blood unit (CBU) is initially kept in a quarantine tank where it will remain in holding until all the results of tests done on the CBU and maternal samples qualify the unit for clinical use. These tests may include, but are not limited to human immunodeficiency virus, hepatitis B, hepatitis C, HTLV1, syphilis and cytomegalovirus tests from maternal samples, as well as hemoglobinopathy testing, blood grouping, sterility tests and colony assays of the CBU. At the Singapore Cord Blood Bank (SCBB), the mother is contacted again at about six months to a year after the birth of the baby and is interviewed about the baby's health and well-being to clarify any issues with regards to the medical history and also to ask the mother if there is anything further in the family or medical history that she wishes to add. At least three such attempts to contact the donor are made and if the donor cannot be contacted, a note that due diligence was made has to be signed off by the Medical Director with a note to attempt contact of the donor again if the CBU is identified for confirmatory typing (CT) prior to requesting the unit for CBT. When all the testing and efforts to contact the donor have been made, then the CBU, if still qualified, will be transferred to the final storage tank. At the SCBB, cords that fail this requirement are designated for research or are used for quality assurance. The units relegated to research are transferred to the research tank.

Whatever the procedures chosen for UCB processing and cryopreservation, it is important that these procedures are closely governed by protocols for validation and qualification of equipment and supplies, clear procedures for identification and labeling, inventory management, inprocess testing, defined conditions for storage, meticulous record keeping, and proper disposal of waste. The processing procedures should also be verified at each individual UCB bank and training of staff should be meticulously tracked and recorded. A robust system of temperature monitoring should track the temperature of the cryopreserved UCB units and ensure that key personnel are contacted if the temperature of the liquid nitrogen tanks goes above a certain temperature. At the SCBB, short message service (SMS) texts will be sent to key personnel if the temperature of tanks goes above a specified temperature. Furthermore, we have clearly defined systems for ensuring that at least one person responds to and addresses the problem. Oxygen concentration should be monitored in the cryopreservation room where there could potentially be fatal leakage of nitrogen. Other useful environmental monitoring systems for the UCB laboratory include those which monitor humidity, air particle count, room temperature, and microbial counts; though some of these parameters may not be tracked in every UCB bank.

SELECTION AND RELEASE

UCB units are selected for transplantation through a close communication between the UCB bank and the transplant center, as well as any third party donor search registries. The transplant center typically sends a search request directly to the UCB bank or indirectly through donor search registries. If the UCB bank receives such a request, it may use its own validated search algorithm to find a list of potentially suitable UCB units, listed according to HLA match and cell dose, which it will send to the transplant center. Transplant centers may also search UCB banks through large donor search registries like the NMDP and BMDW (see above). In this case, the search programs of these registries help to sift through the units from all donors (BM and UCB) to find the most suitable matches for patients in search of suitable HSCs for transplantation. Only when the transplant center has identified that the UCB unit is potentially suitable for the patient and would like to perform CT on the UCB unit, will the UCB bank be contacted. The advantage of going through large donor search registries is that patients will have clearer access to all potential donors, BM or UCB, from many different registries and countries at the same time.

When a suitable unit of UCB is identified, the transplant center will usually request a sample of the CBU for CT. This is preferably from an attached segment of the stored UCB unit and the CT serves several purposes. Firstly, it verifies that there was no mix-up of CBU during or after the processing and cryopreservation. Secondly, it verifies that the HLA typing of the CBU was correct. Thirdly, it also allows transplant centers to look at the HLA typing results at a higher resolution to help with decision making for optimal CBU selection. Besides CT, transplant centers may also request that viability testing, CD34 cell counts and CFU assays be performed on the attached segment of CBU in order to verify the quality of the frozen cell population in the CBU for transplantation. In addition, if further testing needs to be performed on the CBU, either because of new regulatory requirements or the need for additional testing to be performed by testing labs certified by different countries, cryovials or other samples of red blood cells, plasma, unprocessed UCB may be sent to appropriate testing laboratories for further testing, which can usually be performed in a matter of days.

If testing results meet the requirements of the transplant center, the transplant center could send in a request for shipment of the CBU to the transplant center. The CBU is usually expected to arrive at the transplant center before the patient starts on the pre-transplant preparative regimen, which often involves chemotherapy with or without radiotherapy. This is to assure that there will not be any delay or mishaps during shipment of the CBU, as once the patient starts on the preparative regimen, often termed the "conditioning" regimen, the transplant will have to proceed or the patient could die of BM aplasia.

The shipment is usually carried out using a specialized cryoshipper that will maintain and track the temperature of the CBU throughout the shipment with an attached data logger which will track the internal temperature of the dry shipper. The third edition of the standards of the Foundation for the Accreditation of Cellular Therapy (FACT) mandates that "Cryopreserved UCB units stored at a temperature below −135°C shall be transported in a liquid nitrogen-cooled 'dry shipper' that contains adequate absorbed liquid nitrogen and has been validated to maintain temperature at least 48 hours beyond the expected time of arrival at the receiving facility." Furthermore, the time of receipt, internal temperature of the shipper, integrity of product and integrity of the shipper should be recorded and acknowledged by the receiving party. FACT also requires that "Transport records shall document the identity of the courier; the date and time of packaging; the date and time the package left the facility; and the date and time of receipt of the package." A list of items shipped with the CBU should be carefully tracked and the internal temperature of the shipper where the CBU is kept should be carefully tracked by the data logger probe throughout the shipment process. Clinical outcome data of the recipient should be obtained from the transplant center and this should form a basis for statistical analysis and quality assurance programs. At the SCBB we request transplant centers to fill out standard 100-day clinical outcome data forms similar to the transplant essential data (TED) forms used by the IBMTR.

QUALITY ASSURANCE, ACCREDITATION AND LICENSING

Quality assurance of UCB banks is essential to ensure that all processes and systems in UCB banks produce clinical quality CBUs that are safe and effective for transplantation. These processes cover all aspects of collection, processing, cryopreservation, storage, disposal, selection, release and shipping of the CBUs. An overall quality management plan, detailed organizational structure, defined roles and responsibilities, a facility plan, an audit plan and a safety program are needed. A system for agreement review, process control, design control, preventive action, corrective action, validation, qualification, training, inventory management and outcome analysis is essential. Proper labeling, identification and confidentiality of the CBU, donor files and test samples have to be maintained. A Business Continuity Plan (BCP) in event of unexpected interruption of operations should be in place to preserve the inventory under any adverse incident/condition that may occur to the physical facility and to protect the precious investment in each CBU. As such, process and efforts need to be taken to protect the tanks, freezers and related documents, to ensure the ability to release units when a search request finds a proper match, and to resume work when the conditions are resolved. Without adequate preservation of its inventory and related documentation (either in hard or soft copy) the ability of the cord blood bank to succeed in the mission will be diminished, or halted completely. Maintenance of these systems may require a full time or part time quality manager. Proper handling of adverse events, deviations, accidents, errors and complaints, as well as regular internal audits and standard operating procedure (SOP) revisions, is part of the continual responsibilities of such a quality manager. It is beyond the scope of this chapter to delve into all the aspects of quality assurance, and issues specific to it are probably best addressed by the standards issued by the various organizations dealing with accreditation and licensing of UCB banks (see below). The standards provided by these organizations are comprehensive and the pre-accreditation as well as post-accreditation support provided by them has proven to be very useful to many UCB banks.

AABB (http://www.aabb.org/content)

The American Association of Blood Banks (AABB) offers an accreditation program for cellular therapy products and services which include UCB banks. They also provide independent consultancy services for accreditation.

FACT (http://www.factwebsite.org)

The Foundation for the Accreditation of Cellular Therapy (FACT) establishes standards for medical and laboratory practice in cellular therapies, and has a partnership with NETCORD to develop international standards for UCB collection, processing, testing, banking, selection and releases.

FDA (http://www.fda.gov/)

The Food and Drug Administration (FDA) regulates the UCB banking industry, and all UCB banks in the USA need to register with the FDA. Banks also have to notify the FDA of specific adverse reactions in the stem cells that they process, and to allow for FDA inspections.

TGA (http://www.tga.gov.au/)

The Therapeutic Goods Administration of Australia (TGA) also regulates and licenses UCB banks.

FUTURE FOR CORD BLOOD BANKING

CBT currently accounts for a sizeable fraction of all unrelated HSC transplants. We expect an increasing trend of UCB usage in the future, due to better long term prognosis, better methods of overcoming cell dose limitations, larger inventories, and improved collaboration between UCB banks and the transplant centers. The importance of CBT is now widely recognized, and few question the relevance for UCB banking and transplantations.

Work is underway to hasten hematopoietic recovery after CBT by co-infusing CD34+ cells from third-party donors.[39,40] UCB CD34+ cells and unfractionated cells have also been expanded *ex vivo* in attempts to overcome the cell number limitation in a single cord unit, but engraftment was not enhanced in patients who were transplanted with expanded cells. Future expansion efforts may require the optimization of cytokine and stroma to recreate the HSC niche, and better techniques to manipulate cell signaling pathways to expand UCB stem without differentiation.[40]

Additionally, cellular engineering techniques to generate tumor-specific cytotoxic T-lymphocytes from naïve UCB T-cells may find potential in immunotherapy post-transplant to treat leukemia relapse.[41–43]

Much research is ongoing to investigate the potential use of UCB stem cells in regenerative medicine. Clonal lines of multipotent cells (called the multilineage progenitor cell, MLPC) have been established from full term UCB, which can expand and differentiate into cells representing all three germinal layers.[44] Recently, it has been shown that human unrestricted somatic stem cells (USSCs) from umbilical cord blood represent pluripotent, neonatal, non-hematopoietic stem cells with the potential to differentiate into osteoblasts, chondroblasts, adipocytes, and hematopoietic and neural cells.[45] Another group has isolated UCB stem cells with embryonic cell characteristics, and shown that they can differentiate into insulin-producing cells that eliminated hyperglycemia after transplant into a diabetic mouse.[46] *In vivo* animal work has also demonstrated that UCB stem cells can differentiate into the neuronal lineage and have benefit in heatstroke injury, stroke, and spinal cord injury,[47–49] and into the cardiac lineage for treatment of acute myocardial infarction.[50,51] Recently, a clinical trial using UCB-derived mesenchymal stem cells for the treatment of Buerger's disease showed improvement of peripheral circulation, pain, and healing of necrotic skin lesions.[52] It could therefore be envisioned that UCB may one day be used beyond hematologic regeneration.

Although the favorable clinical experience with UCB stem cells has improved adoption of public banking of UCB stem cells in hospitals, public banks still see their growth hindered by insufficient funding to support basic operations of collection and storage, because revenues from the sale of UCB units for transplant are seldom sufficient to finance its operations. Additionally, UCB banks perform collections at a limited number of hospitals throughout the world, and efforts to expand the collection base to hospitals where the maternal population is representative of the general population can improve the diversity of UCB units banked.

Acknowledgments

The authors would like to thank the Singapore Cord Blood Bank and the SingHealth Foundation, as well as the Singapore Biomedical Research

Council and National Medical Research Council for technical and research support. Many thanks to Steve Sobak of the Singapore Cord Blood Bank as well as Mary Halet and Ellen Church of the US National Marrow Donor Program for critical advice with the manuscript.

References

1. Gluckman E, Broxmeyer HA, Auerbach AD, Friedman HS, Douglas GW, Devergie A, *et al.* Hematopoietic reconstitution in a patient with Fanconi's anemia by means of umbilical-cord blood from an HLA-identical sibling. *N Engl J Med* 1989 Oct 26;321(17):1174–8.
2. Wagner JE, Kernan NA, Steinbuch M, Broxmeyer HE, Gluckman E. Allogeneic sibling umbilical-cord-blood transplantation in children with malignant and non-malignant disease. *Lancet* 1995 Jul 22;346(8969):214–9.
3. Barker JN, Krepski TP, DeFor TE, Davies SM, Wagner JE, Weisdorf DJ. Searching for unrelated donor hematopoietic stem cells: availability and speed of umbilical cord blood versus bone marrow. *Biol Blood Marrow Transplant* 2002;8(5):257–60.
4. Dalle JH, Duval M, Moghrabi A, Wagner E, Vachon MF, Barrette S, *et al.* Results of an unrelated transplant search strategy using partially HLA-mismatched cord blood as an immediate alternative to HLA-matched bone marrow. *Bone Marrow Transplant* 2004 Mar;33(6):605–11.
5. Eapen M, Rubinstein P, Zhang MJ, Stevens C, Kurtzberg J, Scaradavou A, *et al.* Outcomes of transplantation of unrelated donor umbilical cord blood and bone marrow in children with acute leukaemia: a comparison study. *Lancet* 2007 Jun 9;369(9577):1947–54.
6. Brunstein CG, Barker JN, Weisdorf DJ, DeFor TE, Miller JS, Blazar BR, *et al.* Umbilical cord blood transplantation after nonmyeloablative conditioning: impact on transplantation outcomes in 110 adults with hematologic disease. *Blood* 2007 Oct 15;110(8):3064–70.
7. Barker JN, Weisdorf DJ, DeFor TE, Blazar BR, McGlave PB, Miller JS, *et al.* Transplantation of 2 partially HLA-matched umbilical cord blood units to enhance engraftment in adults with hematologic malignancy. *Blood* 2005 Feb 1;105(3):1343–7.
8. Loiseau P, Busson M, Balere ML, Dormoy A, Bignon JD, Gagne K, *et al.* HLA association with hematopoietic stem cell transplantation outcome: the number

of mismatches at HLA-A, -B, -C, -DRB1, or -DQB1 is strongly associated with overall survival. *Biol Blood Marrow Transplant* 2007 Aug;13(8):965–74.

9. Howard DH, Maiers M, Kollman C, Logan B, Loren G, Setterholm M. A cost-benefit analysis of increasing cord blood inventory levels. In: Meyer EA, Hanna K, Gebbie K. (eds.) *Cord Blood: Establishing a National Hematopoietic Stem Cell Bank Program*. Institute of Medicine of the National Academies, 2005.

10. Sonnenberg FA, Eckman MH, Pauker SG. Bone marrow donor registries: the relation between registry size and probability of finding complete and partial matches. *Blood* 1989 Nov 15;74(7):2569–78.

11. Beatty PG, Mori M, Milford E. Impact of racial genetic polymorphism on the probability of finding an HLA-matched donor. *Transplantation* 1995 Oct 27;60(8):778–83.

12. Kollman C, Abella E, Baitty RL, Beatty PG, Chakraborty R, Christiansen CL, *et al.* Assessment of optimal size and composition of the U.S. National Registry of hematopoietic stem cell donors. *Transplantation* 2004 Jul 15; 78(1):89–95.

13. Samuel GN, Kerridge IH, Vowels M, Trickett A, Chapman J, Dobbins T. Ethnicity, equity and public benefit: a critical evaluation of public umbilical cord blood banking in Australia. *Bone Marrow Transplant* 2007 Oct;40(8):729–34.

14. Brown J, Poles A, Brown CJ, Contreras M, Navarrete CV. HLA-A, -B and -DR antigen frequencies of the London Cord Blood Bank units differ from those found in established bone marrow donor registries. *Bone Marrow Transplant* 2000 Mar;25(5):475–81.

15. Ballen KK, Kurtzberg J, Lane TA, Lindgren BR, Miller JP, Nagan D, *et al.* Racial diversity with high nucleated cell counts and CD34 counts achieved in a national network of cord blood banks. *Biol Blood Marrow Transplant* 2004 Apr;10(4):269–75.

16. Devolder K. Preimplantation HLA typing: having children to save our loved ones. *J Med Ethics* 2005 Oct;31(10):582–6.

17. Rubinstein P, Rosenfield RE, Adamson JW, Stevens CE. Stored placental blood for unrelated bone marrow reconstitution. *Blood* 1993 Apr 1;81(7): 1679–90.

18. Rubinstein P, Dobrila L, Rosenfield RE, Adamson JW, Migliaccio G, Migliaccio AR, *et al.* Processing and cryopreservation of placental/umbilical

cord blood for unrelated bone marrow reconstitution. *Proc Natl Acad Sci USA* 1995 Oct 24;92(22):10119–22.

19. Kurtzberg J, Laughlin M, Graham ML, Smith C, Olson JF, Halperin EC, *et al.* Placental blood as a source of hematopoietic stem cells for transplantation into unrelated recipients. *N Engl J Med* 1996 Jul 18;335(3):157–66.

20. Gluckman E. Transplantation recent advances in transplantation immunology. *Curr Opin Immunol* 2006 Oct;18(5):556–8.

21. Stevens CE, Rubinstein P, Scaradavou A. HLA matching in cord blood transplantation: clinical outcome and implications for cord blood unit selection and inventory size and ethnic composition [abstract]. *Blood* 2006;108:855a.

22. Majhail NS, Brunstein CG, Wagner JE. Double umbilical cord blood transplantation. *Curr Opin Immunol* 2006 Oct;18(5):571–5.

23. Rocha V, Wagner JE, Jr, Sobocinski KA, Klein JP, Zhang MJ, Horowitz MM, *et al.* Graft-versus-host disease in children who have received a cord-blood or bone marrow transplant from an HLA-identical sibling. Eurocord and International Bone Marrow Transplant Registry Working Committee on Alternative Donor and Stem Cell Sources. *N Engl J Med* 2000 Jun 22;342(25):1846–54.

24. Smythe J, Armitage S, McDonald D, Pamphilon D, Guttridge M, Brown J, *et al.* Directed sibling cord blood banking for transplantation: the 10-year experience in the national blood service in England. *Stem Cells* 2007 Aug;25(8):2087–93.

25. Walters MC, Quirolo L, Trachtenberg ET, Edwards S, Hale L, Lee J, *et al.* Sibling donor cord blood transplantation for thalassemia major: experience of the Sibling Donor Cord Blood Program. *Ann N Y Acad Sci* 2005;1054: 206–13.

26. Reed W, Smith R, Dekovic F, Lee JY, Saba JD, Trachtenberg E, *et al.* Comprehensive banking of sibling donor cord blood for children with malignant and nonmalignant disease. *Blood* 2003 Jan 1;101(1):351–7.

27. Annas GJ. Waste and longing — the legal status of placental-blood banking. *N Engl J Med* 1999 May 13;340(19):1521–4.

28. Johnson FL. Placental blood transplantation and autologous banking — caveat emptor. *J Pediatr Hematol Oncol* 1997 May–Jun;19(3):183–6.

29. Miura M, Gronthos S, Zhao M, Lu B, Fisher LW, Robey PG, *et al.* SHED: stem cells from human exfoliated deciduous teeth. *Proc Natl Acad Sci USA* 2003 May 13;100(10):5807–12.

30. Takahashi K, Tanabe K, Ohnuki M, Narita M, Ichisaka T, Tomoda K, *et al.* Induction of pluripotent stem cells from adult human fibroblasts by defined factors. *Cell* 2007 Nov 30;131(5):861–72.

31. Mayor S. World's first public-private cord blood bank launched in United Kingdom. *Br Med J* 2007 Feb 3;334(7587):229.

32. American Academy of Pediatrics: Work Group on Cord Blood Banking. Cord blood banking for potential future transplantation: subject review. *Pediatrics* 1999 Jul;104(1 Pt 1):116–8.

33. Tamburini A, Malerba C, Mancinelli F, Spagnoli A, Ballatore G, Bruno A, *et al.* Evaluation of biological features of cord blood units collected with different methods after cesarean section. *Transplant Proc* 2006 May;38(4):1171–3.

34. Bornstein R, Flores AI, Montalban MA, del Rey MJ, de la Serna J, Gilsanz F. A modified cord blood collection method achieves sufficient cell levels for transplantation in most adult patients. *Stem Cells* 2005 Mar;23(3):324–34.

35. Tsagias N, Kouzi-Koliakos K, Hamidi-Alamdari D, Karagiannis V, Kostidou E, Koliakos G. Cell recovery sufficient for adult transplantation by additional cord blood collection from placenta. *Transplant Proc* 2007 Dec;39(10):3380–4.

36. Belvedere O, Feruglio C, Malangone W, Bonora ML, Minisini AM, Spizzo R, *et al.* Increased blood volume and CD34(+)CD38(−) progenitor cell recovery using a novel umbilical cord blood collection system. *Stem Cells* 2000;18(4):245–51.

37. Grisaru D, Deutsch V, Pick M, Fait G, Lessing JB, Dollberg S, *et al.* Placing the newborn on the maternal abdomen after delivery increases the volume and CD34 cell content in the umbilical cord blood collected: an old maneuver with new applications. *Am J Obstet Gynecol* 1999 May;180(5):1240–3.

38. Eichler H, Kern S, Beck C, Zieger W, Kluter H. Engraftment capacity of umbilical cord blood cells processed by either whole blood preparation or filtration. *Stem Cells* 2003;21(2):208–16.

39. Lapierre V, Pellegrini N, Bardey I, Malugani C, Saas P, Garnache F, *et al.* Cord blood volume reduction using an automated system (Sepax) vs. a semi-automated system (Optipress II) and a manual method (hydroxyethyl starch sedimentation) for routine cord blood banking: a comparative study. *Cytotherapy* 2007;9(2):165–9.

40. Hofmeister CC, Zhang J, Knight KL, Le P, Stiff PJ. *Ex vivo* expansion of umbilical cord blood stem cells for transplantation: growing knowledge from the hematopoietic niche. *Bone Marrow Transplant* 2007 Jan;39(1):11–23.

41. Introna M, Franceschetti M, Ciocca A, Borleri G, Conti E, Golay J, *et al.* Rapid and massive expansion of cord blood-derived cytokine-induced killer cells: an innovative proposal for the treatment of leukemia relapse after cord blood transplantation. *Bone Marrow Transplant* 2006 Nov;38(9): 621–7.

42. Robinson KL, Ayello J, Hughes R, van de Ven C, Issitt L, Kurtzberg J, *et al. Ex vivo* expansion, maturation, and activation of umbilical cord blood-derived T lymphocytes with IL-2, IL-12, anti-CD3, and IL-7. Potential for adoptive cellular immunotherapy post-umbilical cord blood transplantation. *Exp Hematol* 2002 Mar;30(3):245–51.

43. Hagihara M, Gansuvd B, Ueda Y, Tsuchiya T, Masui A, Tazume K, *et al.* Killing activity of human umbilical cord blood-derived TCRValpha24(+) NKT cells against normal and malignant hematological cells *in vitro*: a comparative study with NK cells or OKT3 activated T lymphocytes or with adult peripheral blood NKT cells. *Cancer Immunol Immunother* 2002 Mar;51(1):1–8.

44. Berger MJ, Adams SD, Tigges BM, Sprague SL, Wang XJ, Collins DP, *et al.* Differentiation of umbilical cord blood-derived multilineage progenitor cells into respiratory epithelial cells. *Cytotherapy* 2006;8(5):480–7.

45. Kogler G, Sensken S, Airey JA, Trapp T, Muschen M, Feldhahn N, *et al.* A new human somatic stem cell from placental cord blood with intrinsic pluripotent differentiation potential. *J Exp Med* 2004 Jul 19;200(2): 123–35.

46. Zhao Y, Wang H, Mazzone T. Identification of stem cells from human umbilical cord blood with embryonic and hematopoietic characteristics. *Exp Cell Res* 2006 Aug 1;312(13):2454–64.

47. Chen SH, Chang FM, Tsai YC, Huang KF, Lin MT. Resuscitation from experimental heatstroke by transplantation of human umbilical cord blood cells. *Crit Care Med* 2005 Jun;33(6):1377–83.

48. Taguchi A, Soma T, Tanaka H, Kanda T, Nishimura H, Yoshikawa H, *et al.* Administration of CD34+ cells after stroke enhances neurogenesis via angiogenesis in a mouse model. *J Clin Invest* 2004 Aug;114(3):330–8.

49. Nishio Y, Koda M, Kamada T, Someya Y, Yoshinaga K, Okada S, *et al.* The use of hemopoietic stem cells derived from human umbilical cord blood to promote restoration of spinal cord tissue and recovery of hindlimb function in adult rats. *J Neurosurg Spine* 2006 Nov;5(5):424–33.

50. Hirata Y, Sata M, Motomura N, Takanashi M, Suematsu Y, Ono M, *et al.* Human umbilical cord blood cells improve cardiac function after myocardial infarction. *Biochem Biophys Res Commun* 2005 Feb 11;327(2):609–14.

51. Chen HK, Hung HF, Shyu KG, Wang BW, Sheu JR, Liang YJ, *et al.* Combined cord blood stem cells and gene therapy enhances angiogenesis and improves cardiac performance in mouse after acute myocardial infarction. *Eur J Clin Invest* 2005 Nov;35(11):677–86.

52. Kim SW, Han H, Chae GT, Lee SH, Bo S, Yoon JH, *et al.* Successful stem cell therapy using umbilical cord blood-derived multipotent stem cells for Buerger's disease and ischemic limb disease animal model. *Stem Cells* 2006 Jun;24(6):1620–6.

14

Commercialization of Cord Blood

Vivek M. Tanavde

ABSTRACT

Banking cord blood is increasingly becoming more popular in many countries in the world. Cord blood banks are set up as public banks or commercial entities that provide fee-based services. This chapter explores the commercial aspects of cord blood banking including the necessity of private banks, their business models, hurdles faced by private banks and the factors that make private cord blood banking a successful enterprise. The differences between public and private banks are also discussed. The business models of these banks will also be impacted by the technological advances in the field of cord blood stem cells. As the use of cord blood cells for treating non-hematological disorders develops, the scale and requirements of banking cord blood will also change and private banks need to adapt themselves to this changing environment. In summary, this chapter discusses unique issues faced in the commercialization of cord blood banking and how private cord blood banks are tackling these issues.

INTRODUCTION

Cord blood is a rich source of stem cells that is often discarded. Since cord blood can only be collected at the time around delivery, it has to be "banked" or cryopreserved for future use. Cord blood banks are either funded by public funds, where the cord blood is stored at no cost to the donor or are run by private companies where the donor pays for the processing and storage of cord blood units. Parents deciding to store cord blood must choose between public or private cord blood banks. Since public banks are funded by public money, they are limited by the number of units they can successfully store. Public banks accept cord blood units as donations to be used for anyone in need. Once the blood is donated, it loses all identifying information after a short period of initial testing, so that families will not be able to retrieve their own blood unit later. Since many parents may want to reserve their child's cord blood for his/her own use or for their siblings, they prefer to use a private cord blood bank and pay for banking their baby's cord blood. This need has given rise to a number of private cord blood banks in different parts of the world and many of them are getting incorporated and publicly listed on various stock exchanges (Table 1). Classically, cord blood stem cells have been used to treat hematologic disorders, but the possibility that cord blood stem cells could also be used for non-hematological disorders makes storing cord blood even more attractive. There are also a few companies trying to commercialize cord blood stem cells, or their derivatives, directly. This chapter does not cover cord blood banking in detail since this has been reviewed in the previous chapter. This chapter mainly reviews the economics of cord blood banking and the specific issues facing private cord blood banks and companies trying to commercialize cord blood stem cells.

IS PRIVATE BANKING NECESSARY?

The larger obstacle facing banks funded by public money are the costs required to run and maintain them. This has prevented more than a handful from opening worldwide. Because public banks do not charge storage fees, many medical centers do not have the funds required to establish and maintain banking operations. Due to donation patterns, different racial

Table 1. Partial list of private cord blood banks in the world.

Company	Year started banking	Accreditation	No. of transplants	Collection fee (US$)
USA				
Cord Blood Registry	1995	AABB	108	1875
Cryobanks International	1994	AABB	18	1745
Cryo-Cell International	1992	AABB	16	1595
LifeBankUSA	1998	AABB	26	1950
Viacord	1995	AABB	125	2050
UK				
Smart Cells International	2001	MHRA+HTA	5	1460
UK Health Solutions	2007	MHRA	3	1470
Virgin Health Bank	2007	MHRA	3	2200
Europe				
Cryo-Save/Belgium	2000	ISO 17025	2	1800
Eurocord/Slovakia	2001	—	5	900
SmartBank/Italy	2005	MHRA+ISO 9001:2000	3	2800
Asia				
CordLife/Singapore	2001	AABB	1	1000
Madifreeze/Israel	2005		12	1560
StemLife/Malaysia	2002		4	850
Reliance Life Sciences/India	2001	AABB	0	1000
LifeCell/India	2005	AABB/ISO 9001:2008	0	1100
Healthbaby/HongKong	2006	AABB/ISO 9001:2000	2	1500
Australia & New Zealand				
Cryosite Ltd./Australia	2001	TGA	0	2200
BioCell/Australia	2004	Partial TGA	0	1450
Cord Bank/New Zealand	2004		1	1300

Note: A comprehensive list of public and private cord blood banks in different countries is available at http://parentsguidecordblood.org/content/usa/banklists/index.shtml?navid=9. The collection and storage fee ranges between US$850–2800 with or without storage. AABB: American Association of Blood Banks; MHRA: Medicines and Health care products Regulatory Agency; HTA: Human Tissue Authority; TGA: Therapeutic Goods Administration.

groups have different likelihood of finding a match through a public cord blood bank. For example in the New York Blood Center, the largest bank in the US, Caucasians find a match 70% of the time (5–6/6 HLA match),[1] while the probability of finding a match is considerably lower for ethnic minorities. Public bank advocacy groups are therefore trying to particularly encourage donations by members of non-Caucasian racial groups.

A number of private for-profit companies have been established that encourage parents to bank their children's cord blood for their own autologous use or by a family member, should the need arise. Parents have been encouraged to bank their children's cord blood as a form of "biological insurance." With increasing reports on the isolation of mesenchymal stem cells from cord blood, applications beyond hematological reconstitution, such as bone reconstitution and other potential applications could be envisioned.

Physicians, employees, and/or consultants of such companies may have potential conflicts of interest in recruiting patients because of their own financial benefit. Annual disclosure of the financial interest and potential conflicts of interest must be made to institutional review boards that are charged with the responsibility of mitigation of these disclosures and risks. Families may be vulnerable to the emotional effects of marketing for cord blood banking at the time of birth of a child and may look to their physicians for advice. No accurate estimates exist of the likelihood of children to need their own stored cord blood stem cells in the future. Available estimates range from one in 1000 to more than one in 200,000.[2] Moreover, currently, the strategy for children receiving their own cord blood stem cells for future autologous use is controversial.[2] There is also no evidence for the efficacy of autologous cord blood stem cell transplantation for the treatment of malignant neoplasms later in life.[2] On the contrary, there is evidence demonstrating the presence of DNA mutations in cord blood obtained from children who subsequently develop leukemia.[3] Thus, an autologous cord blood transplantation might be contraindicated in the treatment of a child who develops leukemia. Many organizations like the American Academy of Pediatrics explicitly discourage private cord blood banking if there is no family history of genetic disorders or familial cancers.

The guidelines issued by the American Academy of Pediatrics[4] state that when consulted by prospective parents who are interested in donating cord blood to a philanthropic bank or paying to have cord blood stored in a for-profit bank, the physician should provide the following information:

1. Although preliminary data show encouraging results in cord blood stem cell transplantation for a variety of genetic, hematologic, and oncologic diseases, the procedure at this time is considered investigational.
2. The indications for autologous transplantation are limited, and the potential for future expansion is unlikely.
3. Given the difficulty of making an accurate estimate of the need for autologous transplantation and the ready availability of allogeneic transplantation, private storage of cord blood as "biological insurance" is unwise. However, banking should be considered if there is a family member with a current or potential need to undergo stem cell transplantation.
4. Conditions such as leukemia or severe hemoglobinopathy may indicate the need for directed-donor cord blood banking for sibling cord blood transplantation.
5. Philanthropic donation of cord blood for banking at no cost for allogeneic transplantation is encouraged. In such instances, the parents should be informed of the appropriate operational principles recommended for the bank listed herein.[5]

However in spite of these issues, private cord blood banks are being established in many countries. Most parents who preserve their child's cord blood see this as an investment in their child's future. Although currently cord blood cells are therapeutically used only for hematological disorders, there is potential for cord blood cells to be used for non-hematological disorders in the future. Autologous transplantation of cord blood cells may be used therapeutically in the future for diseases like Parkinson's and osteoporosis. Since cord blood cells can only be collected at the time around birth, banking these cells at the time of delivery would be the only way these cells would be available in the future. Since public banks cannot guarantee that the child's cord blood will be preserved for autologous use, parents who can afford cord blood banking will demand the opportunity to bank their child's cord blood in a private bank. As the price of banking comes down, more parents would be able to afford banking cord blood. Lower cost of banking cord blood would also reduce the financial burden of this procedure on the family.

Therefore although private banking seems dispensable today, the expanding therapeutic uses of cord blood will likely make them necessary in the future and private cord blood banks will continue to coexist with public cord blood banks.

CAN CORD BLOOD BANKS GENERATE PROFITS?

Private banks can generally be classified according to the business model they use. These are:

1. Pay per user model: this is the oldest private cord blood banking model, where donors pay a fee for collecting and storing their child's cord blood.
2. Semi private bank model where storage is subsidized, but transplanted units are charged heavily to cover storage costs of untransplanted storage units.
3. Split public/private model: This is an innovative model set up by the Virgin Health Bank in the UK. In this model, the bank stores the cord blood in two parts. 20% of the blood is stored for the donor's own use, whereas 80% is registered on a public registry and is publicly available. The bank releases the unit free of charge and the recipient only pays the transportation costs. This is based on the premise that most recipients will require cord blood cells for regenerative purposes rather than full transplants and in the future cord blood stem cells can be expanded for therapeutic use enabling the use of grafts with low stem cell numbers.

Although cord blood banks are expected to generate revenues based on these models, there are a number of factors that need to be considered to determine if a cord blood bank can generate a net profit. These are:

1. *High capital investment.* The equipment and infrastructure needed for setting up a cord blood bank requires high capital investments. The

operational expense is also relatively high especially if the bank is accredited and compliant with guidelines set by the American Association of Blood Banks (AABB) or Foundation for Accreditation of Cellular Therapy (FACT). For example, the half yearly operating expenses for Viacord stood at US$41.5 million with an operating loss of US$11.3 million in June 2007.[6] Cryo-Cell spent US$2.9 million in operating expenses in 2006.[7] In 2004, Cord Blood Registry (CBR) committed US$10 million to its new laboratory and is planning to invest US$75 million more in this facility over the next seven years. The new 80,000-square-foot facility is designed to handle the exponential growth the company expects in the future. CBR announced increased sales, with client contracts up 107% over the same period in 2003.[8]

2. *Long incubation periods before revenues can be generated.* Further compounding the problem is that cord blood banks have to invest the high capital investment upfront before a single cord blood unit is banked. Therefore there is a significant lag between capital expenditure and realization of profits. Often it takes more than five years for a cord blood bank to turn profitable.

3. *Intense competition in developed markets.* In the developed markets, there is intense competition between private cord blood banks offering storage of cord blood. In the US, there are 24 private banks, whereas the UK has nine private cord blood banks. Given the current paucity of donors, the relatively high cost of storage and the ease of transportation of samples within a country, it is unlikely that such a high number of cord blood banks will survive. There is a high likelihood of some form of consolidation of cord blood banks in the future. There is however tremendous growth potential for cord blood banks in developing countries. In the current globalized world it is possible to shift technological expertise, infrastructure and even storage of cord blood units from the developed to developing countries. Since operating costs in such countries tend to be lower, the overall investment required to start a cord blood bank would be significantly lower in these countries. However, cord blood banks in developing countries will have to deal with relatively poor infrastructure, shortage of trained manpower, and backlash against such "outsourcing" from their customers. In such a scenario the safe bet for a cord blood

banking company is to set up a subsidiary bank in a developing country and offer its services locally. Once the subsidiary cord blood bank
achieves similar operating standards and accreditation as the parent
bank the company can think of storing cord blood units "off shore" to
reduce its operating costs.

Another way to tackle some of these problems is for tissue storage
companies to start cord blood banks as an add-on to their existing services. Some companies, like the California Cryobank (CCB) are tissue
banks that have entered the cord blood banking market. CCB was initially
a sperm bank offering semen and embryo storage, and then added
cord blood to its portfolio. The advantage of this approach is that these
banks are able to leverage their extensive experience and infrastructure of
tissue cryopreservation for successful cord blood banking. The disadvan-
tage is that since different tissues require different conditions for
cryopreservation, experience in one tissue may not always be transferable
to other tissues. Given the requirement of having separate processing and
storage facilities for different types of tissue, it is unclear if this strategy
lowers the expense of processing and storing cord blood. While it may
lower marketing and administrative costs, it is unlikely to significantly
lower operating expenses which form the bulk of the cost of banking a
cord blood unit.

As many cord blood banks are privately held, it is difficult to obtain
precise financial information about these banks. Among the publicly
listed cord blood banks, very few banks have been successful in making
a profit based on cord blood banking alone. In fact the largest publicly
traded cord blood bank, Cryo Cell Inc., with operations in Europe, Latin
America and Asia has only made a profit in two of the four years since it
has been listed on the NASDAQ stock exchange. Currently, most investment in cord blood banks is with the hope that the rise of stem cell-based
therapies in the future will produce a large demand for cord blood stem
cells and therefore will turn profitable in the long term. For example,
CBR reported record revenue and sales for the quarter ending June 30,
2006 and a compounded annual growth rate of 69% for the 24-month
period ending June 30, 2006.[9] Whether this is hope or merely hype
remains to be seen.

WHAT DOES IT TAKE TO BE A SUCCESSFUL CORD BLOOD BANK?

A successful private cord blood bank would offer a comprehensive cord blood storage option. Like any business, it should focus on product quality, responsive customer service, customer education and responsible marketing. The cord blood bank should also realize that revenue generated from cord blood storage alone will not be sufficient to be profitable. In the future, cord blood banks are likely to offer additional services such as cord blood expansion, tailor-made grafts of cells differentiated from cord blood stem cells for treating various disorders, and also have extensive R&D plans to provide a pipeline of new services.

HURDLES TO PRIVATE BANKING

1. *High costs for customers.* One of the highest hurdles to the adoption of cord blood banking for private use remains the relatively high costs. The one time fee of approximately US$1800 followed by a recurring storage fee is still too expensive for most people. This currently prevents cord blood banking from becoming a mainstream consumer phenomenon, which would however reduce the costs of storage. The cost of storage is especially acute in developing countries in Asia and Africa, where the majority of the world population lives and most of the potential cord blood units are generated. The high cost coupled with lack of knowledge about the use of cord blood stem cells is the single largest deterrent to widespread adoption of cord blood banking in these countries. The high expense of setting up a cord blood bank is often cited as a justification for the high initial cost. Given the fact that most banks have been in existence for less than a decade, the early consumers are bearing the brunt of this high capital and operating expenditure. However many private cord blood banks are coming up with innovative ways to deal with this issue. For example, Viacord offers a six months interest free payment plan. They also offer a 60 months installment plan with the first installment deferred until the baby is four years old. Companies like CBR have arrangements with GE Finance to provide payment plans with

monthly installments. In Asia, the cost of banking cord blood is significantly cheaper compared to the US and reflects the lower labor costs. In India, Reliance Life Sciences charges under US$1000 for their Relicord S. Program, and LifeCell India, the Indian subsidiary of the US-based LifeCell charges around US$700–900, for banking cord blood. Singapore-based CordLife International, charges US$900 for banking of cord blood and storage for the first year. These lower costs coupled with the rising disposable incomes of Asian families and flexible payment plans are making cord blood banking more affordable to many more families in Asia. Also having a larger client base will serve as an incentive for other investors to enter the field and should make financing of cord blood banking more competitive and consequently cheaper. However, extending credit to qualified couples is not always an easy task especially in developing countries. Many countries do not maintain credit histories or reports, thereby making it difficult to assess the credit risk of applicants. Recovering payment for services rendered is also a troublesome and lengthy procedure in these countries. So many private cord blood banks charge registration or collection fees upfront to cover the costs of cord blood collection and transport.

2. *Operational issues*. Private cord blood banks face some unique operational issues that are not faced by public banks. The biggest decision is when to bank a cord blood unit. Public banks have reasonably well defined criteria for determining if a cord blood unit is bankable or not and units which do not meet these criteria are simply discarded or are used for research. However, for most private banks, this decision is usually taken by the client. Sometimes even though some banks may counsel their clients that the unit is unsuitable for banking due to low volume or nucleated cell count, the client may insist on banking the cord blood unit. This issue is further compounded by the promise of clinically suitable expansion protocols being developed in the future, enabling the transplantation of cord blood units with low cell counts. Also many cord blood banks face a direct conflict of interest since adequate counseling of clients may result in loss of business for the cord blood bank. In addition, adequate information and time may not be available to parents before taking these decisions.

3. *Legal and ethical issues*. There are also some unique medical and legal issues facing private cord blood banks. The biggest issue is the

ownership of the cord blood unit. Do the parents, the child or the cord blood bank own the cord blood unit? In case the parents are unable to pay the storage fees does ownership of the cord blood unit pass to the cord blood bank or the finance company? Also, in case the parents default, the underlying asset i.e., the stored cord blood may be of little value to any one else. If the parents divorce later, who retains the rights to their child's cord blood? Moreover, in an increasing globalized environment, the cord blood unit may be collected in one country, stored in a different country, and the potential recipient may require it in a third country. This presents some unique legal challenges in the event of a dispute as to which law would be applicable. Also, many cord blood banks ask their clients to indemnify the bank in case of any dispute and although clients sign these contracts, their validity can be questioned especially in situations where the client did not have the necessary expertise or option to change the terms and conditions of the contract. And finally, what are the rights of the clients in case of total loss, or loss of quality of their cord blood unit due to poor storage, considering the life-saving value of cord blood stem cells, particularly with foreseeable expansion of cord blood use in non-hematological applications?

4. *Long term storage.* Another major challenge for private cord blood banks is designing fail safe storage systems for the cord blood units. Since cord blood units cannot be duplicated and backed up, the storage facility should have multiple redundancies built into their systems to ensure uninterrupted functioning of the facility over the long term. The other problem that clients face is the fate of their cord blood units if the company goes out of business. Since the financial bases of many cord blood companies are not very healthy, the probability of this happening is fairly high and the client should have the option to transfer the unit to another bank at comparable costs.

OTHER COMMERCIAL USES OF CORD BLOOD STEM CELLS

The scope of using cord blood stem cells is expanding beyond treating hematological disorders. Unrestricted somatic stem cells (USSCs)[10] from cord blood have been used for hematopoietic stem cell expansion,[11] endodermal[12] and neuronal[13] differentiation and cytokine production.[14] Many cord blood banks today are either part of or have associations with companies developing the use of cord blood for non-hematological diseases. This will have a two-fold impact on their business. These companies can then offer additional services based on therapeutic usage of cord blood stem cells. This will lead to additional revenues for the company. Such companies by expanding the scope of diseases treatable by cord blood stem cell transplants will be able to better justify the need for banking cord blood to their potential customers and will make their core banking process easier to market. This trend is already observable at companies such as Viacord and Cryo-Cell. Viacell, Viacord's parent company acquired Kourion Therapeutics in 2003, a company specializing in the development of Unrestricted Somatic Stem Cells (USSCs) from cord blood. Viacell is currently exploring the applications of USSCs in the areas of cardiac disease, cancer and diabetes. Similarly, Cryo-Cell has a controlling stake in Saneron Therapeutics, a company that focuses on the use of cord blood cells for treating neuronal disorders. Similarly, BioE is commercializing the use of Multilineage Progenitor Cells (MLPC™) from cord blood. These cells can differentiate into a variety of tissues including epithelial cells found in the lungs.[15] Another area for commercialization of cord blood has been the use of cord blood-derived cells for research. AllCells, a California-based company and Lonza are leading players in this area. BioE also sells its MLPC™ cell line for research and is developing the use of MLPC™ as a platform for screening drugs. However, the therapeutic use of cord blood stem cells for non-hematological applications is still largely experimental and commercial exploitation of cord blood stem cells by mainstream biotechnology companies for cell-based or cell-derived products still remains a distant possibility.

WHY COMMERCIALIZATION OF CORD BLOOD MAY BECOME FEASIBLE IN THE FUTURE

The biggest growth in cord blood banking results from the prospects of using cord blood cells for treating non-hematological disorders in the future. Cord blood is an attractive source of mesenchymal stem cells (MSCs), although MSC purification protocols from cord blood have to be considerably improved. Cord blood-derived MSCs could thus be envisioned for regeneration of bone and as immune modulatory cells in the context of allogeneic transplantation (Chapters 5 to 9). If cotransplantation of cord blood hematopoietic stem cells (HSCs) with MSCs is able to increase the success of haploidentical or completely mismatched transplants, this will significantly increase the probability of finding an allogenic match in a cord blood bank. This will increase the turnover of cord blood units in the bank, resulting in increased profit margins for the bank. MSCs are also attractive targets for gene therapy and could be used as vectors for gene transfer. Providing genetically modified cord blood stem cells suitable for therapeutic use could be another service provided by an integrated cord blood banking and cell processing company.

All these technologies will lead to greater numbers of cord blood units being processed and banked either by companies or parents of potential users. This should lead to better economies of scale, making cord blood banking affordable to many people. Also, increased competition will lead to eventual consolidation ensuring that only large companies with good infrastructure, array of applications and new products in the pipeline will survive. This consolidation will lead to harmonization of standards and practices leading to lower costs. Thus, cord blood products and services may become affordable to most people in the near future.

References

1. Schoemans H, Theunissen K, Maertens J, Boogaerts M, Verfaillie C, Wagner J. Adult umbilical cord blood transplantation: a comprehensive review. *Bone Marrow Transplant* 2006 Jul;38(2):83–93.
2. Johnson FL. Placental blood transplantation and autologous banking — caveat emptor. *J Pediatr Hematol Oncol* 1997 May–Jun;19(3):183–6.
3. Rowley JD. Backtracking leukemia to birth. *Nat Med* 1998 Feb;4(2):150–1.
4. AAP. *Frequently Asked Questions about Cord Blood Banking* 2007. Available from: http://www.aap.org/advocacy/releases/jan07cordbloodfaq.htm.
5. Lubin BH, Shearer WT. Cord blood banking for potential future transplantation. *Pediatrics* 2007 Jan;119(1):165–70.

6. Viacord. *ViaCell Reports Second Quarter 2007 Financial Results* (News Release). Cambridge, MA, August 9, 2007.

7. Company Financials: CCEL CRYO-CELL INTL INC OTCBB [database on the Internet]. *EDGAR Online 2007*. Available from: http://www.nasdaq.com/asp/ExtendFund.asp?page=full&symbol=ccel&selected=ccel&rpage=fundamentals&.

8. CBR. *Cord Blood Registry Reports Record Growth in Revenue* 2005. Available from: http://www.cordblood.com/cord_blood_news/media/press_releases/article_financial.asp.

9. CBR. *Cord Blood Banking Market Leader Reports Continued Growth in Consumer Demand for New Technology*. San Bruno, CA: Cord Blood Registry, 2006. Available from: http://www.cordblood.com/cord_blood_news/media/press_releases/cbr_high_growth.asp.

10. Kogler G, Sensken S, Airey JA, Trapp T, Muschen M, Feldhahn N, *et al*. A new human somatic stem cell from placental cord blood with intrinsic pluripotent differentiation potential. *J Exp Med* 2004 Jul 19;200(2):123–35.

11. Chan SL, Choi M, Wnendt S, Kraus M, Teng E, Leong HF, *et al*. Enhanced *in vivo* homing of uncultured and selectively amplified cord blood CD34$^+$ cells by cotransplantation with cord blood-derived unrestricted somatic stem cells. *Stem Cells* 2007 Feb;25(2):529–36.

12. Sensken S, Waclawczyk S, Knaupp AS, Trapp T, Enczmann J, Wernet P, *et al*. *In vitro* differentiation of human cord blood-derived unrestricted somatic stem cells towards an endodermal pathway. *Cytotherapy* 2007; 9(4):362–78.

13. Fallahi-Sichani M, Soleimani M, Najafi SM, Kiani J, Arefian E, Atashi A. *In vitro* differentiation of cord blood unrestricted somatic stem cells expressing dopamine-associated genes into neuron-like cells. *Cell Biol Int* 2007 Mar;31(3):299–303.

14. Kogler G, Radke TF, Lefort A, Sensken S, Fischer J, Sorg RV, *et al*. Cytokine production and hematopoiesis supporting activity of cord blood-derived unrestricted somatic stem cells. *Exp Hematol* 2005 May;33(5):573–83.

15. Berger MJ, Adams SD, Tigges BM, Sprague SL, Wang XJ, Collins DP, *et al*. Differentiation of umbilical cord blood-derived multilineage progenitor cells into respiratory epithelial cells. *Cytotherapy* 2006;8(5):480–7.

Closing Remarks

Cord Blood Stem Cells: A Future for Regenerative Medicine?

Suzanne Kadereit and Gerald Udolph

While this book's chapters focuses mainly on giving an overview on basic research with cord blood stem cells, the idea was to communicate to the reader the exiting possibilities lying in this abundant stem cell source. Indeed, every minute around 150 babies are born worldwide, rendering umbilical cord blood the most abundant source of human cell material. Currently, this source of human cells is largely underutilized and still most umbilical cords and with it valuable cells are discarded. Nevertheless, around 400,000 cord blood units have been banked worldwide,[1] with a strongly increasing trend. One could thus envision for the future banks that cover most HLA haplotypes represented in the world's population and thus provide matching stem cells for almost everybody.

In fact, despite the advances in translational research, and success in animal models of diseases, translation of cell therapies into the human setting suffers from insufficient availability of human cells. Patients waiting for organ transplants are still dying while on waiting lists,[2] and only about 30% of patients needing a HSC transplant have a matched donor in bone marrow registries.[3] Also, regeneration of organs with cellular therapy is only

335

taking off very slowly, again mainly for lack of matched tissues and cells. Accordingly, large hopes are set on a technology that emerged surprisingly rapidly in the last four years: cellular reprogramming and generation of induced pluripotent stem cells (iPS cells).[4] Pluripotent stem cells can theoretically be propagated in culture indefinitely without loosing their potential to make most cells of the body, and could thus provide clinically relevant numbers of cells. In particular, the iPS technology allows generation of patient-specific cells, and as such the problem of generating autologous cells relevant for therapies, which was believed to be the most challenging aspect of stem cell applications, has been alleviated.

As discussed in this book, the culture amplification of HSCs is still a challenge, and MSCs can only be propagated for limited numbers of passages. And although cord blood is abundant, amplification of stem cells contained in it is still a problem, and cellular therapies derived from umbilical cord likely are still far in the future. Transplantation of tissue derived from iPS cells made from a patient's own skin or bone marrow therefore has become a realistic possibility. Such cells would presumably not be rejected by the patient as the immune system would not recognize such cells as foreign. However, in general, at the age one begins to need regenerative medicine, both somatic cells and stem cells are on the decline and the DNA has accumulated mutations. Moreover, some treatments (e.g. chemotherapy) are genotoxic, provoking DNA lesions in surviving cells, and skin fibroblasts (currently used for iPS derivation) are often exposed to damaging UV irradiation. Whether such cells would make attractive transplants after reprogramming, culture expansion and differentiation, is questionable. Moreover, many patients need transplant material urgently and would benefit from off-the-shelf therapy.

When we started assembling this book, the question mark after the title was appropriate. At that point in time umbilical cord was already extremely promising, but clinical applications, beyond hematopoietic reconstitution, were far out in the future. But only within the last year, a relatively small step (within the gigantic advance of iPS technology) has totally jumbled the dices, and all bets are now on. Several groups have derived iPS cells from umbilical cord blood stem and precursor cells, thus opening undreamed-off possibilities. iPS cells derived from endothelial precursor cells, CD34+ or CD133+ cells isolated from cord blood could be differentiated into beating cardiomyocytes, dopaminergic neurons and also hematopoietic cells.[1,5–7] Cord blood stem cells appeared more amenable to reprogramming than fibroblasts or keratinocytes and were karyotypically stable. Importantly, however, two of the groups also derived iPS cells from frozen cord blood. That, and the fact that all banked units are stored with their immunological information, adds another great advantage to cord blood as a source for iPS derivation. One could prospectively identify units with commonly represented HLA haplotypes and generate iPS cell banks. With the large HLA haplotype representation within the cord donor pool, iPS cells for many patients could be derived. Bank size could be further reduced by identifying donors homozygous for common haplotypes.[1] And for future generations, one could envision storage of a few cells at birth for later usage, provided the newborn has no genetic defects precluding derivation and transplantation of own material.

However, until realization of such potential, many hurdles remain, most prominent the not yet fully optimized differentiation protocols to obtain clinically relevant amounts of pure differentiated cells from stem cells including pluripotent stem cells. While enormous advances have been made in these last years, differentiation of clinically relevant cells from pluripotent stem cells is still a major challenge. Differentiation of transplantable HSCs from ESCs for example has not yet been achieved. Also, the reprogramming process has to be further optimized, to eliminate retro/lentiviral vectors, and teratoma formation and other degeneration of transplanted cells needs to be thoroughly assessed after long-term follow-up.

Nevertheless, one undeniable enthusiasm for cord blood, apart from it being a good source for iPS cell derivation, is its shear unlimited supply and its genetic 'youth'. So we may see the question mark behind the title of this book disappear in a few years.

References

1. Giorgetti A, Montserrat N, Aasen T, Gonzalez F, Rodríguez-Pizà I, Vassena R, *et al.* Generation of induced pluripotent stem cells from human cord blood using OCT4 and SOX2. *Cell Stem Cell* 2009;5(4):353–7.

2. Lechler RI, Sykes M, Thomson AW, Turka LA. Organ transplantation —
    how much of the promise has been realized? *Nat Med* 2005;11(6):605–13.
3. Tse W, Bunting KD, Laughlin MJ. New insights into cord blood stem cell
    transplantation. *Curr Opin Hematol* 2008;15(4):279–84.
4. Okita K, Yamanaka S. Induction of pluripotency by defined factors. *Exp Cell
    Res* 2010;316(16):2565–70.
5. Haase A, Olmer R, Schwanke K, Wunderlich S, Merkert S, Hess C, *et al.*
    Generation of induced pluripotent stem cells from human cord blood. *Cell
    Stem Cell* 2009;5(4): 434–41.
6. Ye Z, Zhan H, Mali P, Dowey S, Williams DM, Jang Y-Y, *et al.* Human-
    induced pluripotent stem cells from blood cells of healthy donors and
    patients with acquired blood disorders. *Blood* 2009;114(27):5473–80.
7. Takenaka C, Nishishita N, Takada N, Jakt LM, Kawamata S. Effective
    generation of iPS cells from CD34+ cord blood cells by inhibition of p53.
    Exp Hematol 2010;38(2):154–62.

Index